COMMERCIAL
AVIATION SAFETY

ABOUT THE AUTHORS

CLARENCE C. RODRIGUES, PH.D., PE, CSP, CPE, is the Manager for Health, Safety and Environment and also holds a faculty appointment in the Department of Mechanical Engineering at the Petroleum Institute in Abu Dhabi, UAE. Before joining the Petroleum Institute, Dr. Rodrigues was a tenured full-professor in the College of Aviation and was also the program chair for the BS in Safety degree at Embry-Riddle Aeronautical University (ERAU) in Daytona Beach, Florida. Before joining ERAU, Dr. Rodrigues was on the safety sciences' faculty at the Indiana University of Pennsylvania (IUP) and was also an OSHA consultant for the State of Pennsylvania. Prior to joining IUP, Dr. Rodrigues was a worldwide engineering manager at the Campbell Soup Company for 7 years. While at Campbell Soup, he also held an adjunct faculty appointment at The University of Pennsylvania's Systems Engineering department. Dr. Rodrigues has consulted for industry and for government, has made several national and international presentations, and has authored or coauthored numerous technical publications. Dr. Rodrigues holds a Ph.D. in industrial engineering and a master's degree in civil engineering from Texas A&M University. He also holds a bachelor's degree in civil engineering, and his master's degree is in environmental engineering from the Indian Institute of Technology (IIT), Mumbai, India.

STEPHEN K. CUSICK, J.D., is an Associate Professor for the College of Aeronautics at the Florida Institute of Technology in Melbourne, Florida, and a member of the university graduate faculty for its Masters Degree program in Applied Aviation Safety. Before joining Florida Tech, Dr. Cusick was a corporate attorney for Harris Corporation, Intersil Corporation, and DRS Technologies. Following law school, he served as an attorney with the Navy Office of General Counsel, Naval Air Systems Command. A Naval Aviator, he retired in the rank of Captain, USN, and is an active commercial pilot with multiengine, instrument and helicopter FAA ratings. He is a member of the Florida Bar and serves on its Aviation Law Committee. Dr. Cusick holds a Juris Doctor degree and a bachelor of science degree from the University of Louisville, and a master's of public administration degree from the University of West Florida.

COMMERCIAL AVIATION SAFETY

CLARENCE C. RODRIGUES
STEPHEN K. CUSICK

FIFTH EDITION

New York   Chicago   San Francisco   Lisbon   London   Madrid
Mexico City   Milan   New Delhi   San Juan   Seoul
Singapore   Sydney   Toronto

The **McGraw·Hill** Companies

Cataloging-in-Publication Data is on file with the Library of Congress

McGraw-Hill books are available at special quantity discounts to use as premiums and sales promotions, or for use in corporate training programs. To contact a representative please e-mail us at bulksales@mcgraw-hill.com.

Commercial Aviation Safety, Fifth Edition

1 2 3 4 5 6 7 8 9 0   DOC/DOC   1 9 8 7 6 5 4 3 2 1

ISBN 978-0-07-176305-9
MHID 0-07-176305-8

This book is printed on acid-free paper.

Sponsoring Editor
Larry S. Hager

Proofreader
Manisha Sinha

Acquisitions Coordinator
Bridget Thoreson

Indexer
Robert Swanson

Editorial Supervisor
David E. Fogarty

Production Supervisor
Richard C. Ruzycka

Project Manager
Neha Rathor, Neuetype

Composition
Neuetype

Copy Editor
Monoleena Misra

Art Director, Cover
Jeff Weeks

CONTENTS

CHAPTER THREE. THE NATIONAL TRANSPORTATION SAFETY BOARD

61

CHAPTER FOUR. RECORDING AND REPORTING OF SAFETY DATA

83

CHAPTER FIVE. REVIEW OF SAFETY STATISTICS

109

CHAPTER SIX. ACCIDENT CAUSATION MODELS

CHAPTER SEVEN. HUMAN FACTORS IN AVIATION SAFETY

CHAPTER TEN. AIRPORT SAFETY   243

CHAPTER ELEVEN. AVIATION SECURITY   269

CHAPTER TWELVE. AIRLINE SAFETY   301

CHAPTER THIRTEEN. AVIATION SAFETY MANAGEMENT SYSTEMS

333

PREFACE

This text is dedicated to studying the principles and regulatory practices of commercial aviation safety in the United States and worldwide community in the twenty-first century. In addition to being a major rewrite of previous editions, this fifth edition introduces coauthor Dr. Stephen Cusick, who is a member of Florida Institute of Technology's College of Aeronautics faculty in Melbourne, Florida.

This edition updates, revises, and makes current the aviation safety and security information contained in previous editions; establishes new changes in the format and content of the chapters to make the flow of information progressive and logical; and broadens the field of study to include regulatory information on ICAO and Safety Management Systems (SMS) that is essential to the practicing aviation safety professional.

Today's aviation safety practitioner has to contend with more than just the safety dictates of the FAA, NTSB, OSHA, and EPA. Commercial aviation safety is an international subject, heavily regulated by these agencies as well as their international counterparts, including the Standards and Recommended Practices of ICAO. The introduction of SMS principles by ICAO throughout the world has created a paradigm shift in the study of aviation safety which is explored in this book.

This text is intended for individuals with a limited background in the subject and serves as a foundation for further study in the field of aviation safety and security. This book will benefit students enrolled in aviation safety and/or management programs and college or university flight programs. This text also serves as a resource for the practicing safety professional. Course instructors should readily be able to supplement text topics based on specific course needs, as several references are provided throughout the text.

FEATURES OF THE NEW FIFTH EDITION

- Chapter 1, covering the regulatory framework, has been expanded to include the background and chronology of key ICAO developments and discuss how it fits into the international regulatory structure. Additional information on the importance of the Colgan air crash and aftermath, as well as the NextGen Air Transportation System, has been added.

- Chapter 2 discusses the organizational structure and the rulemaking process of ICAO, FAA, OSHA, and EPA. All relevant information from previous editions has been updated. The organizational structure of various departments within

ICAO and FAA, and their missions and responsibilities, has been added. Additional information on the organizational structure, rulemaking process, and selected regulatory issues of OSHA and EPA has been updated in this edition.

- Chapter 3 discusses the structure, function, and workings of the NTSB; and ICAO's role in the international accident investigation process. It was deemed appropriate to review the organizations involved in safety before any discussion on safety measures and statistics occurred. To this chapter was added information on NTSB's most wanted aviation safety improvements.

- Chapter 4 deals with the reporting and recording of safety data. The ICAO five basic traits of an effective safety reporting system have been added. Information from previous editions has been revised and updated, and additional information on LOSA and AQP has been added. Also, ICAO and NTSB accident and incident definitions are reviewed here. Finally, information on OSHA and EPA accident and incident definitions and reporting requirements has been updated.

- Chapter 5 of this edition contains revised and updated information on safety statistics and accidents of major U.S. and international aviation operators from 1980 until 2010. Safety statistics of accident rates, fatalities, and other information on commercial jet accidents as compiled by ICAO, NTSB, and the Boeing Company are included. A discussion on occupational accidents from the U.S. Bureau of Labor Statistics has been added to this chapter.

- Chapter 6 on accident causation models reformulates the information on this topic from previous editions. Information on Dr. James Reason's "Swiss Cheese" model of accident causation has been expanded and updated from the latest ICAO safety documentation. The Shell model, another widely used conceptual tool, has been added.

- Chapter 7 on human factors has been substantially revised to focus on the latest concepts in this critical area. The Human Factors Analysis and Classification System is introduced, followed by modern FAA aeronautical decision making and human error management techniques such as CRM, LOA, and TEM. The section on control strategies to manage threats and errors has been revised and updated.

- Chapter 8 on Air Traffic Control (ATC) safety systems has been completely revised. The National Air Space System evolution to a GPS-based Next Generation Air Transportation System (NextGen) has been added to reflect modern developments in ATC safety.

- Chapter 9 on aircraft safety systems has been revised to include recent developments in jet engine design and new cockpit enhancements from Boeing and Airbus on their latest aircraft models. Information on glass cockpit design and crew alerting systems has been revised and updated.

- Airport safety regulations have been revised by the FAA, and Chapter 10 of this edition updates the airport inspection and compliance requirements of FAR Part 139. Likewise, the information on runway incursions has been changed to reflect recent ICAO and FAA improvements in this important area.

- Aviation security has significantly evolved since 9/11, so Chapter 11 updates current developments in the international and domestic areas. A new section on the 9/11 Commission has been added, and the role of intelligence and counterterrorism has been expanded.

- Airline safety is in the midst of a paradigm shift to Safety Management Systems (SMS) following ICAO's lead and worldwide acceptance of this new safety concept. Chapter 12 has been rewritten to reflect this change, focusing on the four basic components (pillars) of SMS. The chapter concludes with a discussion of the successful implementation of SMS at a major air carrier.

- Chapter 13 on Aviation Safety Management Systems is new. The chapter discusses the evolution of SMS principles, and explains safety risk management and safety assurance as the heart of an effective SMS organization. The chapter concludes with a brief discussion of the future of the SMS process in commercial aviation safety.

- Tables, figures, statistics, key terms, review questions, and references contained in this text have been updated and revised.

- Numerous Web sites have been included to help students and instructors utilize the vast amount of information available on the World Wide Web.

- Each chapter contains a number of features that are designed to facilitate student learning. These features include:
  - *Chapter outlines.* Each chapter opens with an outline of the major topics.
  - *Learning objectives.* The objectives of the chapter are included so students know exactly what is to be accomplished after completing the material.
  - *Relevance.* All examples, applications, and theories are current as of this writing.
  - *Figures and tables.* Figures and tables are drawn from sources such as ICAO, NTSB, Airbus, Boeing, and other current Web sites.
  - *Logical organization and frequent headings.* Itemized bulleted lists are used as frequently as possible to enhance reading.
  - *Key terms.* Each chapter concludes with a list of key terms used in the test.
  - *Review questions.* Review questions at the end of each chapter cover all of the important points.
  - *References.* A list of references is included at the end of each chapter for students who wish to pursue the material in greater depth.

ACKNOWLEDGMENTS

Publication of a milestone such as the Fifth Edition of a textbook that has endured over two decades in such a dynamic area as Commercial Aviation Safety truly humbles the current authors. While acknowledgments are too numerous to mention, we wish to express our sincere thanks and gratitude to Dr. Alexander T. Wells, who was the original source of inspiration and contribution to the text. Dr. Wells, who is Emeritus Professor of Aviation at Embry-Riddle Aeronautical University, is a prolific writer whose vision and contributions to the field of aviation will remain a lasting legacy for generations to come. As always, we are sincerely appreciative of the many public and private institutions that have provided resource material for this edition. We are particularly indebted to the International Civil Aviation Organization, Federal Aviation Administration, National Transportation Safety Board, Department of Homeland Security, Occupational Safety and Health Administration, Environmental Protection Agency, US Airways, the Boeing Company, and other industry partners for their numerous publications. A special thanks for the dedicated support from faculty, administration, and support staff of the Florida Institute of Technology and The Petroleum Institute in Abu Dhabi, UAE.

Every book has one individual who anchors the team through meticulous professional preparation of the manuscript, figures and tables. We have been truly fortunate to have Kim Wise of Jacksonville, Florida, as our outstanding executive assistant working with the authors and publisher. We are also grateful to the editor(s) at McGraw-Hill for their contribution, especially Larry Hager, Bridget Thoreson, David Fogarty, and Pamela Pelton.

Finally, we gratefully thank our families for their patience and encouragement in this project, especially our wives Nicola and Jean, who made this effort possible.

CLARENCE RODRIGUES, PH.D.
The Petroleum Institute

STEPHEN K. CUSICK, J.D.
Florida Institute of Technology

THE REGULATORY FRAMEWORK

LEARNING OBJECTIVES

After completing this chapter, you should be able to

- Describe some of the key ICAO developments that helped shape international aviation.
- Identify some of the important results of the Chicago Convention of 1944.
- Describe some of the early federal legislation that helped shape the airline industry in its formative years.
- Discuss some of the factors that led to the passage of the Federal Aviation Act of 1958.

- Identify some of the safety provisions of the FAA Act of 1958.
- Recognize the important legislation that followed airline deregulation in 1978.
- Distinguish between FAR Part 121 and 135 air carriers.
- Discuss the importance of the Colgan Air crash and its aftermath.
- Describe the important features of the Airline Safety and FAA Act of 2010.
- Discuss the evolution of EPA.
- List and discuss major EPA laws that are of importance to aviation operations.
- Discuss the evolution of OSHA.
- List and discuss major OSHA standards that are of importance to aviation operations.

This chapter reviews the development and regulatory framework of the four major safety-related agencies that regulate today's commercial aviation sector. The International Civil Aviation Organization is discussed first, followed by the Federal Aviation Administration next, followed by the Environmental Protection Agency and the Occupational Safety and Health Administration.

THE INTERNATIONAL CIVIL AVIATION ORGANIZATION (ICAO)

OVERVIEW

The International Civil Aviation Organization is a United Nations specialized agency, which serves as the global forum for international civil aviation. The ICAO was established by the Chicago Conference in December 1944, and was originally known as the Provisional International Civil Aviation Organization (PICAO). It underwent a name change to ICAO in 1947, and Montreal, Canada, became its permanent headquarters. The ICAO's vision as mentioned on its Web site is to achieve "safe, secure, and sustainable development of civil aviation through cooperation amongst its member states." To implement this vision, ICAO has the following strategic objectives for 2005–2010:

- *Safety*. Enhance global civil aviation safety
- *Security*. Enhance global civil aviation security
- *Environmental Protection*. Minimize the adverse effect of global aviation on the environment
- *Efficiency*. Enhance the efficiency of aviation operations
- *Continuity*. Maintain the continuity of aviation operations
- *Rule of Law*. Strengthen law governing international civil aviation

BACKGROUND

The ICAO set forth aims and objectives at the Chicago Convention of 1944 to develop "the principles and techniques of international air navigation and to foster the planning and development of international air transport." These aims and objectives are reproduced from the ICAO Web site from Article 44 of the convention to

- Ensure the safe and orderly growth of international civil aviation throughout the world
- Encourage the arts of aircraft design and operation for peaceful purposes
- Encourage the development of airways, airports, and air navigation facilities for international civil aviation
- Meet the needs of the peoples of the world for safe, regular, efficient, and economical air transport
- Prevent economic waste caused by unreasonable competition
- Ensure that the rights of Contracting States are fully respected and that every Contracting State has a fair opportunity to operate international airlines
- Avoid discrimination between Contracting States
- Promote safety of flight in international air navigation
- Promote generally the development of all aspects of international civil aeronautics

CHRONOLOGY OF KEY ICAO DEVELOPMENTS

The roots of ICAO date back before World War I to a period shortly after the Wright Brothers' historic flight in December 1903. The development of ICAO and its history is well-documented on the ICAO Web site; therefore, only the major milestones will be discussed here.

THE PARIS CONVENTION OF 1910.   At the invitation of France, the first important conference to establish an international air law code was convened in Paris in 1910. Eighteen states from Europe attended to begin the long path of this international aviation body.

The Paris Convention of 1919 establishes the International Commission for Air Navigation (ICAN). Of course World War I completely changed the development path of aviation, so it is not surprising that international aviation was on the agenda in Versailles, France, at the proceedings to end the war in 1919. Once again at the invitation of France, the postwar Allied powers convened in Paris to deal with the technical, operational, and organizational aspect of civil aviation. This 1919 Paris Convention was ultimately ratified by 38 states and consisted of 43 articles that served as the major starting point for the regulation of air navigation. This marked the beginning of the International Commission for Air Navigation, which remained headquartered in Paris for the next 25 years.

CONVENTIONS AND REGIONALISM BETWEEN THE WORLD WARS. Although ICAN had great prospects for aviation global development and cooperation, international meetings between the world wars focused on regional, not global issues. These international meetings focused on various political and regional differences with the *Paris Convention* of 1910 (ICAN). Major meetings and their topics during this period are as follows:

1926 The Madrid Convention (Equality of States)

1927 The Hague Conference (First Air Mail provision)

1928 The Havana Convention (Mutual Freedom of Air Passage)

1928 The International Civil Aeronautics Conference, Washington, D.C. (Celebration—25 years after Kitty Hawk)

1929 Warsaw Convention (International Liability)

1937 Pan American Conference—Lima (Final meeting of series of Pan American conferences held in the United States and South America 1916–1937)

THE CHICAGO CONVENTION OF 1944—THE BIRTH OF THE ICAO. Toward the end of World War II, the leaders of the United States and Great Britain began studies of probable postwar civil aviation problems with a view toward promotion of economic development to heal the wounds of war. These studies resulted in U.S. invitations to 55 states to attend an International Civil Aviation Conference in Chicago. One of the goals of this conference was to obtain uniformity in international regulations and standards that was lacking under the split between ICAN and the Havana Convention. Important work was accomplished in the technical field because the Chicago Convention paved the way for a common air navigation system throughout the world. The major accomplishments of the Chicago Convention include the following:

- The Convention on International Civil Aviation provided the basis for a complete modernization of the basic public international law of the air.

- Twelve technical annexes were drafted to cover the technical and operational aspects of international civil aviation such as airworthiness of aircraft, air traffic control, telecommunications, and air navigation services.

- Regions and regional offices were established in specific areas where operating conditional and other relevant parameters were comparable.

- The Provisional International Civil Aviation Organization (PICAO) was established until the permanent ICAO organization came into force in April 1947.

- ICAN was dissolved when the Chicago Convention came into force with ICAO, and the Havana Convention was superceded. At last, the international community of states had adopted a single unifying body.

THE GROWTH OF ICAO—POST WORLD WAR II DEVELOPMENTS AND NEW CHALLENGES.   As aircraft grew in size, speed, and dependability following the war, it was clear that air transportation was quickly becoming a major economic force. Under U.S. leadership, civilian and military aircraft aviation operations were blended into a common system of air navigation and communication procedures. Montreal, Canada, was chosen as the new ICAO headquarters and Dr. Edward Warner of the United States was chosen as ICAO President. He served in this capacity for 10 years.

During this postwar period, several major technical decisions were reached regarding air navigation and communication matters, such as

- The standard aid for aircraft approaches and landing was the Instrument Landing System (ILS).
- The standard short-range navigation aids were chosen as the VHF omni-directional radio (VOR) supplemented with Distance Measuring Equipment (DME); while standard long-range navigation was chosen as the Long Range Air Navigation (LORAN) radio system.
- VHF voice communication was chosen as the primary means of air to ground communication.
- Air traffic corridors were adopted to ensure safe aircraft separation. This concept would evolve into our modern airway system.
- As the speed of jet aircraft increased, more air traffic cooperation was required between states and ICAO regional offices. Under the new ICAO organization, congested air routes were efficiently managed.

ICAO—FUTURE SAFETY CHALLENGES.   Today ICAO has grown to encompass 190 countries on a global basis. At its 37th triennial assembly held in Montreal in October 2010, ICAO has listed the following safety challenges:

- To proactively improve safety in a complex operating environment with a wide range of technologies, from older and latest generation (NextGen) aircraft flying in the same airspace to the progressive introduction of remote-controlled airborne vehicles
- To focus on these regions of the world with the highest level of safety risks

ICAO safety strategy will be based on the following three areas considered by the assembly:

- Transparency and sharing of safety information
- Greater involvement of regional safety organization
- Increased cooperation between regulators and industry stakeholders

THE FEDERAL AVIATION ADMINISTRATION (FAA)

OVERVIEW

The Federal Aviation Administration is a U.S. government agency with primary responsibility for the safety of civil aviation. The FAA was established by the *Federal Aviation Act* of 1958 and was originally known as the Federal Aviation Agency. It underwent a name change (Agency changed to Administration) in 1967 when the FAA became part of the Department of Transportation. The FAA's major roles and responsibilities as stated on its Web site include

- Regulating civil aviation to promote safety
- Encouraging and developing civil aeronautics, including new aviation technology
- Developing and operating a common system of air traffic control (ATC) and navigation for both civil and military aircraft
- Researching and developing the National Airspace System and civil aeronautics
- Developing and carrying out programs to control aircraft noise and other environmental effects of civil aviation
- Regulating U.S. commercial space transportation

BACKGROUND

The FAA is responsible for performing several activities that are in support of its above-mentioned functions. These activities as reproduced from the FAA Web site include

- *Safety regulation.* The FAA issues and enforces regulations and minimum standards covering the manufacture, operation, and maintenance of aircraft. The agency is responsible for the rating and certification of airmen and for certification of airports serving air carriers.
- *Airspace and air traffic management.* The safe and efficient utilization of the navigable airspace is a primary objective of the FAA. The agency operates a network of airport towers, air route traffic control centers, and flight service stations. It develops air traffic rules, assigns the use of airspace, and controls.
- *Air navigation facilities.* The FAA is responsible for the construction and installation of visual and electronic aids to air navigation, and for the maintenance, operation, and quality assurance of these facilities. Other systems maintained in support of air navigation and air traffic control include voice/data communications equipment, radar facilities, computer systems, and visual display equipment at flight service stations.
- *Civil aviation abroad.* The FAA promotes aviation safety and encourages civil aviation abroad. Activities include exchanging aeronautical information with

foreign authorities; certifying foreign repair shops, airmen, and mechanics; providing technical aid and training; negotiating bilateral airworthiness agreements; and taking part in international conferences.

- *Commercial space transportation.* The agency regulates and encourages the U.S. commercial space transportation industry. It licenses commercial space launch facilities and private sector launching of space payloads on expendable launch vehicles.

- *Research, engineering, and development.* The FAA does research on and develops the systems and procedures needed for a safe and efficient system of air navigation and air traffic control. The agency supports development of improved aircraft, engines, and equipment. It also does aeromedical research and conducts tests and evaluations of specified items such as aviation systems, devices, materials, and procedures.

- *Other programs.* The FAA provides a system for registering aircraft and recording documents affecting title or interest in aircraft and their components. Among other activities, the agency administers an aviation insurance program; develops specifications for aeronautical charts; and publishes information on airways air-port services, and other technical subjects in aeronautics.

CHRONOLOGY OF MAJOR FAA REGULATIONS

The roots of today's aviation safety programs extend back to the early days of commercial aviation following World War I. Many returning pilots bought surplus war aircraft and went into business. These happy-go-lucky barnstormers toured the country, putting on shows and giving rides to local townsfolk. By the mid-1920s, uses of aircraft included advertising, aerial photography, crop dusting, and carrying illegal shipments of liquor during prohibition. Initial efforts to establish scheduled passenger service were short-lived, as service catered primarily to wealthy east coast tourists. This service was expensive compared to the country's well-developed rail and water travel networks.

AIR MAIL SERVICE. Growth of commercial aviation was greatly stimulated by the establishment of the U.S. Air Mail Service in the early 1920s. Regulations established by the Post Office Department required its pilots to be tested and to have at least 500 hours of flying experience. The Post Office set up aircraft inspection and preventive maintenance programs for the pilots. These early regulatory requirements improved air mail carrier safety. In 1924, commercial flyers experienced one fatality every 13,500 miles, while the Air Mail Service had one fatality every 463,000 miles.

In 1925, Congress enacted the *Air Mail Act* of 1925, authorizing the Post Office Department to transfer air mail service to private operators. Twelve carriers, some of which evolved into today's major airlines, began air mail operations in 1926 and 1927. These carriers offered limited passenger service, which was much less profitable than carrying mail. Initially, air mail contractors were paid a percentage of postage revenues. In 1926, however, an amendment to the Air Mail Act required payment by weight carried. Small independent operators, using Ford and Fokker trimotor airplanes, handled most of the passenger service in the late 1920s, the forerunners of today's commuter airlines and air taxis.

EARLY SAFETY LEGISLATION.   No federal safety program existed, which prompted a number of states to pass legislation requiring aircraft licensing and registration. In addition, local governments enacted ordinances regulating flight operations and pilots, creating a patchwork of safety-related requirements and layers of authority. Despite strong industry support for federal legislation, Congress was unable to reach agreement on the scope and substance of a statute until 1926, when the *Air Commerce Act* was passed. Key issues debated by Congress included whether to separate military and civil aviation activities, what responsibilities should be left to state and local governments, and how to provide federal support for airports.

The new law gave the Department of Commerce regulatory authority over commercial aviation and responsibilities aimed at promoting the fledgling industry. The major provisions of the act authorized the regulation of aircraft and pilots in interstate and foreign commerce; provided federal support for charting and lighting airways, maintaining emergency fields, and making weather information available to pilots; authorized aeronautical research and development programs; and provided for the investigation of aviation accidents. Local governments were left with jurisdiction over airport control.

Within the Department of Commerce, a new Aeronautics Branch, comprising existing offices already engaged in aviation activities, was formed to oversee the implementation of the new law. Nine district offices of the Regulatory Division of the Aeronautics Branch were established to conduct inspections and checks of aircraft, pilots, mechanics, and facilities. District offices also shared licensing and certification responsibilities with the Washington, D.C., office. The basic allocation of responsibilities survives to this day, although the Department of Commerce responsibilities now rest with the Department of Transportation (DOT) and its branch, the FAA.

The first set of regulations was drafted with substantial input from aircraft manufacturers, air transport operators, and the insurance industry. Compared with current standards, pilot requirements were minimal. In addition to written and flight tests, transport pilots were required to have 100 hours of solo flight experience, while industrial pilots needed only 50 hours. Current procedures for certifying aircraft and engines also originated under these early regulatory programs. Aircraft manufacturers were required to comply with minimum engineering standards issued by the Department of Commerce in 1927, and one aircraft of each type was subject to flight testing to obtain an airworthiness certificate.

The Aeronautics Branch also collected and analyzed data from aircraft inspection reports, pilot records, and accident investigations. Data were made accessible to the insurance industry, which allowed for the development of actuarial statistics. A direct consequence of this step was a significant reduction in insurance rates for many carriers. However, the Department of Commerce, cognizant of its role to promote the aviation industry, was reluctant to make public disclosures about the results of individual accident investigations despite a provision in the Air Commerce Act directing it to do so. Eventually, in 1934, the Air Commerce Act was amended, giving the Secretary of Commerce extensive powers to investigate accidents, including a mandate to issue public reports of its findings. This congressional policy put safety considerations ahead of protecting the industry's image.

As additional regulations to improve safety were implemented, accidents involving passenger carriers and private aircraft decreased significantly. Between 1930 and 1932, the fatality rate per 100 million passenger-miles declined by 50 percent. Updated regulations established more stringent requirements for pilots flying aircraft in scheduled interstate passenger service, including flight-time limitations.

Pilots were restricted to flying 100 hours per month, 1000 hours during any 12-month period, 30 hours for any 7-day period, and 8 hours for any 24-hour period. A 24-hour rest period was also required for every 7-day period. These requirements, which were established in 1934 and are virtually the same today, upgraded earlier restrictions that limited pilots to 110 hours of flight time per month. In addition, a waiver of the 8-hour limitation for a 24-hour period could be granted by the Department of Commerce. The 8-hour waiver rule was ultimately eliminated following a fatal accident involving a pilot who had exceeded 8 hours of flight and pressure from the Air Line Pilots Association.

Other requirements specified the composition of flightcrews, established standards for flight schools, improved takeoff and landing procedures, set minimum flight altitudes and weather restrictions, and required multiengine aircraft to be capable of flying with one inoperative engine. In addition, certification of carriers providing scheduled passenger service in interstate commerce began in 1930.

EARLY ECONOMIC LEGISLATION. During the 1930s, industry expansion and the development of aircraft and communication technologies required continuous improvements of regulations, airways, and airports. However, budget constraints prevented the Department of Commerce from conducting sufficient inspections and keeping up with airway development needs. Moreover, a series of fatal accidents in 1935, 1936, and 1937, including one in New Mexico that killed a New Mexico senator, called into question the adequacy of existing regulations. The fatality rate rose from 4.78 per 100 million passenger-miles in 1935 to 10.1 per 100 million passenger-miles in 1936.

The *Civil Aeronautics Act* of 1938 marked the beginning of economic regulation. It required airlines with or without mail contracts to obtain certificates authorizing service on specified routes if the routes passed a test of public convenience and necessity.

The Civil Aeronautics Act created the *Civil Aeronautics Authority* (CAA), which was responsible for safety programs and economic regulations that included route certificates, airline tariffs, and air mail rates. Within the CAA, a separate Administrator's Office, answering directly to the president, was responsible for civil airways, navigation facilities, and air traffic control. Increasing air traffic between Newark, Cleveland, and Chicago prompted a group of airlines to establish an air traffic control system in 1934. By 1936, however, the Department of Commerce assumed control of the system and issued new regulations for instrument flight.

However, in June 1940, under the Reorganization Act of 1939, the CAA was transferred back to the Department of Commerce. The *Civil Aeronautics Board (CAB)* was created. The CAB was responsible for regulatory and investigatory matters.

Federal responsibilities for airway and airport development grew tremendously during World War II, leading to passage of the *Federal Airport Act* of 1946. Federal financial assistance to states and municipalities was also initiated at this time. The federal government assumed responsibility for ATC. However, the inspector force could not keep pace with the rapidly increasing numbers of new airplanes, pilots, and aviation-related facilities. As early as 1940, the CAA had designated certain parts of the certification process to industry. For example, flight instructors were permitted to certificate pilots, and a certificated airplane repaired by an approved mechanic could fly for 30 days until it was checked by an available CAA inspector. After the war, the CAA limited its aircraft certification and inspection role to planes, engines, and propellers. Manufacturers were responsible for ensuring that other aircraft parts met CAA standards.

Regulatory and organizational changes also took place during and after the war. Regional offices of the CAA, reduced in number to seven in 1938, became more autonomous in 1945. Regional officials became directly responsible for operations in their regions, although technical standards and policies were still developed in Washington, D.C. Except for a brief return to more centralized management in the late 1950s, regional autonomy with the FAA has persisted to this day.

Fatal crashes in the late 1940s and early 1950s prompted revised standards setting minimum acceptable performance requirements that were designed to ensure continued safe flight and landing in the event of failure of key aircraft components. These standards also distinguished small and large airplanes based on existing airplane and power plant design considerations. Small airplanes were those with a maximum certificated takeoff weight of 12,500 pounds or less; airplanes above 12,500 pounds were defined as large. This distinction is still applied by the FAA today despite significant changes in aircraft design.

INDUSTRY GROWTH AFTER WORLD WAR II.   Surplus war transport airplanes and a new supply of pilots led to the development of the nonscheduled operator, or air taxi. Exempt from economic regulation by the Civil Aeronautics Act of 1938, these operators transported people or property over short distances in small airplanes, often to locations not serviced by the certificated airlines. The CAA, at the time sympathetic to private and small operators, applied less-stringent safety regulations to air taxis. In 1952, exemption from economic regulation became permanent, even for carriers using small aircraft to provide scheduled service.

The decade following World War II witnessed enormous industry growth. Pressurized aircraft traveling at greater speeds and carrying more passengers were introduced. Initially, Lockheed produced the Constellation, which carried 60 passengers and was 70 miles per hour faster than the DC-4. To compete with Lockheed, Douglas developed the DC-6. Subsequently, upgraded versions of each aircraft, the DC-7 and the Super Constellation, were introduced.

In addition to scheduled passenger service, air freight operations expanded when the CAB granted temporary certificates of public convenience and necessity to four all-cargo airlines in 1949. The four carriers were Air News, Flying Tigers, Slick, and U.S. Airlines. Certification and operating rules for commercial operators— those offering air service for compensation or hire—were also adopted in 1949.

However, despite continuing increases in air traffic and the need for better airports to accommodate larger and faster aircraft, federal support for ATC facilities, airport development, and airway modernization was insufficient. The CAA, faced with budget reductions in the early 1950s, was forced to abandon control towers in 18 small cities and numerous communications facilities, postpone jet development and navigation improvements, and curtail research efforts. In addition, the number of CAA regional offices was reduced from seven to four, 13 safety inspection field offices were eliminated, and the industry designee program was expanded.

THE FEDERAL AVIATION AGENCY.   The impending introduction of jet aircraft and a 1956 midair collision over the Grand Canyon involving a DC-7 and a Super Constellation helped promote congressional authorization of increased levels of safety-related research and more federal inspectors. In 1958, Congress passed the *Federal Aviation Act,* which established a new aviation organization, the Federal Aviation Agency. Assuming many of the duties and functions of the CAA and the CAB, the agency was responsible for fostering air commerce, regulating safety, all future ATC and navigation systems, and airspace allocation and policy. The CAB was continued as a separate agency responsible for economic regulation and accident investigations. However, the Federal Aviation Agency Administrator was authorized to play an appropriate role in accident investigations. In practice, the Federal Aviation Agency routinely checked into accidents for rule violations, equipment failures, and pilot errors. Moreover, the Civil Aeronautics Board delegated the responsibility to investigate nonfatal accidents involving fixed-wing aircraft weighing less than 12,500 pounds to the Federal Aviation Agency.

The safety provisions of the 1958 act, restating earlier aviation statutes, empowered the Agency to promote flight safety of civil aircraft commerce by prescribing

- Minimum standards for the design, materials, workmanship, construction, and performance of aircraft, aircraft engines, propellers, and appliances.

- Reasonable rules and regulations and minimum standards for inspections, servicing, and overhauls of aircraft, aircraft engines, propellers, and appliances, including equipment and facilities used for such activities. The agency was also authorized to specify the timing and manner of inspections, servicing, and overhauls and to allow qualified private persons to conduct examinations and make reports in lieu of agency officers and employees.

- Reasonable rules and regulations governing the reserve supply of aircraft, aircraft engines, propellers, appliances, and aircraft fuel and oil, including fuel and oil supplies carried in flight.
- Reasonable rules and regulations for maximum hours or periods of service of pilots and other employees of air carriers.
- Other reasonable rules, regulations, or minimum standards governing other practices, methods, and procedures necessary to provide adequately for national security and safety of air commerce.

In addition, the act explicitly provided for certification of pilots, aircraft, air carriers, air navigation facilities, flying schools, maintenance and repair facilities, and airports. In the years following creation of the agency, federal safety regulations governing training and equipment were strengthened despite intense opposition from industry organizations. The number of staff members also grew in the early 1960s, and inspection activities were stepped up, including en route pilot checks and reviews of carrier maintenance operations and organizations. The FAA staff grew from 30,000 in 1959 to 40,000 in 1961.

In 1966, the Federal Aviation Agency became the *Federal Aviation Administration* (*FAA*), when it was transferred to the newly formed Department of Transportation (DOT). The *National Transportation Safety Board* (*NTSB*) was also established to determine and report the cause of transportation accidents and conduct special studies related to safety and accident prevention. Accident investigation responsibilities of the CAB were moved to the NTSB.

Renewed support for improvements to airports, ATC, and navigation systems was also provided by the *Airport and Airway Development Act of 1970*. The act established the Airport and Airway Trust Fund, which was financed in part by taxes imposed on airline tickets and aviation fuel. This act has been reauthorized in subsequent years.

AIRLINE DEREGULATION.   Prompted by widespread dissatisfaction with CAB policies and the belief that increased competition would enhance passenger service and reduce commercial airline fares, Congress enacted the *Airline Deregulation Act of 1978*. Congress believed that fares would drop based on the record of intrastate airlines, where fares were 50 to 70 percent of the Civil Aeronautics Board– regulated fares over the same distance. In addition, the Civil Aeronautics Board had already reduced restrictions on fare competition in 1976 and 1977 and allowed more airlines to operate in many city-pair markets.

Specifically, in a 6-year period, the act phased out CAB control over carrier entry and exit, routes, and fares. In 1984, the remaining functions of the CAB were transferred to DOT. These functions include performing carrier fitness evaluations and issuing operating certificates, collecting and disseminating financial data on carriers, and providing consumer protection against unfair and deceptive practices.

During the 60-year history of federal oversight, federal regulatory and safety surveillance functions have been frequently reorganized and redefined. Moreover, public concerns about how the FAA carries out its basic functions have remained remarkably constant despite a steadily improving aviation safety record.

The *Airport and Airway Improvement Act of 1982* reestablished the operation of the Airport and Airway Trust Fund with a slightly revised schedule of user taxes. The act authorized a new capital grant program, called the Airport Improvement Program (AIP). In basic philosophy, the AIP was similar to the previous Airport Development Aid Program (ADAP). It was intended to support a national system of integrated airports that recognizes the role of large and small airports together in a national air transportation system. Maximized joint use of underutilized, nonstrategic U.S. military fields was also encouraged. The 1982 act also contained a provision to make funds available for noise compatibility planning and to carry out noise compatibility programs as authorized by the *Noise Abatement Act* of 1979.

The *Aviation Safety and Capacity Expansion Act of 1990* authorized a passenger facility charge (PFC) program to provide funds to finance airport-related projects that preserve or enhance safety, capacity, or security; reduce noise from an airport that is part of such a system; or furnish opportunities for enhanced competition between or among air carriers by local imposition of a charge per enplaned passenger. This act also established a Military Airport Program for current and former military airfields, which should help improve the capacity of the national transportation system by enhancement of airport and ATC systems in major metropolitan areas.

COMMERCIAL AVIATION EVOLVES. The Commuter Safety Initiative of 1995 (Commuter Rule). An *air carrier* is a commercial operator or company that has been certificated by the FAA under *FAR Part 121* or *FAR Part 135* to provide air transportation of passengers or cargo. These operators possess operations specifications and an air carrier certificate, which is a document that describes the conditions, authorizations, and limitations under which the air carrier operates.

Before December 14, 1995, FAR Part 121 included the regulations that govern air carriers in multiengine airplanes with more than 30 seats or 7500-pound payload. As authorized by the operations specifications, FAR Part 121 operations can be *domestic air carrier* (scheduled passenger service, generally within the United States), *flag air carrier* (scheduled passenger service in international operations), or *supplemental air carrier* (all cargo and charter operations).

As of December 14, 1995, all airplanes with 10 or more passenger seats and all turbojets operated in scheduled passenger service had to operate under FAR Part 121. Commuter operations with 9 or fewer seats and on-demand air taxi airplanes with 30 or fewer seats and all rotorcraft still operated under FAR Part 135. To operate under FAR Part 121, the aircraft had to meet additional standards involving operational and airplane certification and equipment and performance upgrades.

The Commuter Rule required the regional (commuter) air carriers to comply with nearly all FAR Part 121 operational and safety requirements, just like the large, mainstream airlines. During the years after 1995, the regional air carriers grew rapidly using codeshare agreements with their large airline partners to provide one face to the traveling public. Unfortunately, the regional air carriers did not enjoy the same safety record as the major air carriers as shown by future events.

THE NEW CENTURY.   As the FAA began the 21st century, it was quickly apparent that modernization of its organization and infrastructure was required to bring the agency up to date. In April 2000, the *Wendell H. Ford Aviation Investment and Reform Act* for the 21st century had two major provisions. The first created the FAA's Air Traffic Organization (ATO), which provided a new business structure that focused on efficient operation of the air traffic control system under a new ATO Chief Operating Officer. The second major provision of this act set up an FAA Whistleblower Protection Program to protect employees who reported safety problems in the field.

THE NEXT GENERATION AIR TRANSPORTATION SYSTEM (NEXTGEN).   In December 2003, this ambitious new law was signed to create an integrated program to improve the FAA air traffic technical infrastructure which was woefully out of date. Among the goals of the NextGen legislation are

- Improve the level of safety, security, efficiency, quality, and affordability of the National Airspace System and aviation services and
- Take advantage of data from emerging ground-based and space-based communications, navigation, and surveillance technologies.

This latter goal emphasizes use of the Department of Defense created Global Positioning System (GPS) cluster of satellites for navigation. NextGen will be more fully discussed in Chapter 8, Air Traffic Safety Systems.

THE COLGAN AIR CRASH AND ITS AFTERMATH.   In early 2009, a regional commuter flight from Newark Liberty International Airport to Buffalo crashed while on final approach, causing 50 fatalities (Colgan Air Flight 3407). The details of this accident will be further explored in Chapter 5 entitled Review of Safety Statistics. This single accident has had a profound effect on commercial aviation safety though sweeping new legislation passed in its wake.

The Airline Safety and Federal Aviation Administration Act of 2010 passed August 1, 2010, mandates broad changes in Airline Safety and Pilot Training Improvement. Title II of the Act requires the FAA to issue new regulations in a variety of areas:

- Establishment of a comprehensive FAA Pilot Records database to ensure that airlines have access to all relevant pilot training information.

- Establishment of a special FAA Task Force on Air Carrier Safety and pilot training to evaluate best practices in the industry and provide recommendations.

- High-level review of FAR Part 121 air carriers by the Department of Transportation Inspector General.

- Establishment of a flight crewmember mentoring, professional development, and leadership program to emphasize the highest standards of professional performance with special focus on strict compliance with the "sterile cockpit rule" to eliminate nonessential voice communications between crewmembers during the critical phases of flight.

- A study of aviation industry best practices with regard to flight crewmember pairing, crew resource management, and pilot commuting to and from airline hubs from their place of residence.

- Implementation of specific Colgan crash NTSB remedial training recommendations regarding aircraft stall and recovery training, stick pusher, and weather event training including icing conditions, microburst, and wind shear weather events.

- Establishment of a multidisciplinary panel to recommend best practices in FAR Parts 121 and 135 training methods including initial and recurrent testing requirements for pilots; classroom instruction requirements; the best methods to allow specific academic training courses to be credited toward an Airline Transport Pilot Certification, among other priorities.

- Random, onsite safety inspections of Regional Air Carriers by the FAA at least once each year.

- Special emphasis on combating pilot fatigue through specific limitations on the hours of flight and duty time allowed for pilots. Part 121 air carriers must now submit a "Fatigue Risk Management Plan" and update it every 2 years.

- Bolstering of four industry voluntary safety programs in the following areas:
  - Aviation Safety Action Plan (ASAP)
  - Flight Operational Quality Assurance Program (FOQA)
  - Line Operations Safety Audit (LOSA)
  - Advanced Qualification Program (AQP)

- Specific development and implementation of ASAP and FOQA programs is now required by FAR Part 121 air carriers.

- Safety Management Systems (SMS) are now legislatively required for all FAR Part 121 air carriers. This is indeed a revolutionary change which will have a profound effect on the commercial aviation industry. Under this section of the Act, each carrier's SMS program shall consider ASAP, FOQA, LOSA, and AQP concepts.

- Enhanced Air Transport Pilot (ATP) Certificate requirements. Under the act, all flight crewmembers must hold an ATP certification which is the highest pilot rating available under FAA regulations. This requirement is also a major change for the regional (commuter) airlines which previously allowed a pilot with a commercial rating to serve as first officer on their aircraft. The Act provides a default clause which requires an ATP certificate for all FAR Part 121 pilots by 3 years from enactment of this Act (July 31, 2013).

The minimum requirements for a pilot to be qualified to receive an ATP are proposed as at least 1500 flight hours and flight training, academic training, or operations experience to

- Function effectively in a multipilot environment
- Function effectively in adverse weather conditions, including icing
- Function effectively during high altitude operations
- Adhere to the highest professional standards
- Function effectively in an air carrier operational environment

An expert panel has been established to access and recommend credit for specific academic training courses which may enhance safety more than requiring the pilot to fully comply with the 1500 flight hour requirement. The FAA must issue final rules within 3 years of August 1, 2010 to ensure compliance with the sweeping legislation.

THE ENVIRONMENTAL PROTECTION AGENCY (EPA)

OVERVIEW

On December 2, 1970, the *Environmental Protection Agency (EPA)* was established in the executive branch of government as an independent agency pursuant to President Nixon's Reorganization Plan No. 3 of July 9, 1970. The EPA was created to enable coordinated and effective government action on behalf of the environment. The agency strives to abate and control pollution systematically, by integrating a variety of research, monitoring, standard-setting, and enforcement activities. To complement its other activities, EPA coordinates and supports research and antipollution activities by state and local governments, private and public groups, individuals, and educational institutions. The EPA also reinforces efforts among other federal agencies with respect to the impact of their operations on the environment. The EPA is specifically charged with making public its written comments on environmental impact assessment proposals and lobbying against proposal/projects that are unsatisfactory from the standpoint of public health, community welfare, or environmental quality. The EPA's mandate is to serve as the nation's advocate for a quality livable environment.

BACKGROUND

Before the establishment of the EPA there was no single integrated agency to systematically address air, water, and land pollution, an issue that affected the daily lives of every person in the United States. The agencies that existed at that time were segregated primarily along pollution medium domains of air, water, and land. The sources of air, water, and land pollution, however, are interrelated and interchangeable and overlap more than one medium. For example, a single source such as a power generating station may pollute the air with smoke and chemicals, the land with solid wastes, and a river or lake with chemical and other wastes. Also, controlling air pollution may produce solid wastes, which then pollute the land or water. Conversely, treating wastewater effluents may generate solid wastes, which must be disposed of on land or burned in an incinerator, which could cause air pollution problems. Finally, some pollutants such as radiation and pesticides can appear in all three media at once. Controlling this multifaceted problem required the coordinated efforts of a variety of separate agencies and departments which often resulted in ineffective regulatory action. Thus, the need to integrate the various departments into a single entity, the EPA, was established. Per the dictates of Reorganization Plan No. 3, the following agencies' functions were consolidated in the Environmental Protection Agency:

- The National Air Pollution Control Administration which resided within the Department of Health, Education, and Welfare.
- The Department of the Interior's Federal Water Quality Administration.
- The Council on Environmental Quality, which has the authority to perform studies relating to ecological systems.
- The Food and Drug Administration's pesticides division, which was part of the Department of Health, Education, and Welfare.
- The Department of the Interior's pesticide studies' division.
- The Department of Agriculture's Research Service group dealing with pesticides registration and related activities.
- The Bureau of Solid Waste Management and the Bureau of Water Hygiene. Also included here were duties carried out by the Bureau of Radiological Health of the Environmental Control Administration, which was part of the Department of Health, Education, and Welfare.
- The Atomic Energy Commission and the Federal Radiation Council's radiation criteria and standards division.

CHRONOLOGY OF MAJOR ENVIRONMENTAL LAWS AFFECTING AVIATION OPERATIONS

NATIONAL ENVIRONMENTAL POLICY ACT (NEPA).   The *National Environmental Policy Act of 1969* was significant in that it was among the first laws to establish a broad national framework for protecting the environment. NEPA's fundamental policy was to make certain that all branches of government gave careful consideration to the environment before any major federal undertaking such as airports, military complexes, highways, parkland purchases, and other federal activities that are proposed. The Act requires *Environmental Assessments* (*EAs*) and *Environmental Impact Statements* (*EISs*) on the impact of all major undertakings and alternative courses of action on the environment.

CLEAN AIR ACT (CAA).   *The Clean Air Act of 1970* now 40 years old, regulates air emissions from area, stationary, and mobile sources. By virtue of this law the Environmental Protection Agency is authorized to establish National Ambient Air Quality Standards (NAAQSs) to protect public health and the environment. The goal of the Act was to achieve NAAQSs in every state in the nation by 1975. The Act set upper limits for pollution levels and at the same time required the states to develop specific plans (also called SIPs) for controlling pollution from each industrial pollution source within the state. The act was amended in 1977 to give the states more time to comply with the NAAQS requirements. In 1990, the Clean Air Act was amended again to address special pollution problems of acid rain, ground-level ozone, stratospheric ozone depletion, and air toxics. Sample aviation operations coming under the jurisdiction of this law would include exhaust emissions of smoke from aircraft engines and venting of fuel emissions into the atmosphere.

In recent years, global warming has become a significant issue, and aircraft emissions have become important to the environmental movement. Aircraft are producers of "greenhouse gases," i.e., compounds that trap the sun's heat within the earth's climate. The EPA has existing authority to regulate greenhouse gas emissions under the Clean Air Act, or Congress could address aviation legislation through cap-and-trade or carbon tax proposals. Foreign countries, particularly the European Union, may attempt to control aviation emissions as well. In this era of "green energy" production, aviation greenhouse gas emissions will remain an issue for the foreseeable future.

CLEAN WATER ACT (CWA).   First established as the Federal Water Pollution Control Act Amendments of 1972 to address public concerns for controlling water pollution, this act was amended in 1977 and came to be known as the *Clean Water Act*. The Act gave the EPA the authority to implement pollution control programs such as setting wastewater standards for industry and water quality standards for surface waters. Under the Act, it was unlawful for anyone to discharge pollutants into navigable waters, unless a permit was obtained. It also established grants to fund the construction of sewage treatment plants. In 1981 the municipal construction grants process was streamlined, and treatment plants were required to have enhanced capabilities if funded under the program. In 1987 the funding strategy changed to build on the partnerships developed between EPA and the state. This caused the construction grants program to be phased out and replaced with the State Water Pollution Control Revolving Fund, more commonly known as the Clean Water State Revolving Fund.

RESOURCE CONSERVATION AND RECOVERY ACT (RCRA).   This Act by far is the most far-reaching of all and is of major importance to the aviation industry. By virtue of the RCRA of 1976, the EPA has the authority to control hazardous waste from "cradle to grave." This cradle-to-grave approach governs all phases of the waste from generation and transportation through treatment, storage, and disposal. Hazardous wastes can be solids, liquids, or contained gaseous materials that could pollute the environment. For the purposes of this regulation, wastes are categorized into two groups: listed wastes and characteristics wastes. Over 400 specific substances are regulated under the "listed" category. Wastes that are ignitable, corrosive, reactive, or toxic are regulated under the "characteristics" category. Sources of hazardous wastes in aviation operations include these:

- Painting, degreasing, and cleaning of aircraft generate paint wastes, phenols, organic solvents, acids, and alkalis
- Plating, stripping, rust prevention, and stain removal generate cyanides, chromium, and other toxic metals
- Spills and leaks from fuel systems and storage tanks generate fuels, oils, and grease
- Spent or leaking batteries from aircraft, ATC tower backup, and other power supply sources generate toxic (lead, lithium, nickel, and cadmium) and reactive (acid) wastes
- Miscellaneous wastes include glycol used for deicing and other detergents

RCRA focuses only on active and future facilities and does not address abandoned or historical sites. Historical and abandoned sites are addressed by CERCLA.

HAZARDOUS AND SOLID WASTE AMENDMENTS (HSWA).   The federal *Hazardous and Solid Waste Amendments of 1984* added major changes to RCRA. These amendments required the phasing out of land disposal of hazardous waste, tightened restrictions on waste recycling and pollutant releases from old abandoned facilities, increased enforcement authority for EPA, and required more stringent hazardous waste management standards. HSWA, under Subtitle I of RCRA, also comprehensively addressed environmental issues of underground storage tanks (USTs) that stored petroleum and other hazardous substances. The greatest potential hazard of USTs is that their contents could leak, seep into the soil, and contaminate groundwater. Leaking USTs can cause fire and explosions in addition to posing other health and environmental risks. HSWA also addressed management of hazardous wastes in aboveground storage tanks and their transportation. In addition, RCRA references the Department of Transportation's Hazardous Materials Transportation Act (HMTA), which governs the packaging, labeling, and transportation of hazardous materials by air, water, rail, or highway. Fuel storage facilities and ramp fueling operations are sample aviation operations governed by this regulation.

TOXIC SUBSTANCES CONTROL ACT (TSCA).   In 1976, the *Toxic Substances Control Act* was enacted by Congress to give the EPA the ability to track industrial chemicals produced or imported into the United States and to ban their manufacture and import if those chemicals pose an unreasonable risk. In addition, EPA incorporated the mandate and infrastructure to track the thousands of new chemicals developed by industry each year that may have unknown or dangerous characteristics. With this ability the EPA could monitor and control these chemicals as needed to protect human health and the environment. TSCA supplements the Clean Air Act and the Toxic Release Inventory under EPCRA. Aircraft manufacturing and assembly processes, discharges, and effluents are controlled by this regulation.

COMPREHENSIVE ENVIRONMENTAL RESPONSE, COMPENSATION, AND LIABILITY ACT (CERCLA) OR "SUPERFUND." *Comprehensive Environmental Response, Compensation, and Liability Act* of 1980 gave to federal authority broad powers to respond directly to releases or potential releases of hazardous chemicals that posed risks to public health or the environment. The Act established requirements concerning closed and abandoned hazardous waste sites and provided for liability of individuals responsible for contaminating sites with hazardous wastes. The Act also gave EPA the power to seek out the parties that were responsible for the pollution and force them to clean up. Finally, this Act created a tax on the chemical and petroleum industries to generate funds to provide for cleanup when a responsible party could not be identified. Hence this Act is also known as the *Superfund*. This fund generated about $1.6 billion in its first 5 years of existence and was mainly used to cleanup abandoned hazardous waste sites. Sample applications of this law would be to aircraft burial (retirement) sites and old (discontinued) fuel dumping/storage facilities involving underground fuel storage.

SUPERFUND AMENDMENTS AND REAUTHORIZATION ACT (SARA). As a result of the complex administering requirements of Superfund, changes and additions were made to the program which resulted in the enactment of the *Superfund Amendments and Reauthorization Act of 1986*. Several site-specific amendments, definitions, clarifications, and technical requirements were added to the legislation, including additional enforcement authorities. The size of the fund was increased to $8.5 billion to pay for cleanup activities around the country. Some of the other salient features of the Act were that it required states to get involved in all phases of the Superfund program, stressed human health concerns posed by hazardous waste sites, and encouraged greater community involvement in the site cleanup decision-making process. Title III of SARA also authorized the *Emergency Planning and Community Right-to-Know Act* (EPCRA) and is discussed next. Most businesses in the United States are affected by SARA Title III.

EMERGENCY PLANNING AND COMMUNITY RIGHT-TO-KNOW ACT (EPCRA). Also known as Title III of SARA, EPCRA was passed by Congress in 1986. This law was enacted to help local communities protect public health, safety, and the environment from chemical hazards. As required by the Act, each state was required to appoint a state emergency response commission (SERC), divide the state into emergency planning districts, and name a local emergency planning committee (LEPC) for each district. These planning committees were required to have representation from all groups that were considered essential to managing emergencies. Groups to be given due consideration include firefighters, health officials, government and media representatives, community groups, industrial facilities, and emergency managers. Aircraft rescue and fire fighting (ARFF) and airport fires unrelated to aircraft are governed by this regulation.

THE OIL POLLUTION ACT (OPA).   The *Oil Pollution Act of 1990* was adopted after the Exxon Valdez oil spill and was passed to strengthen EPA's ability to prevent and respond to catastrophic oil spills. A trust fund, financed by a tax on oil, was established to clean up spills when the responsible party was unable or unwilling to do so. Oil storage facilities and transport vessel operators are required to submit detailed plans to the federal government on how they will manage large-scale inadvertent discharges. The EPA also has specific requirements for aboveground storage facilities. This regulation applies to aviation fueling and storage operations. EPA has jurisdiction for inland oil spills and the U.S. Coast Guard responds in coastal waters.

Of course, the Oil Pollution Act became very important to the U.S. Government in the wake of the British Petroleum Oil Rig disaster that occurred in the Gulf of Mexico near the mouth of the Mississippi River in Louisiana on April 20, 2010. The Oil Pollution Act of 1990 provides that the responsible party must pay for oil spill damages, and British Petroleum agreed to establish a $20 billion trust fund to pay for these damages. The cataclysmic nature of the BP oil spill to the Gulf Coast region will have an impact on subsequent aviation oil spills for years to come. New safety standards for oil production will certainly be forthcoming in the near future.

THE NOISE CONTROL ACT.   By virtue of the *Noise Control Act of 1972*, which created the Office of Noise Abatement and Control (ONAC), the EPA was required to submit to the FAA proposed aircraft noise control regulations that promoted and protected public health and welfare. The FAA would then be required to publish the proposed regulations in a notice of proposed rulemaking and invite community comments through public hearings. The *Quiet Communities Act* of 1978 expanded the scope of the Noise Control Act of 1972 to include setting specific decibel limits for civil aircraft; tightening noise emission standards; coordinating federal noise abatement research; working with industry, state, and local regulators to develop consensus standards; and sponsoring research on the effects of noise and abatement strategies. The Act also established a nationwide Quiet Communities Program. The *Aviation Safety and Noise Abatement Act* of 1979 authorized the FAA, under the Airport Improvement Program (AIP), to award noise mitigation and control grants to state and local governments. In 1981, however, Congress, while keeping the Noise Control Act intact, ceased funding for the ONAC. As a result, the EPA was unable to stay on top of its noise abatement and control regulatory effort despite the advancement of relevant science and technology that demonstrated a better understanding on how noise affected people. This led to outdated EPA emission and labeling standards. *The Federal Aviation Reauthorization Act* of 1996 required the FAA to appoint an aviation noise ombudsman to serve as a liaison with the public on issues regarding aircraft noise and to be consulted when the FAA administrator proposed changes in aircraft routes that included flying over populated areas.

Legislative interest in noise abatement and control of aircraft noise resurfaced in the 106th Congress through the introduction of several bills. Three bills of particular interest included the mitigation of noise levels from aircraft flights over national parks, the banning of commercial operation of certain types of supersonic aircraft due to the high noise levels, and the reduction of noise resulting from takeoffs and landings at airports in metropolitan areas. The Reform Act for the 21st century proposed an increase in the amount of funding set aside for noise mitigation from 31 to 34 percent. Public Law 106-81 set aside nearly $207 million for fiscal year 2000 for AIP noise mitigation. Among other things, this law directed the FAA to develop more stringent aircraft noise standards and develop (with the National Parks Service) air tour management plans for national parks and monuments, to minimize the effects of aircraft noise on the natural environment. In addition to the above legislation, a bill has been introduced to reestablish EPA's Office of Noise Abatement and Control and support its activities with an annual funding of $21 million for fiscal years 2000 through 2004. The important functions of this reestablished office would be to provide states and local communities with technical assistance and grants to develop noise control programs and to fund research on the health effects of noise on humans.

OCCUPATIONAL SAFETY AND HEALTH ADMINISTRATION (OSHA)

OVERVIEW

The *Occupational Safety and Health Administration Act* (*OSHA Act*) was signed into law by President Nixon on January 29, 1970 (took effect April 28, 1971), to ensure safe and healthful working conditions for working men and women. This was to be accomplished by authorizing enforcement of the standards developed under the Act, assisting and encouraging the states in their efforts to ensure safe and healthful working conditions, and providing for research, information, education, and training in the field of occupational safety and health. The Act established the Occupational Safety and Health Administration, the regulatory body that promulgates and enforces safety and health regulations in the United States. The Act also created the *National Institute for Occupational Safety and Health* (*NIOSH*) and the *Occupational Safety and Health Review Commission* (*OSHRC*). NIOSH is responsible for conducting research and making recommendations for the prevention of work-related disease and injury. The function of the OSHRC, which is comprised of three judges appointed by the President and confirmed by the Senate, is to resolve disputes between OSHA and industry before the issues run through the judicial system. Although NIOSH and

OSHA were created by the same Act of Congress, they are two distinct agencies with separate responsibilities. NIOSH is in the U.S. Department of Health and Human Services and is a research agency. OSHA is in the U.S. Department of Labor and is responsible for creating and enforcing workplace safety and health regulations. NIOSH and OSHA often work together toward the common goal of protecting worker safety and health.

The OSHA Act covers all employers and employees who do business in the United States except workplaces already protected by other federal agencies under other federal statutes. This means that OSHA's jurisdiction does not extend into the aircraft, but would apply to all ground, ramp, and airport operations. While state and federal agencies are not covered by the Act, Order 3900.19B requires the FAA to establish and maintain an agencywide occupational safety and health program that is consistent with the OSHA Act of 1970. In a separate agreement, OSHA and FAA signed a *memorandum of understanding (MOU)* on August 7, 2000, in which they pledge to work together to improve the working conditions of flight attendants while aircraft are in operation. By virtue of this MOU, a joint FAA/OSHA team will review the application of OSHA standards to flight attendant safety and health on issues of recordkeeping, bloodborne pathogens, noise, sanitation, hazard communication, and access to employee exposure and medical records. The team will also address whistleblower protections. FAA will issue a new policy statement and request public comment on the team's recommendations. OSHA, for its part, will consult with the FAA before proposing a standard that could compromise aviation safety. Under the MOU, OSHA will continue to enforce its standards for other aviation industry employees, such as maintenance and ground support personnel, while the FAA will continue to cover the flight deck crew, pilots, and copilots.

BACKGROUND

Before OSHA was established, the responsibility for occupational safety and health rested mainly with individual states. Regulations and enforcement were sporadic, varied by state, and were ineffective for the most part. The few federal laws that did exist at that time were limited in scope and application. For example, the *Walsh-Healey Public Contract Act of 1936*, the forerunner of the OSHA Act, applied only to federal contracts in excess of $10,000. The industrial sector was rampant with exposures ranging from exposed equipment and unguarded moving parts that maimed or killed workers to electrocutions and blocked/locked exits that trapped individuals during emergencies. Most notable of all was New York City's Triangle Factory fire of 1911 that killed 146 textile workers as they were prevented from escaping because of exits that had been locked to prevent theft. The Pittsburgh Survey of 1907, a one-year study of industrial accidents in Allegheny County, Pennsylvania, found that there were almost two deaths for each day of the entire year of the study. The above are two of the many commonly cited examples that mobilized the nation to call for federal intervention to protect citizens in the workplace. Common law doctrine and lack of any worker's compensation laws left few incentives for employers to improve working conditions as survivors and/or their families had the burden of carrying medical costs and losses due to wages. As a result of this failed system, the annual occupational death rate leading up to the Act reached 14,300 with disabling injuries totaling 2.2 million.

CHRONOLOGY OF MAJOR OSHA STANDARDS AFFECTING AVIATION OPERATIONS

During OSHA's 40-year history, the agency has helped to save many thousands of lives and reduce occupational injury and illness rates by more than half. OSHA's regulatory development and enforcement process has been credited with this feat. What follows is an aviation-related subset of OSHA's regulations that were promulgated over the past 40 years to improve working conditions for employees in the United States.

- *May 29, 1971.* Comprehensive standards were first adopted to provide a baseline for safety and health protection in occupational environments. A majority of standards were incorporated from other standard-setting organizations such as the American National Standards Institute (ANSI), National Fire Protection Agency (NFPA), and American Conference of Governmental Industrial Hygienists (ACGIH). All private sector employees are covered by the OSHA Act. Sample aviation operations that are regulated include aircraft manufacturing and assembly, hangar and other maintenance shop operations, painting and stripping, ramp and flight line operations, baggage handling, cleaning crew activities, and airport operations.

- *November 14, 1978.* The lead standard was introduced to protect workers from occupational exposure to lead, an element that is known to cause damage to human nervous, urinary, and reproductive systems. It was estimated that about 835,000 workers were exposed to lead in its various forms and that this standard would reduce lead exposure risk by 75 percent. Aviation applications of this standard include battery maintenance and aircraft painting and stripping.

- *May 23, 1980.* By virtue of the medical and exposure records standard being finalized, workers and OSHA would be able to access employer-maintained medical and toxic exposure records. Records may include, but are not limited to, personal and area exposure monitoring, baseline measurements, preplacement physicals, and postexposure testing. This regulation would apply to baseline monitoring and audiometric testing for noise exposures for ramp operators and manufacturing and assembly workers, and testing for blood and other potentially infectious material (OPIM) exposures for flight attendants and aircraft cleaning crew.

- *September 12, 1980.* The fire protection standard was updated to include specific procedures for fire brigades that were responsible for extinguishing major workplace fires. Typical applications in aviation would be aircraft manufacturing and maintenance facilities where compressed gases, flammables, and solvents are stored; where welding and brazing operations occur; and where testing and maintenance of sprinkler systems are required.

- *January 16, 1981.* The electrical standards were updated to make compliance easier and adopt a performance-based approach. These standards apply to most businesses. Applications in aviation would include aircraft manufacturing and assembly, and hangar and other maintenance shop activities.

- *November 25, 1983.* The hazard communication standard was passed to require information, training, and labeling of toxic materials handled by employers and employees. Any employee who has the potential of coming in contact with a chemical used at the workplace must be trained in its safe use, handling, and decontamination procedures, if exposed to the chemical. All chemicals must have a material safety data sheet (MSDS) on site within easy access, and all containers must be labeled with their contents. Aviation applications would include aircraft manufacturing and assembly jobs, cleaning crew tasks, and hangar and other maintenance shop activities.

- *March 6, 1989.* The hazardous waste operations and emergency response standard (HAZWOPER) was promulgated to protect 1.75 million public and private sector workers exposed to toxic wastes from spills or at hazardous waste sites. Aircraft refueling, battery maintenance and disposal, deicing operations, and manufacturing process discharges are covered by this standard.

- *September 1, 1989.* The control of hazardous energy sources standard, also known as the *lockout/tagout (LOTO)* standard, was promulgated to protect workers (especially during maintenance) from unexpected energizing or start-up of equipment. It was estimated that by implementing this standard 120 deaths and 50,000 injuries would be prevented annually. Individuals who maintain baggage and cargo handling equipment and manufacturing and assembly equipment are covered by this standard.

- *December 6, 1991.* The bloodborne pathogens standard was introduced to prevent occupational exposure to AIDS, hepatitis B, and other infectious diseases. Flight attendants, the members of the go-team, baggage and cargo handlers, aircraft cleanup crew, and high-exposure-potential manufacturing and assembly jobs are covered by this standard. At the very least, employees in these positions require basic training on exposure prevention to blood and other potentially infectious material (OPIM).

- *January 14, 1993.* The confined spaces (and permit-required confined spaces) standard was promulgated to prevent more than 50 deaths and more than 5000 serious injuries annually. Employees who perform maintenance and fabrication work in elevators, bulkheads, and cargo holds are covered by this standard. Manufacturing facilities requiring employees to enter and work in spaces that have a limited means of entry, can engulf the employee, and are not designed for normal continuous work are covered by this standard.

- *November 6, 2000.* The Needlestick Safety and Prevention Act required OSHA to revise its Bloodborne Pathogens standard to include new exposure control plans, an injury log, and safer medical devices to protect employees from inadvertent injury from sharp objects.

- *November 14, 2000.* The ergonomics program standard was initiated to prevent a painful and debilitating category of musculoskeletal injuries that affect more than 102 million workers. These injuries develop from jobs requiring excessive repetitive motion and/or high forceful applications and/or awkward postures. Sample aviation jobs that could lead to repetitive-motion injuries are those performed by flight attendants, baggage handlers, data entry personnel, and aircraft assembly workers. Although this standard was repealed by Congress in March 2001, OSHA's "Effective Ergonomic Strategies for Success" program remains viable in today's workplace.

KEY TERMS

International Civil Aviation Organization (ICAO)

Provisional International Civil Aviation Organization (PICAO)

Chicago Convention of 1944

Paris Convention of 1910 and 1919

International Commission for Air Navigation (ICAN)

The Havana Convention of 1928

Next Generation Air Transportation System (NextGen)

Air carrier

Air Mail Act of 1925

Air Commerce Act of 1926

Air taxi operators

Civil Aeronautics Act of 1938

Civil Aeronautics Board (CAB)

Federal Airport Act of 1946

Federal Aviation Act of 1958

Federal Aviation Administration (FAA)

Department of Transportation (DOT)

National Transportation Safety Board (NTSB)

Airport and Airway Development Act of 1970

Airline Deregulation Act of 1978

Airport and Airway Improvement Act of 1982

Aviation Safety and Capacity Expansion Act of 1990

Commuter Safety Initiative of 1995 (Commuter Rule)

FAR Part 119

FAR Part 121

FAR Part 135

Domestic air carrier

Flag air carrier

Supplemental air carrier

Commuter air carrier

Regional air carrier (Commuter)

Wendell H. Ford Aviation Investment and Reform Act for the 21st Century

FAA Air Traffic Organization (ATO)

Global Positioning System (GPS)

Airline Safety and Federal Aviation Administration Act of 2010

Aviation Safety Action Plan (ASAP)

Flight Operational Quality Assurance Program (FOQA)

Line Operations Safety Audit (LOSA)

Advanced Qualification Program (AQP)

Safety Management Systems (SMS)

Air Transport Pilot (ATP)

National Environmental Policy Act

Clean Air Act (CAA)

Clean Water Act (CWA)

Resource Conservation and Recovery Act (RCRA)

Hazardous and Solid Waste Amendments (HSWA)

Toxic Substances Control Act (TSCA)

Comprehensive Environmental Response, Compensation, and Liability Act (Superfund)

Superfund Amendments and Reauthorization Act (SARA)

Emergency Planning and Community Right-to-Know Act (EPCRA)

Oil Pollution Act

Noise Control Act

OSHA the Agency

National Institute for Occupational Safety and Health (NIOSH)

Occupational Safety and Health Review Commission (OSHRC)

Needlestick Safety and Prevention Act

REVIEW QUESTIONS

1. List three strategic objectives of ICAO and discuss their importance in international aviation.

2. Discuss the significance of The Paris Convention of 1919.

3. What were the important accomplishments of the Chicago Convention of 1944?

4. List three major functions of the FAA and discuss some of the activities that support these functions.

5. What was the significance of the Air Mail Act of 1925?

6. Describe the major provisions of the Air Commerce Act of 1926, and discuss the role of the aeronautics branch.

7. Distinguish between the Civil Aeronautics Authority (CAA) and the Civil Aeronautics Board (CAB).

8. Discuss some of the factors that led to the passage of the Federal Aviation Act of 1958, and identify several of the safety provisions of this Act.

9. What was the primary reason for the passage of the Airline Deregulation Act of 1978?

10. Describe the major provisions of the Airline Safety and Federal Aviation Administration Act of 2010.

11. List the four industry voluntary safety programs described in the Airline Safety and FAA Act of 2010.

12. Explain the evolution of the EPA.

13. Highlight the major environmental acts relevant to aviation, giving examples of each.

14. Explain the evolution of OSHA.

15. Discuss the major OSHA acts that are relevant to aviation, giving examples of each.

REFERENCES

Briddon, Arnold E., Ellmore A. Champie, and Peter A. Marraine. 1974. *FAA Historical Fact Book: A Chronology 1926–1971.* DOT/FAA. Washington, D.C.: U.S. Government Printing Office.

Davies, R. E. G. 1972. *Airlines of the United States Since 1914.* Washington, D.C.: Smithsonian Institution Press.

Freer, Duane W. 1986. Chicago Conference (1944)—Despite uncertainty, the spirit of internationalism soars. ICAO Bulletin, 41(9), 32–33.

Jenkins, Darryl (ed.). 1995. ALPA's One Level of Safety (Chapter 73). *Handbook of Airline Economics.* New York: McGraw-Hill.

Komons, Nick A. 1978. *Bonfires to Beacons.* DOT/FAA. Washington, D.C.: U.S. Government Printing Office.

Krieger, G. R., and J. F. Montgomery. 1997. *Accident Prevention Manual for Business and Industry— Engineering and Technology,* 11th ed. Itasca, Ill.: National Safety Council.

McCarthy, James E. 2020. Aviation and Climate Change. Washington, D.C.: Congressional Research Service.

Rochester, Stuart I. 1976. *Takeoff at Mid-Century: Federal Aviation Policy in the Eisenhower Years, 1953–1961.* Washington, D.C.: U.S. Government Printing Office.

U.S. Congress, Office of Technology Assessment. 1988. *Safe Skies for Tomorrow: Aviation Safety in a Competitive Environment,* OTA-SET-38. Washington, D.C.: U.S. Government Printing Office.

Wells, Alexander T. 2003. *Air Transportation: A Management Perspective,* 5th ed. Belmont, Calif.: Wadsworth Publishing Company.

———. 2004. *Airport Planning and Management,* 5th ed. New York: McGraw-Hill.

Wilson, John R. M. 1979. *Turbulence Aloft: The Civil Aeronautics Administration amid Wars and Rumors of Wars, 1938–1953.* DOT/FAA. Washington, D.C.: U.S. Government Printing Office.

WEB REFERENCES

http://www.icao.int/
http://www.faa.gov/
http://www.epa.gov/
http://www.osha.gov/

REGULATORY ORGANIZATION AND RULEMAKING

LEARNING OBJECTIVES

After completing this chapter, you should be able to

- Recognize the organizational structure of the ICAO and describe its three governing bodies.

- Recognize the organizational structure of the FAA and describe the functions of its major functional offices.

- Describe the organizational structure of the FAA Air Traffic and Aviation Safety Organizations.

- Explain why inspector workload has been greatly affected since airline deregulation.

- Explain the rulemaking process of the ICAO, FAA, OSHA, and EPA.

- Recognize the organizational structure of OSHA and describe the functions of its nine directorate offices.

- Recognize the organizational structure of the EPA and describe the functions of its 13 major offices.

To carry out their diverse mandates, regulatory organizations have evolved into organizational structures that best support their missions. The ICAO and FAA's organization are reviewed together with detailed discussions on offices that are relevant to safety and rulemaking. This topic is followed by discussions on the organizational structure and rulemaking process of OSHA and the EPA.

ICAO ORGANIZATION AND RULEMAKING

As previously stated in Chapter 1, ICAO was formed at the Chicago Convention in December 1944, which enacted the Convention on International Civil Aviation. This document serves as the constitution of ICAO, and each of the 190 Contracting States is a party to this convention. An overview of ICAO's organizational structure is outlined, followed by more detailed discussions on office functions that are relevant to the focus of this text. This section concludes with the international rulemaking process, otherwise known as the ICAO Standards or "Standards and Recommended Practices" (SARPs), in the language of ICAO.

ICAO ORGANIZATION

ICAO has its headquarters in Montreal, Canada, with seven regional offices throughout the world. These regional offices are very important to coordinate international aviation policy and standards in third world and developing nations which do not have a modern aviation infrastructure.

The seven ICAO regional offices are located as follows:

- Bangkok, Thailand—Asia & Pacific office
- Cairo, Egypt—Middle East office
- Dakar, Senegal—Western & Central African office
- Lima, Peru—South American office
- Mexico City—North American, Central American and Caribbean office
- Nairobi, Kenya—Eastern & Southern African office
- Paris, France—European & North Atlantic office

According to the Chicago Convention, ICAO is made up of three governing bodies: an Assembly, a Council of limited membership with various subordinate bodies, and a Secretariat. The chief officers of ICAO are the President of the Council and the Secretary General.

THE ASSEMBLY.   This is the first, sovereign governing body of ICAO. It meets every 3 years and recently concluded its 37th session on October 8, 2010. ICAO's 190 member States and a large number of international organizations were invited to the Assembly which reviewed the complete work of the organization in the economic, legal, and technical fields. Each state is entitled to one vote, and decision of the Assembly are decided by a majority of votes cast except when otherwise provided for in the Convention.

THE COUNCIL.   This is the permanent, second governing body of ICAO, which is elected by the Assembly and is composed of 36 Contracting States. In accordance with Article 50 of the Chicago Convention, the Assembly elects the council member States using three criteria:

• States of chief importance in air transport

• States which make the largest contribution to the provision of facilities for civil air navigation

• States whose designation will ensure that all major areas of the world are represented

As its primary governing body, the Council gives continuing direction to the work of ICAO. The primary focus is to adopt and incorporate ICAO Standards and Recommended Practices (SARPs) as Annexes to the Convention on International Civil Aviation. To develop SARPs, the Council is assisted by the Air Navigation Commission in technical matters, the Air Transport Committee in economic matters, and the Committee on Unlawful Interference in aviation security matters.

The primary work of the Air Navigation Commission is to advise the Council on aviation navigation issues using international experts with appropriate qualifications and experience in this area. It is important to note that these experts are expected to serve worldwide aviation interests and function independently, and not as representatives of their member States.

THE SECRETARIAT.   This third governing body of ICAO is headed by the Secretary General and is divided into five main divisions or bureaus. An organization chart for the ICAO Secretariat is shown in Fig. 2-1.

The five bureaus are

• Air Navigation Bureau

• Air Transport Bureau

• Technical Cooperation Bureau

• Legal Bureau

• Administration and Services Bureau

Of these five bureaus, the Air Navigation Bureau (ANB) is by far the most important for the purposes of this text. The Standards (Rulemaking) process of ICAO

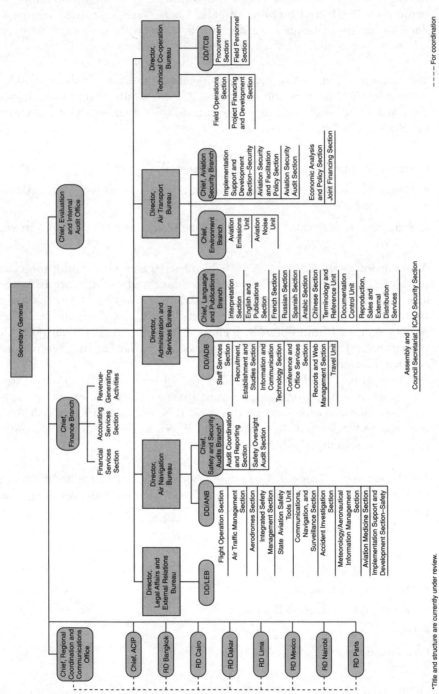

FIGURE 2-1 ICAO organization structure. (*Source: www.icao.int*)

*Title and structure are currently under review.

---- For coordination

area kept current by ANB through updates to the Annexes of the Convention. Sixteen out of 18 annexes to the Convention are of a technical nature, and therefore fall within the responsibilities of ANB and its subordinate offices.

ICAO RULEMAKING

As clearly indicated on the ICAO Web site, its rulemaking has a language and process with an international focus and lexicon all its own. ICAO standards and other provisions are primarily developed in the following categories:

- Standards and Recommended Practices (SARPs)
- Procedures for Air Navigation Services (PANS)
- Regional Supplementary Procedures (SUPPS)

Some definitions are helpful at this point to distinguish between these three categories. The ICAO Web site has an excellent discussion in the area of SARPs and their development.

A *Standard* is defined as any specification for physical characteristics, configuration, material, performance, personnel, or procedure, the uniform application of which is recognized as necessary for the safety or regularity of international air navigation and to which Contracting States must conform in accordance with the Convention; in the event of impossibility of compliance, notification to the Council is compulsory under Article 38 of the ICAO Convention.

A *Recommended Practice* is any specification for physical characteristics, configuration, material, performance, personnel, or procedure, the uniform application of which is recognized as desirable in the interest of safety, regularity, or efficiency of international air navigation, and to which Contracting States will endeavor to conform in accordance with the Convention. Please note, States are merely invited to inform the Council of noncompliance.

SARPs are formulated in broad terms and restricted to essential requirements. For complex systems such as communications equipment, SARPs material is constructed in two sections: core SARPs—material of a fundamental regulatory nature contained within the main body of the Annexes, and detailed technical specifications placed either in Appendices to Annexes or in manuals.

Procedures for Air Navigation Services (or PANS) comprise operating practices and material too detailed for Standards or Recommended Practices—they often amplify the basic principles in the corresponding Standards and Recommended Practices. To qualify for PANS status, the material should be suitable for application on a worldwide basis.

Regional Supplementary Procedures (or SUPPs) have application in the respective ICAO regions. Although the material in Regional Supplementary Procedures is similar to that in the Procedures for Air Navigation Services, SUPPs do not have the worldwide applicability of PANS, and are adapted to suit the particular region of the world, through the ICAO regional offices.

Additional Guidance Material is also produced to supplement the SARPs and PANS and to facilitate their implementation. Guidance material is issued as Attachments to Annexes or in separate documents such manuals, circulars, and lists of designators/addresses. Usually such guidance is approved at the same time as the related SARPs are adopted.

RULEMAKING PROCESS.   Figure 2.2 provides a comprehensive flow chart on the ICAO rulemaking process. The formulation of new or revised SARPs begins with a proposal for action from ICAO itself or from its Contracting Member States. Proposals also may be submitted by international organizations. The SARP process has been very effective in recent years to ensure the safe, efficient, and orderly growth of international civil aviation. The reason for success in this area lies in the four "C's"of ICAO international aviation: Cooperation, Consensus, Compliance, and Commitment; i.e., cooperation by member States in the formulation of SARPs, consensus in their approval, compliance in their application, and commitment of adherence to the on-going process.

As indicated previously, for highly technical SARPs, proposals are analyzed first by the Air Navigation Commission (ANC). Depending on the nature of the proposal, the Commission may assign its review to specialized, expert working groups or panels. As previously stated, these experts act in their individual capacity and not as representatives of their member States. ICAO Council technical committees may also be established to deal with problems involving technical, economic, social and legal aspects. Less complex issues may be assigned to the Secretariat for further examination, perhaps with the assistance of an air navigation study group.

These various groups report back to the Air Navigation Commission in the form of a technical proposal either for revisions to SARPs or for new SARPs, for preliminary review. This review is normally limited to consideration of controversial issues which, in the opinion of the Secretariat or the Commission, require examination before the recommendations are circulated to member States for comment. States are normally given 3 months to comment on the proposals.

Standards developed by other recognized international organizations can also be considered, provided they have been subject to adequate verification and validation. The comments of States and international organizations are analyzed by the Secretariat and a working paper detailing the comments and the Secretariat proposals for action is prepared. The Commission undertakes the final review of the recommendations and establishes the final texts of the proposed amendments to SARPs, PANS, and associated attachments. The amendments to

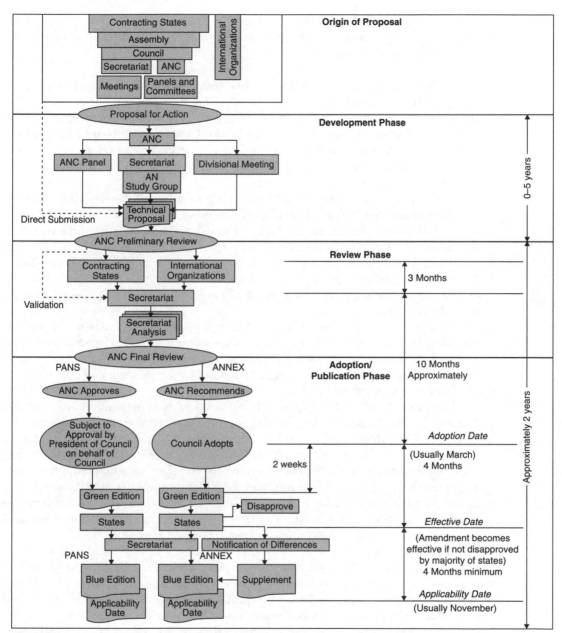

FIGURE 2-2   The ICAO rulemaking process. (*Source: ICAO http://www.icao.int*)

Annexes recommended by the Commission are presented to the Council for adoption under cover of a "Report to Council by the President of the Air Navigation Commission."

NOTIFICATION AND PUBLICATION OF REPORT AND ADMINISTRATIVE PROCESS. The Council reviews the proposals of the Air Navigation Commission and adopts the amendment to the Annex if two-thirds of the members are in favor. Within 2 weeks of the adoption of an Annex amendment by the Council, an interim edition of the amendment, referred to as the "Green Edition," is dispatched to States with a covering explanatory letter. This covering letter also gives the various dates associated with the introduction of the amendment.

ICAO policy prescribes that Contracting member States be allowed three months to indicate disapproval of adopted amendments to SARPs. A further period of 1 month is provided for preparation and transit time, making the Effective Date approximately 4 months after adoption by Council. There should be a period of 4 months between an amendment's Effective Date and its Applicability Date; however, this can be longer or shorter as the situation requires. Provided a majority of States have not registered disapproval, the amendment will become effective on the Effective Date.

On the Notification Date, which is 1 month prior to the Applicability Date, the member States must notify the Secretariat of any differences that will exist between their national aviation regulations and the provision of the Standard as amended. The reported differences are then published as regional supplementary procedures in supplements to Annexes (SUPPs).

Immediately after the Effective Date, a letter is sent announcing that the amendment has become effective and the Secretariat takes action to issue the "Blue Edition" which is the form of the amendment suitable for incorporation in the Annex to the Convention or PANS. On the Applicability Date, member States must implement the amendments unless, of course, they have notified differences. To limit the frequency of Annex and PANS amendments, the Council has established one common applicability date for each year, usually in November. The result of this adoption procedure is that the new or amended Standards and Recommended Practices become part of the relevant Annex.

It takes on average 2 years from the Preliminary Review by the ANC to the applicability date. Although this process may seem lengthy at first glance, it provides for repeated consultation and extensive participation of member States and international organizations in producing a consensus based on logic and experience. Cooperation and consensus have thus provided international aviation with the vital infrastructure for safe and efficient air transport. The third "C," compliance, brings this comprehensive regulatory system to life.

Attachments to Annexes, although they are developed in the same manner as Standards and Recommended Practices, are approved by the Council rather than adopted. Regional Supplementary Procedures, because of their regional application, do not have the same development as the previously mentioned amendments; they also must be approved by the Council.

The proposed amendments to PANS are approved by the Air Navigation Commission, under power delegated to it by the Council, subject to the approval by the President of the Council after their circulation to the Representatives of the Council for comment.

IMPLEMENTATION OF SARPs/UNIVERSAL SAFETY OVERSIGHT AUDIT PROGRAM.   Under the Convention on International Civil Aviation, the implementation of SARPs lies with Contracting member States. To help them in the area of safety, ICAO established in 1999 a Universal Safety Oversight Audit Program. The Program consists of regular, mandatory, systematic, and harmonized safety audits carried out by ICAO in all Contracting States.

The objective of the Safety Oversight Audit is to promote global aviation safety by determining the status of implementation of relevant ICAO SARPs, associated procedures, and safety-related practices. The audits are conducted within the context of critical elements of a State's safety oversight system. These include the appropriate legislative and regulatory framework; a sound organizational structure; technical guidance; qualified personnel; licensing and certification procedures; continued surveillance; and the resolution of identified safety concerns.

FAA ORGANIZATION AND RULEMAKING

The FAA is organized into 15 headquarters offices, nine field and regional offices and two primary technical centers. An overview of FAA's organizational structure is outlined, including more detailed discussions on office functions that are of relevance to this textbook. The FAA current organization chart is provided in Fig. 2-3. This section concludes with an overview of the rulemaking processes of FAA.

FAA ORGANIZATION

The largest agency of the U.S. Department of Transportation (DOT), the FAA, has authority to regulate and oversee all aspects of civil aviation with an annual budget of over $16 billion. The FAA stated mission is to provide the safest, most efficient aerospace system in the world.

HEADQUARTERS OFFICES

OFFICE OF THE ADMINISTRATOR.   This office provides overall direction and supervision of the Agency and reports directly to the Secretary of Transportation (DOT), a cabinet level position. Within this office is a Deputy Administration and Chief of Staff.

AIR TRAFFIC ORGANIZATION (ATO).   This is a new organization formed in November 2003 to serve as the operations branch of the FAA. Authorized by the Wendell H. Ford

FEDERAL AVIATION ADMINISTRATION

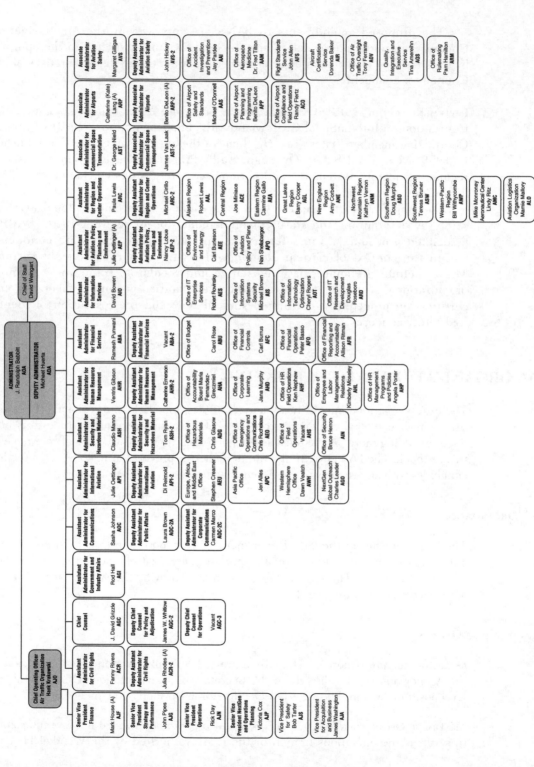

FIGURE 2-3 FAA organization chart, Oct. 2010. (*Source: http://www.faa.gov*)

Aviation Investment and Reform Act for the 21st century, ATO is a new "performance-based" business structure unlike other government agencies. It focuses on safe, secure, and efficient air traffic management services under the guidance of a Chief Operating Officer.

ATO OPERATIONS   As provided on the FAA Web site, there are four operations service units within this branch to provide ATC and aeronautical information to customers in the national airspace system:

- En Route and Oceanic Services
- System Operations Services
- Technical Operations Services
- Terminal Services

NEXT GENERATION (NEXTGEN) AND OPERATIONS PLANNING   This department of ATO sets forth the strategy and planning required for the FAA to transition to the NextGen air transportation system of the future. This transition involves moving from today's ground-based communications, navigation, and surveillance system to a modern next generation system based on satellite navigation, digital communications, and advanced networking. The FAA NextGen system will be fully explained in Chapter 8 entitled Air Traffic Safety Systems.

STRATEGY AND PERFORMANCE   Since the ATO is a performance-based organization, this office provides a framework for executing and integrating ATO plans, programs, and activities to make sure that performance stays on track.

FINANCE   This ATO office is in charge of financial metrics, comparative analysis productivity measures, business case evaluation, and competitive sourcing.

OFFICE OF SAFETY   Established in 2004, this office works with FAA operational service units in the safety policy arena to manage risks, assure quality standards, instill an open culture of disclosure, educate employees, and provide continuous improvement.

Responsible for identifying and mitigating aircraft collision risks during the delivery of air traffic separation services, the Office of Safety is the focal point for

- Applying the agency's Safety Management System principles (See Chapter 13)
- Auditing safety, quality assurance and quality control in the ATO and reporting findings to improve safety performance
- Integrating the functions and information of risk reduction, investigations, evaluations,independent operational testing and evaluation, safety risk management, runway safety and operational services, in order to identify collision risks and influence their resolution

- Providing information on assessments of operational and safety performance within the National Airspace system (NAS)
- Working with the Associate Administrator for Aviation Safety and other external entities undertaking special projects in support of increasing the safety of the NAS

AVIATION SAFETY.   This is another important FAA organization that encompasses a wide range of Agency functions and responsibilities. In addition to certification of aircraft, pilots, mechanics and others in safety-related positions, AVS is also responsible for the certification and safety of approximately 7,300 U.S. commercial airlines and air operators. As stated on the FAA Web site, AVS consists of over 7,000 personnel located in the following headquarters and field offices:

- Office of Accident Investigation & Prevention. This office works closely with the National Transportation Safety Board (NTSB). Detailed information on the NTSB will be provided in Chapter 3.
- Office of Aerospace Medicine
- Air Traffic Oversight Service. This office establishes safety standards and provides independent oversight of the Air Traffic organization.
- Aircraft Certification Service. This office is responsible for the design and production approval, airworthiness certification, and continued airworthiness programs of all U.S. civil aviation products.
- Flight Standards Service (AFS). This is another important office responsible for the certification and oversight of airmen, air operators, air agencies, and their designees. In addition to its Federal civilian workforce, FAA uses over 11,000 designees to perform selected safety oversight duties. Some of the key functions performed by AFS are the following:
  - Certificate Management Office for the certification, surveillance and inspection of all major air carriers.
  - Civil Aviation Registry—This office, primarily located in Oklahoma City, is responsible for maintaining and operating all the U.S. national programs for the registration of civil aircraft and certification of airmen. The staff of approximately 240 people interact with over 1 million FAA customers each year.
  - Air Transportation Division
  - Aircraft Maintenance Division
  - Flight Technologies and Procedures Division
  - Flight Standards Service Regional Divisions—Eight regional offices throughout the U.S. which supervise:
    - Flight Standards District Offices (FSDO) Located in approximately 82 locations in cities throughout the United States, these local FAA offices concentrate on enforcement of the Federal Aviation Regulations (FAR) in areas such as aircraft maintenance, licensing operational issues, permits, certification, and accident reporting. The local FSDO aviation safety inspector is the FAA's "first line of defense" for the public to report low flying aircraft and other suspected FAR violations.

Additional FAA Headquarters offices:

- Airports (ARP)
- Aviation Policy Planning and Environment (AEP)
- Chief Counsel (AGC)
- Civil Rights (ACR)
- Commercial Space Transportation (AST)
- Communications (AOC)
- Financial Services (ABA)
- Government and Industry Affairs (AGI)
- Human Resources Management (AHR)
- Information Services (AIO)
- International Aviation (API- handles ICAO matters)
- Security and Hazardous Materials (ASH)

PRIMARY FAA TECHNICAL CENTERS

- Mike Monroney Aeronautical Center, Oklahoma City, OK
  - Home of FAA Academy Training, FAA Logistics Center and other technical programs
  - Reports to FAA Assistant Administrator of Regions and Center Operations (ARC)
- William J. Hughes Technical Center, Atlantic City, NJ
  - Home of FAA research and development, test and evaluation facility, and many other technical initiatives
  - Reports to FAA Chief Operating Officer of Air Traffic Organization (ATO)

Additional information on each of the offices together with their contacts is available on the FAA Web site. Some of the office functions that are important to the scope and intent of this text will be discussed in greater detail throughout the book.

AIR CARRIER RESPONSIBILITY FOR SAFETY.   Section 601(b) of the FA Act specifies, in part, that when prescribing standards and regulations and when issuing certificates, the FAA shall give full consideration "to the duty resting upon air carriers to perform their services with the highest possible degree of safety in the public interest." The FA Act charges the FAA with the responsibility for promulgating and enforcing adequate standards and regulations. At the same time, the FA Act recognizes that the holders of air carrier certificates have a direct responsibility for providing air transportation with the highest possible degree of safety. The meaning of Section 601(b) of the FA Act should be clearly understood. It means that this responsibility rests directly with the air carrier, irrespective of any action taken or not taken by an FAA inspector or the FAA.

Most of the day-to-day inspections, reviews, and sign-offs are performed by the manufacturers, airlines, and airports; the system depends on self-inspections, and it is simply not possible for the FAA to make every inspection on every airplane in every location around the world. This self-inspection, or "designee," concept is startling to many of the general public, but it has worked effectively for many decades. The airlines and the manufacturers have a great concern for the safety of their airplanes and operations; it is in their business interests to place a high priority on safety.

Before certification, the FAA's objective is to make a factual and legal determination that a prospective certificate holder is willing and able to fulfill its duties as set forth by the FA Act and complies with the minimum standards and regulations prescribed by the FAA. This objective continues after certification. If a certificate holder fails to comply with the minimum standards and regulations, Section 609 of the FA Act specifies that the FAA may reexamine any certificate holder or appliance. As a result of an inspection, a certificate may be amended, modified, suspended, or revoked, in whole or in part. Additionally, Section 605(b) generally provides that whenever an inspector finds that any aircraft, aircraft engine, propeller, or appliance used or intended to be used by any air carrier in air transportation is not in a condition for safe operation, the inspector shall notify the air carrier, and the product shall not be used in air transportation until the FAA finds the product has been returned to a safe condition.

The following conditions or situations could indicate that an air carrier's management is unable or unwilling to carry out its duties as set forth by the FA Act:

- Repetitive noncompliance with minimum standards and regulations
- Lack of knowledge and/or understanding of minimum standards and safe practices
- Lack of motivation to exercise operational and/or quality airworthiness control
- Inaccurate record-keeping procedures

The FA Act and the FARs contain the principle that air carriers offering services to the public must be held to higher standards than the general-aviation community. Inspectors must also be aware of the private rights of citizens and air carriers. Since public safety and national security are among the FAA's highest priorities, FAA inspectors must be prepared to take action when any air carrier does not, or cannot, fulfill its duty to perform services with the highest possible degree of safety.

FAA SAFETY INSPECTION PROGRAM.   Aviation safety depends in part on the quality and thoroughness of the airlines' maintenance programs and on oversight and surveillance by safety inspectors of the FAA. Even though the frequency of maintenance-related accidents has not increased since 1978, airline deregulation has been accompanied by increasing concern that maintenance standards might have been lowered at some carriers and that pressures of the marketplace might lead to unsafe operating practices. At the same time, deregulation has increased the stress on FAA inspection programs. The existing regulatory inspection program, with its local and regional structure, does not have sufficient flexibility to adapt to a dynamic industry environment.

The 82 FSDOs nationwide handle the dual functions of safety inspection and advice for airlines. In addition to scheduled airline surveillance, the local offices are responsible for safety inspections of nonscheduled air taxis and other operations, such as flight schools, engine overhaul shops, and private pilots. An air carrier's operating certificate is held at a specific flight standards office, typically the one nearest the carrier's headquarters or primary operations or maintenance base. For each carrier, a principal inspector is assigned to operations (flights, training, and dispatch), airworthiness (maintenance), and avionics (navigation and communications equipment). For large airlines, each of the principal inspectors can have one or two assistants.

Local FSDOs conduct several types of inspections on each airline's operations and maintenance functions. Inspectors periodically conduct maintenance-base inspections, which focus on the records kept by an airline. For example, records demonstrating that an airline has complied with airworthiness directives might be inspected. Inspectors conduct shop inspections to observe maintenance procedures and carry out ramp inspections to observe the airworthiness of aircraft. A similar operations-base inspection focuses on records concerning the hours of training and checkrides given pilots and the rest periods between duty shifts given crews as required by regulations. En route inspections involve observations of actual flight operations, with the inspector riding in the jump seat in the cockpit.

Base inspections are preannounced. There is a tendency to focus on records rather than to probe deeply into the data underlying carrier records concerning maintenance, training, and flightcrew logs. Inspectors at the local level try to work with air carriers to achieve compliance when they find discrepancies. Violations and fines are viewed as a last resort.

Every major airline has a reliability program that monitors maintenance activities and looks for emerging problems. For example, most airlines monitor engine temperatures, oil consumption, and the metal content of oil. They then use these tests to determine when an individual engine needs to be overhauled or repaired. Some airlines also use statistical measures such as the number of engines requiring premature overhaul, engines that are shut down in flight, the number of mechanical discrepancies that are left outstanding at flight time, and the rate at which these discrepancies are cleared. In some companies, analysts search for adverse trends that might indicate, for example, a shop procedure that needs to be revised, both to ensure safety and to reduce maintenance expense. Some FSDOs have taken advantage of these statistical data to monitor the effectiveness of airline maintenance. In some cases, flight standards inspectors have encouraged airlines to set up or expand their statistical reliability programs.

INSPECTOR WORKLOAD.   The inspector workload has been affected greatly by airline deregulation. Every time a new airline has been formed, an airline has placed a new aircraft type into service, or two airlines have merged, flight standards have been obliged to devote resources to certificating the new or changed air carrier. Certification involves page-by-page approval of the airline's operations and maintenance manuals, checkrides for the airline's senior pilots, and final proving runs for the operation as a whole.

Certification is an activity that competes for inspectors' time with the entire safety inspection program. Because it is a potential bottleneck in the establishment of new and changing airlines, requests for certification divert resources from other surveillance activity and usually bring pressure on the FAA to address the certification on a priority basis. After all certification requests have been fulfilled, any remaining inspector time is devoted to surveillance. The surveillance program consequently experiences curtailment and fluctuations as a result of this operating philosophy.

The situation has improved in recent years. However, the large number of airline mergers during the mid-1980s has affected flight standards personnel in a unique way. When airlines merge, the acquired carrier often is kept intact as an operating entity. This merger is temporarily convenient for the merged company due to differences in aircraft types, crew training, and maintenance programs. Consequently, the FAA inspection force assigned to the acquiring carrier might become responsible for the operations of a carrier that is much different and is located in another region.

Because principal inspectors can be far away from their newly assigned airlines, there is a far greater dependence on inspections by the local offices that happen to be nearby. Inspections performed by one local office on another office's carrier are called *geographic* inspections by the FAA. Although the FAA is taking steps to remedy the situation, at present the geographic inspections do not count in an office's workforce staffing standards. Offices that have fewer assigned certificates but greater geographic responsibilities tend to be understaffed.

REEXAMINATION OF AIR CARRIERS. When one carrier is merged with another, Aviation Flight Standards Service local office acquires responsibility for the subsidiary airline, which is for that office a new air carrier certificate, operations program, and maintenance program. However, the airline and its certificates and programs are not new—they have been approved by and have been in operation under supervision of a different local office prior to the merger. In some cases, the new local office might be concerned that the merged operation for which it is responsible is not ready to operate safely in its new configuration. Yet the new local office might feel that it cannot reinspect or reopen the certification of business units that have already been approved but that might not meet its own standards in their present combination.

The ability of airlines to move their operations to different geographic areas is one of the prime contributors to airline competitiveness and efficiency. However, the mobility of airlines has left FAA surveillance behind in certain cases. For example, in one case an airline maintains business offices in one city, while its principal inspectors are relocated to a nearby flight standards local office. However, the airline no longer has flight operations within hundreds of miles of that city. This case can lead to reduced efficiency in the surveillance program and undue dependence on geographic surveillance by other local offices.

PUBLIC COMPLAINTS.   The new FAA Office of Audit and Evaluation is the focal point for public safety complaints and whistleblower contributions. It also handles relations with the U.S. Office of Special Counsel, the Government Accountability Office and the Department of Transportation's Office of Inspector General. The new office improves the FAA's ability to conduct candid self-examinations in aviation safety matters and questionable personnel practices. The office's goal is to manage complaints and audits more effectively, making sure they receive the necessary consideration and coordination and are acted on in a timely manner. The office also helps FAA managers and employees resolve workplace conflicts, but it is not a substitute for existing equal opportunity, grievance or FAA Accountability Board processes. The director of the office reports to the FAA Chief Counsel, but has direct access to the FAA Administrator as well.

FAA RULEMAKING

Over the past 50 years, numerous reports have documented the causes and delays in the FAA rulemaking process. There is no question but that rulemaking is extremely time-consuming and complex, requiring careful consideration of the impact of proposed rules on the public, aviation industry, the economy, and the environment. The U.S. General Accounting Office (now Government Accountability Office) has reported that after formally initiating a rule, FAA took an average of 30 months to complete the rulemaking, and 20 percent of the rules took over 10 years or longer to complete. In recent years, difficult rulemaking issues such as changes to the Helicopter Emergency Management System FAR rules pertaining to hospital MedEvac helicopter flights have taken several years in development in spite of numerous crashes, dozens of fatalities, and NTSB strong public recommendations in this area. Often it takes public outcry and political pressure to move the process forward such as seen in the wake of the Colgan Air Commuter flight crash in Buffalo, New York, in February of 2009.

In general, all government administrative rulemaking is governed under the Administrative Procedures Act (APA), which requires advance notice and opportunity for the public and industry to comment before a rule is issues or amended. Under the Federal Advisory Committee Act, the FAA obtains advice and recommendations from industry concerning the full range of its rulemaking activity including all aviation-related issues regarding such issues as air carrier operations, airman and aircraft certification, airports and noise abatement. The primary committee chartered by the FAA for this purpose is called the Aviation Rulemaking Advisory Committee (ARAC) whose primary purpose is to provide the public an earlier opportunity to participate in the FAA rulemaking process before the formal APA notification period begins.

FAA OFFICE OF RULEMAKING.   This office operates under the FAA Associate Administration for Aviation Safety (AVS), and provides the charter and support for the ARAC process. The ARAC process allows the FAA to obtain direct, first-hand information from industry and other experts who are substantially affected by proposed rules. The ARAC consists of approximately 55 member organizations selected by the FAA as most representative of the various viewpoints of those impacted by FAA organizations. The activities of the ARAC are advisory only, and while given great weight, they do not circumvent the formal APA rulemaking process.

STEPS IN THE FORMAL PROCESS.   The formal FAA rulemaking procedures are provided in Question and Answer format in 14 CFR Part II. A diagram from the GAO Report 01-950T on Aviation Rulemaking outlining the major steps is provided in Fig. 2-4.

The Secretary of Transportation reviews FAA programs for regulatory compliance. The DOT secretary notifies the FAA of schedules for certain projects, expresses general departmental policy issues, and identifies specific areas of concern.

The FAA, through the DOT secretary, submits to the *Office of Management and Budget (OMB)* an annual draft regulatory program. The program covers regulatory policies, goals, and objectives and also provides information concerning

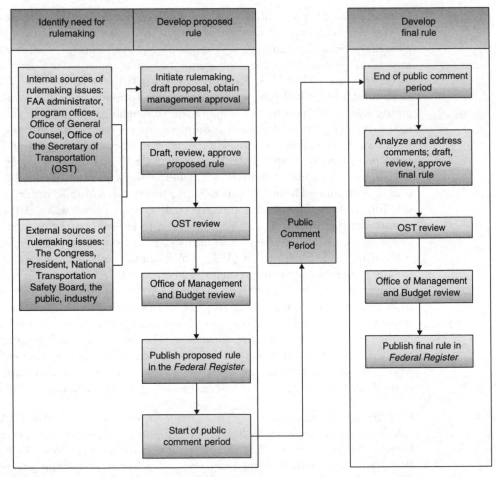

FIGURE 2-4   Rulemaking process. (*Source: www.faa.gov*)

significant regulatory actions or actions that might lead to rulemaking. The OMB reviews the program to ensure that all proposed regulatory actions are consistent with administration principles.

After a rule is cleared to proceed, the appropriate FAA program office organizes a team to manage each rule. This team proceeds with input from the ARAC team and other interested parties. An Advance Notice of Proposed Rulemaking may be prepared and published at this point to obtain additional information, followed by a draft Notice of Proposed Rulemaking (NPRM) which is reviewed for regulatory impact analysis when appropriate. An impact analysis includes potential costs and benefits imposed on society, lower-cost approaches that were not chosen and why, and an explanation of any legal preclusions from cost/benefit criteria. When the team receives the regulatory impact analysis, it develops a new draft and briefs the principals (associate administrators, etc.) of interest. Following the principals' briefing, the team coordinates the package at the branch and division levels. After the branches and divisions concur, the package is coordinated with the principals. Next, the Associate Administrator for Aviation Safety reviews the package and forwards it to the chief counsel, who prepares the package for Office of the Secretary of Transportation and OMB review.

After the FAA review process is complete, the FAA chief counsel submits the NPRM package to the Department of Transportation (DOT).

After they sign off, the package is finally forwarded to the Office of Management and Budget (OMB) for its review.

If anyone at any stage in the process objects, the package is reworked until the problem is resolved. At that point, the FAA recoordinates the package with the appropriate people. This process continues until everyone concurs or at least until all comments are considered.

After the Notice of Proposed Rulemaking is published in the *Federal Register,* the public comments on the rule for a set period. At the close of the comment period, the FAA compiles the comments and considers appropriate changes in the rule. It develops a draft of the final rule and the whole process starts again. The final rule is reviewed by the FAA, the Office of the Secretary of Transportation (OST), and the OMB. After the rule clears the process again, the OMB releases it to the FAA. The administrator issues the final rule and has it published in the *Federal Register.* Once a regulation is completed and has been printed in the *Federal Register* as a final rule, it is "codified" by being published in the Code of Federal Regulations (CFR). The CFR is the official record of all regulations created by the federal government. These regulations are organized into 50 topics called titles or volumes. FAA regulations are found in Title 49 of the CFR.

Most analysts of the FAA rulemaking process agree that the area most in need of improvement is that of timeliness in identifying and responding to safety issues. Critics complain that excessive delays in the rulemaking process complicate and delay new or amended certification programs and make them more costly. They complain of excessive rewriting and divergent rules for different aircraft categories when such rules should be identical. Some critics blame the Office of the Secretary of Transportation and the Office of Management and Budget for much of the delay.

Some critics complain that the OST and the OMB lack the necessary technical background to understand and review FAA rules. Others suggest that rulemaking quality and timeliness would improve if one person were accountable for the whole process. On the other hand, some argue that the OST reduces the delay over the long term by helping the FAA deal with the OMB. Most disagreements between the FAA and either the OST or the OMB concern economic analysis. OST officials often question the quality of the cost/benefit analysis coming from the FAA. FAA officials maintain that lack of understanding hinders the quality of those evaluations.

The actual delay from involvement by three different parties in this process is difficult to determine. Sometimes the parties negotiate while the rule sits in a particular office. Sometimes the OST and the FAA work together to build consensus when they develop the rule. Other times, the OST or the OMB sends a rule back to the FAA for further work. The OST or the OMB might say the responsibility for the delay lies with the FAA because the proposal was incomplete or flawed. The FAA might place blame for the same delay on the OST or the OMB because the FAA thought the package was acceptable and needed no additional analysis.

The problem of rulemaking delay seems to lie with the whole process, not just the DOT, the OMB, or the FAA. In any case, most analysts agree that the rulemaking process needs to be streamlined. The essential steps in developing sound public safety policy must be identified, and the rest eliminated.

OSHA ORGANIZATION STRUCTURE AND RULEMAKING

An agency of the U.S. Department of Labor, OSHA has 9 directorate offices and implements its enforcement and assistance activities through 10 regions across the country. The latest OSHA organization chart is provided as Fig. 2-5.

OSHA ORGANIZATION

ASSISTANT SECRETARY (**OSHA**).   This is the Office of the Assistant Secretary of Labor, which consists of the following offices:

- Office of Equal Employment Opportunity (EEO)
- Office of Communications

DIRECTORATE OF ADMINISTRATIVE PROGRAMS.   This division houses the

- Office of Administrative Services
- Office of Human Resources
- Office of Management Systems and Organization
- Office of Program Budgeting, Planning, and Financial Management

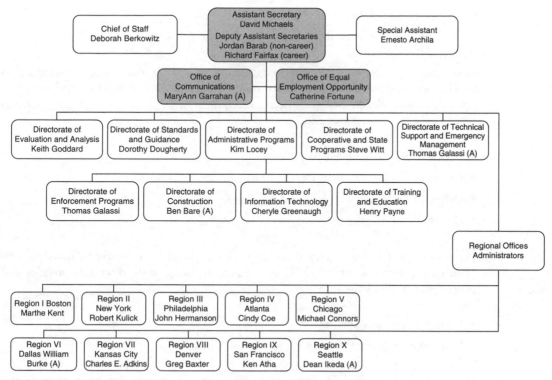

FIGURE 2-5 OSHA organization chart, Oct. 2010. (*Source: www.osha.gov*)

DIRECTORATE OF CONSTRUCTION.   This directorate serves as OSHA's principal source for standards, regulations, policy, programs, and assistance to OSHA, other federal agencies, the construction industry, and the general public with respect to occupational safety and health. Offices within this directorate include

- Office of Construction Services
- Office of Construction Standards and Guidance
- Office of Engineering Services

DIRECTORATE OF ENFORCEMENT PROGRAMS.   This is the compliance enforcement directorate of OSHA. OSHA inspectors enforce employer compliance to the OSHA code of federal regulations and other safety regulations through site audits and inspections and assess citations and fines where appropriate. Offices within this directorate include

- Office of Maritime Enforcement
- Office of Health Enforcement
- Office of General Industry Enforcement

- Office of Federal Agency Programs
- Office of the Whistleblower Protection Program

DIRECTORATE OF COOPERATIVE AND STATE PROGRAMS. This directorate implements OSHA's cooperative programs, coordinates the Agency's compliance and outreach activities, and coordinates the Agency's relations with state governments. Offices within this directorate include

- Office of Small Business Assistance
- Office of Partnership and Recognition
- Office of Outreach Services and Alliances
- Office of State Programs

DIRECTORATE OF TRAINING. This directorate develops, directs, oversees, manages and ensures implementation of OSHA's national training and education policies and procedures. Offices include

- OSHA Training Institute
- Outreach Training Program

DIRECTORATE OF STANDARDS AND GUIDANCE. This directorate develops and promulgates standards. In addition, this office plans, develops, and manages nonregulatory approaches to safety and health guidelines which supplement the agency's rulemaking efforts. Offices within this directorate include

- Office of Engineering Safety
- Office of Physical Hazards
- Office of Chemical Hazards (metal)
- Office of Chemical Hazard (nonmetal)
- Office of Engineering Safety
- Office of Maritime
- Office of Safety Systems

DIRECTORATE OF INFORMATION TECHNOLOGY (DIT). This directorate manages OSHA computer data systems and internet services from its Salt Lake City Technical Center. Offices include

- Office of Management Data Systems
- Internet/Intranet Services

DIRECTORATE OF EVALUATION AND ANALYSIS.   This directorate monitors and measures the agency's performance to track its progress against strategic objectives. In addition, customer service–related functions and on-site audits of regional and area office operations are also conducted. Offices within this directorate include

- Office of Evaluations and Audit Analysis
- Office of Program Review
- Office of Regulatory Analysis
- Office of Statistical Analysis

DIRECTORATE OF TECHNICAL SUPPORT AND EMERGENCY MANAGEMENT.   The mission of the directorate is to provide field and staff support to the rest of OSHA by providing specialized technical expertise and advice in occupational safety and health. Some of these areas include industrial hygiene, ergonomics, occupational health nursing and medicine, and safety engineering along with sample analysis and equipment calibration and repair. Offices within this directorate include

- Office of Emergency Management
- Cincinnati Technical Center
- Office of Ergonomic Support
- Office of Occupational Health Nursing
- Office of Occupational Medicine
- Office of Science and Technology Assessment
- Salt Lake Technical Center
- Office of Technical Programs and Coordination Activities
- Technical Data Center and Docket Office

OSHA Regional Offices

- OSHA region 1 (CT, ME, NH, RI, VT) Boston
- OSHA region 2 (NJ, NY, PR, VI) New York
- OSHA region 3 (DE, DC, MD, PA, VA, WV) Philadelphia
- OSHA region 4 (AL, FL, GA, KY, MS, NC, SC, TN) Atlanta
- OSHA region 5 (IL, IN, MI, MN, OH, WI) Chicago
- OSHA region 6 (AR, LA, NM, OK, TX) Dallas
- OSHA region 7 (IA, KS, MO, NE) Kansas City
- OSHA region 8 (CO, MT, ND, SD, UT, WY) Denver

- OSHA region 9 (AZ, CA, HI, NV) San Francisco
- OSHA region 10 (AK, ID, OR, WA) Seattle

OSHA RULEMAKING

OSHA can initiate new standards on its own or if petitioned by other parties of relevance such as the Secretary of Health and Human Services (HHS), the *National Institute for Occupational Safety and Health (NIOSH)*, state and local governments, any nationally recognized standards-producing organization (e.g., National Fire Protection Association, or NFPA), employer or labor representatives, or any other interested individual or citizen groups.

ADVISORY COMMITTEES AND NIOSH.   If it is determined by OSHA that a specific standard is warranted, one or more of several advisory committees may be asked to work on specific recommendations. OSHA has two standing committees at its disposal, and in addition to these an ad hoc committee may be appointed to examine special concerns. All advisory committees, standing or ad hoc, are required to have representation from management, labor, state agencies, and one or more designees of the Secretary of Health and Human Services (HHS). The two standing advisory committees are

- *National Advisory Committee on Occupational Safety and Health (NACOSH)*, which advises and makes recommendations to the Secretary of Labor on specifics pertaining to the administration of the Act and the Secretary of HHS
- *Advisory Committee on Construction Safety and Health (ACCSH)*, which advises the Secretary of Labor on the development of health and safety standards for the construction industry

NIOSH, the research arm of OSHA, may also make recommendations to modify and/or promulgate standards. This agency conducts research and provides technical assistance to OSHA and industry on various safety and health hazards. During the course of its research, NIOSH may conduct workplace investigations, gather testimony from employers and employees, and require that employers measure and report employee exposure to potentially hazardous materials. NIOSH also may require employers to provide medical examinations and tests to determine the incidence of occupational illness among employees.

STANDARDS ADOPTION.   OSHA publishes its plans to recommend, modify, or reject standards in the *Federal Register* as a Notice of Proposed Rulemaking. Sometimes the earlier option of Advance Notice of Proposed Rulemaking is used when information is needed to draft a proposal. Before publishing any major regulation, OSHA is required to consult with the Office of Management and Budget as mandated by Executive Order. OSHA is also required by the Small Business Regulatory Enforcement and Fairness Act (SBREFA) to consult with small businesses on regulations that significantly affect them.

The Notice of Proposed Rulemaking will include the details of the new rule together with a time frame of at least 30 days (usually 60 days or more) from the date of publication for the public to respond. Concerned parties may submit written comments together with supporting documentation of their concerns. The citizenry also has a right to request a public hearing on the proposal if OSHA has not proposed one. If requested, OSHA must schedule a public hearing and publish the time and place of its occurrence, in advance, in the *Federal Register*. After the close of the comment and public hearing period, the full text of the regulation and the date it becomes effective must be published in the *Federal Register*. OSHA must also include the reasons for implementing the regulation together with the preamble that led up to this regulation and a cost/benefit analysis of the regulation. Once this is done, the regulation is coded and placed in the Code of Federal Regulation. Alternatively, OSHA may decide after due consideration that a standard is not warranted and publish the reasons for such a determination. OSHA Safety and Health Standards for General Industry are placed in 29 CFR Part 1910, and the Construction Standards are located in 29 CFR Part 1926.

EPA ORGANIZATION STRUCTURE AND RULEMAKING

An independent agency of the U.S. Government, EPA has 13 major offices and implements its regulatory authority through 10 regions across the country. The latest EPA organizational chart is provided as Fig. 2-6.

EPA Organizational Structure			
Office of the Administrator 202-564-4700			
Headquarters offices:			
Office of Administration and Resources Management	Office of Air and Radiation	Office of Chemical Safety and Pollution Prevention	Office of the Chief Financial Officer
Office of Enforcement and Compliance Assurance	Office of Environmental Information	Office of General Counsel	Office of Inspector General
Office of International and Tribal Affairs	Office of Research and Development	Office of Solid Waste and Emergency Response	Office of Water
Regional offices around the nation:			
Region 1/Boston	Region 2/New York	Region 3/ Philadelphia	Region 4/Atlanta
Region 5/Chicago	Region 6/Dallas	Region 7/ Kansas City	Region 8/Denver
Region 9/ San Francisco	Region 10/Seattle		

FIGURE 2-6   EPA organization structure, June 2011. (*Source: www.epa.gov*)

EPA ORGANIZATION

OFFICE OF THE ADMINISTRATOR. This office provides overall direction and supervision of the Agency and reports directly to the President of the United States. Divisions within the Office of the Administrator include Administrative Law Judges, Children's Health Protection, Civil Rights, Congressional and Intergovernmental Relations, External Affairs and Environmental Education, Regional Operations, Small Business Programs, Science Advisory Board, Environmental Appeals Board, and Office of Policy, Economics and Innovation.

OFFICE OF ADMINISTRATION AND RESOURCES MANAGEMENT. This office is responsible for national leadership, policy and procedures governing administrative services, and human resources. In addition, this office handles facilities services, information resources management, automated data processing systems, grants, debarment, and acquisition management.

OFFICE OF AIR AND RADIATION. This office develops national programs, policies and regulations for controlling air pollution and radiation exposure. OAR is concerned with pollution prevention and energy efficiency, indoor and outdoor air quality, industrial air pollution, pollution from vehicles and engines, radon, acid rain, stratospheric ozone depletion, climate change, and radiation protection. OAR is responsible for administering the Clean Air Act, the Atomic Energy Act, the Waste Isolation Pilot Plant Land Withdrawal Act, and other applicable environmental laws.

OFFICE OF CHEMICAL SAFETY AND POLLUTION PREVENTION. This office is responsible for protecting the environment from potential risks from pesticides, as well as toxic chemicals and also works to prevent pollution before it begins. It implements and monitors regulatory activities relating to the following laws:

- Federal Insecticide, Fungicide and Rodenticide Act
- Federal Food, Drug and Cosmetic Act
- Toxic Substances Control Act
- Pollution Prevention Act

OFFICE OF ENFORCEMENT AND COMPLIANCE ASSURANCE. This office is the regulatory enforcement wing of the EPA and advises the EPA Administrator on matters concerning administrative, civil, and criminal enforcement; environmental-equity efforts; and compliance monitoring and assurance activities. There is also a strong focus on environmental justice by protecting vulnerable communities.

OFFICE OF THE CHIEF FINANCIAL OFFICER. This office develops, manages, and coordinates EPA's strategic planning and implements the Government Performance and Results Act. In addition, it manages the budget function (including payroll and disbursements), provides resource and financial management services, and manages agencywide internal controls and audit resolution.

OFFICE OF ENVIRONMENTAL INFORMATION.   Headed by the Chief Information Officer, this office manages the life cycle of information to support EPA's mission of protecting human health and the environment. It identifies and implements information technology and information management solutions, ensures the quality of EPA's information, and the efficiency and reliability of EPA's technology, data collection and exchange efforts, and access services.

OFFICE OF INSPECTOR GENERAL.   This independent office audits and investigates EPA programs and operations. It is also responsible for detecting and preventing fraud, waste, and abuse. This office provides hotline services and keeps senior EPA management and Congress appraised of any serious abuses and deficiencies encountered in EPA programs and operations.

OFFICE OF GENERAL COUNSEL.   This office is the chief legal advisor to EPA and provides legal services to all offices within the Agency with respect to the Agency's workings and programs. This office offers legal opinions, legal counsel, and litigation support in federal court and other tribunals to provide legal positions for the formulation and administration of the Agency's policies and programs.

OFFICE OF INTERNATIONAL AND TRIBAL AFFAIRS.   Working with the experts from EPA's other program and regional offices, other government agencies, and other nations and international organizations, this office identifies international environmental issues and helps implement technical and policy options to address them. It also includes the American Indian Environmental Office which leads and coordinates the agencywide effort to strengthen public health and environmental protection in Indian country, with a special emphasis on helping tribes administer their own environmental programs.

OFFICE OF RESEARCH AND DEVELOPMENT.   Science at EPA provides the foundation for credible decision-making to safeguard human health and ecosystems from environmental pollutants. This office is the scientific research arm of EPA, whose leading-edge research helps provide the solid underpinning of science and technology for the Agency. The work at EPA research and development laboratories, research centers, and offices across the country helps improve the quality of air, water, soil, and the way resources are utilized.

OFFICE OF SOLID WASTE AND EMERGENCY RESPONSE.   This office provides policy, guidance and direction for the Agency's emergency response and waste programs, and develops guidelines for the land disposal of hazardous waste and underground storage tanks. It administers the Brownfields program which supports state and local governments in redeveloping and reusing potentially contaminated sites, and manages the Superfund program, which responds to abandoned and active hazardous waste sites and accidental oil and chemical releases.

OFFICE OF WATER.   Ensures drinking water is safe, and restores and maintains oceans, watersheds, and their aquatic ecosystems to protect human health, support economic and recreational activities, and provide healthy habitat for fish, plants and wildlife.

EPA REGIONAL OFFICES (STATES COVERED) AND OFFICE LOCATIONS

- Region 1 (CT, ME, MA, NH, RI, VT) Boston
- Region 2 (NJ, NY, PR, and VI) New York
- Region 3 (DE, MD, PA, VA, WV, DC) Philadelphia
- Region 4 (AL, FL, GA, KY, MS, NC, SC, TN) Atlanta
- Region 5 (IL, IN, MI, MN, OH, WI) Chicago
- Region 6 (AR, LA, NM, OK, TX) Dallas
- Region 7 (IA, KS, MO, NE) Kansas City
- Region 8 (CO, MT, ND, SD, UT, WY) Denver
- Region 9 (AZ, CA, HI, NV, and U.S. Pacific Islands) San Francisco
- Region 10 (AK, ID, OR, WA) Seattle

EPA RULEMAKING

The EPA rulemaking process is similar to the one described under OSHA rulemaking. Initially, an authorized agency such as the EPA decides that a regulation may be needed. The Agency then researches the need for the regulation and proposes it, if needed. The proposal is listed in the *Federal Register,* and comments are invited from members of the public. The Agency reviews all comments, makes changes where appropriate, and issues a final rule. During the standards development process, all information regarding the original proposal, requests for public comment, notices about meetings (time and place) for public discussions, and the text of the final regulation is published in the *Federal Register.* Biannually, the Agency publishes a report that documents its efforts on all the regulations it is working on or has recently finished. These are published in the *Federal Register,* usually in April and October, as the "EPA Semiannual Regulatory Agenda." Environmental regulations appear in Title 40 of the U.S. Code of Federal Regulations (CFR).

KEY TERMS

Standards and Recommended Practices (SARPs)

Procedures for Air Navigation Services (PANS)

Regional Supplementary Procedures (SUPPs)

Air Traffic Organization (ATO)

Aviation Safety (AVS)

Flight Standards Services (AFS)

Flight Standards District Offices (FSDO)

Government Accountability Office (GAO)

Administrative Procedures Act (APA)

Aviation Rulemaking Advisory Committee (ARAC)

Notice of Proposed Rulemaking (NPRM)

National Institute for Occupational Safety and Health (NIOSH)

National Fire Protection Association (NFPA)

National Advisory Committee on Occupational Safety and Health (NACOSH)

Advisory Committee on Construction Safety and Health (ACCSH)

Office of Management and Budget (OMB)

REVIEW QUESTIONS

1. Discuss the organizational structure of the ICAO, and describe the functions of its three governing bodies.

2. Discuss the ICAO Rulemaking Process including the role of SARPs, PANS, and SUPPs.

3. Briefly describe the organizational structure of the FAA together with the function of the Air Traffic Organization and the Aviation Safety office.

4. What does Section 601(b) of the FA Act say about an air carrier's responsibility for safety?

5. Why are most day-to-day inspections, reviews, and sign-offs performed by the airlines, not the FAA? Give examples of how an air carrier might demonstrate its inability or unwillingness to carry out its duties as set forth by the FA Act.

6. How has inspector workload been affected since airline deregulation? Why has it been difficult attracting inspectors to major metropolitan areas?

7. Explain the tendency for inspectors to focus on records during maintenance-base inspections, and discuss how public complaints are handled.

8. What is the major role of the FAA Flight Standards District Office (FSDO) organization?

9. What is the purpose of the Aviation Rulemaking Advisory Committee (ARAC)?

10. Discuss the FAA rulemaking process, and explain the basis for the NPRM being published in the *Federal Register*.

11. Explain the rulemaking process of OSHA and EPA.

12. Discuss the organizational structure of OSHA, and describe the functions of its nine directorate offices.

13. Discuss the organizational structure of the EPA, and describe the functions of its 13 major offices.

REFERENCES

ICAO, Convention on International Civil Aviation (Doc 7300/9). Montreal, Canada.

U.S. Government Printing Office, FAA General Rulemaking Procedures (14 CFR Part II). Washington, D.C.

U.S. Government Printing Office, Federal Administrative Procedures Act (5 USC §553, Rulemaking). Washington, D.C.

WEB REFERENCES

International Civil Aviation Organization, http://www.icao.int

Federal Aviation Administration, http://www.faa.gov

Aviation Rulemaking, U.S. General Accounting Office, http://www.gpo.gov

Occupational Safety and Health Administration, http://www.osha.gov

Environmental Protection Agency, http://www.epa.gov

THE NATIONAL TRANSPORTATION SAFETY BOARD

LEARNING OBJECTIVES

After completing this chapter, you should be able to

- Explain the International Accident Investigation process
- Discuss ICAO's role in international accidents
- Describe the purpose of the National Transportation Safety Board (NTSB) and its organizational structure.

- List the types of aviation accidents investigated by the NTSB.
- Explain the steps involved in investigating a major commercial aviation accident.
- Discuss the composition, function, and working of the go-team, the party system, and the board of inquiry as they relate to accident investigation.
- Summarize the responsibilities of the FAA during an investigation.
- Discuss the NTSB most wanted aviation safety improvements.

INTERNATIONAL ACCIDENT INVESTIGATION

OVERVIEW

As introduced in Chapters 1 and 2, ICAO has played a major part in international aviation matters since its establishment in 1944. Commercial aviation safety has truly become a worldwide concern in the modern era of jet airline transportation, and the ICAO Chicago Convention provides for international cooperation in accident investigations in two documents: Article 26 of the Convention and in its Annex 13 entitled *"Aircraft Accident and Incident Investigation."* This section will set forth ICAO's role in this process and explain how the aircraft accident investigation process is handled on a worldwide basis.

ICAO's ROLE

As stated, the international process for aircraft accident investigation is set forth in Annex 13 of the Chicago Convention. This document provides the following principles:

- The ultimate objective of accident investigation is prevention.
- Responsibility for an investigation belongs to the member state in which the accident or incident occurred (State of Occurrence).
- All ICAO States that may be involved must be promptly notified of the accident or incident occurrence.
- Other member States may participate in an investigation based upon their relationship to the accident such as the State of:
  - Registry
  - Operator
  - Design and Manufacture
- States of Registry, Operator, Design and Manufacture are entitled to appoint an *accredited representative* of that State to take part in the investigation.
- Experts and advisors may also be appointed to assist accredited representatives. The ICAO member State conducting the investigation may call on the best technical expertise available from any source to assist with the investigation.

- The investigation process includes the gathering, recording, and analysis of all relevant information; the determination of the causes of the accident; formulating appropriate safety recommendations and completion of the final report.

REGIONAL AND NATIONAL AUTHORITIES

The following are the primary international authorities empowered to investigate aircraft accidents in their state or region:

- Australia—Australian Transport Safety Bureau
- Canada—Transportation Safety Board of Canada (BST/TSB)
- France—Bureau d'Enquêtes et d'Analyses (BEA) pour la Sécurité de l'Aviation civile
- Mexico—Secretariat of Communications and Transportation (SCT)
- Russia (Commonwealth of Independent States, Former USSR area)—Interstate Aviation Committee (MAK)
- United Kingdom—Air Accidents Investigation Branch (AAIB) of the UK Department for Transport
- United States—National Transportation Safety Board (NTSB)

RECENT INTERNATIONAL INVESTIGATION

AIR FRANCE FLIGHT 447.   On June 1, 2009, an Airbus A330-200 aircraft en route from Rio de Janeiro to Paris—Charles de Gaulle Airport crashed in the Atlantic Ocean causing 228 fatalities. This has been described as the worst accident in French aviation history by the BEA, which is currently investigating the accident under Annex 13 of the ICAO Convention. The NTSB has assigned an "accredited representative" to assist the BEA in this investigation. Additional information on this accident will be provided later in Chapter 5 of this book.

Aviation safety remains a top priority of ICAO and its member States. The recent 37th Assembly of ICAO, which concluded in Montreal in October 2010, strongly endorsed the sharing of safety information, and the United States offered its leadership and support by signing a precedent-setting agreement with ICAO, the European Union and the International Air Transport Association (IATA) to facilitate such sharing using modern Safety Management Systems (SMS) concepts. The NTSB organization and accident investigation process will be explored next.

NATIONAL TRANSPORTATION SAFETY BOARD

The *National Transportation Safety Board (NTSB)* is an independent agency of the U.S. government that determines the probable cause of transportation accidents and promotes transportation safety through the recommendation process. The NTSB also conducts safety studies, evaluates the effectiveness of other government agencies' transportation safety programs, and reviews appeals of adverse actions by the U.S. Department of Transportation (DOT) involving pilot and mariner certificates and licenses.

To help prevent accidents, the NTSB develops and issues safety recommendations to other government agencies, industry, and organizations that are in a position to improve transportation safety. These recommendations are always based on the NTSB's investigations and studies and are the focal point of its efforts to improve safety in U.S. transportation systems.

The NTSB's origins can be found in the Air Commerce Act of 1926, in which Congress charged the Department of Commerce with investigating the causes of aircraft accidents. Later that responsibility was given to the Civil Aeronautics Board's Bureau of Aviation Safety. In 1967, Congress consolidated all transportation agencies into a new Department of Transportation and established the National Transportation Safety Board as an independent agency within the department.

NTSB MISSION.   Every 3 years, the Board issues its Strategic Plan. In its plan for fiscal years 2010 through 2015, the NTSB states its mission is to promote transportation safety by

- Maintaining its congressionally mandated independence and objectivity;
- Conducting objective, precise accident investigations and safety audits;
- Performing fair and objective airman and mariner certification appeals;
- Advocating and promoting safety recommendations;
- Assisting victims of transportation accidents and their families through Transportation Disaster Assistance (TDA).

In creating the NTSB, Congress envisioned that a single agency could develop a higher level of safety than the individual modal agencies working separately. Unlike the Bureau of Safety, the NTSB was to make its recommendations for safety reforms publicly. In summary, the NTSB's mission is to determine the "probable cause" of transportation accidents and to formulate safety recommendations to improve transportation safety.

With the passage of the *Independent Safety Board Act of 1974*, Congress made the NTSB completely independent outside the DOT, because "no Federal agency can properly perform such investigatory functions unless it is totally separate and independent from any other . . . agency of the United States." Because the DOT is charged with both the regulation and the promotion of transportation in the United States, and accidents may suggest deficiencies in the system, the NTSB's independence is necessary for objective oversight.

It is important to note that the NTSB has no authority to regulate, fund, or be directly involved in the operation of any mode of transportation. Therefore, it has the ability to oversee the transportation system, conduct investigations, make recommendations from a totally objective viewpoint, and make recommendations for needed safety improvements. Its effectiveness depends on an ability to make timely and accurate determinations of the cause of accidents, along with comprehensive and well-considered safety recommendations.

The most visible portion of the NTSB involves major accident investigations. Under its accident selection criteria, the NTSB's investigative response depends primarily on

- The need for independent investigative oversight to ensure public confidence in the transportation system
- The need to concentrate on the most significant and life-threatening safety issues
- The need to maintain a database so that trends can be identified and projected

NTSB investigations include the participation of modal agencies and other parties (such as manufacturers, operators, and employee unions). Within the transportation network, each government organization has been established to fulfill a unique role. Each modal agency investigates accidents to varying degrees of depth and with different objectives. As the only federal agency whose sole purpose is promoting transportation safety, the NTSB conducts detailed, open, and thorough accident investigations that often uncover significant systemwide problems that need to be corrected to prevent future similar accidents.

Under the Independent Safety Board Act of 1974, the NTSB investigates hundreds of accidents annually, including

- All accidents involving 49 Code of Federal Regulations (CFR) Parts 121 and 135 air carriers
- Accidents involving public (i.e., government) aircraft (except military accidents);
- Foreign aircraft accidents involving U.S. airlines and/or U.S.-manufactured transport aircraft or major components
- Accidents involving air traffic control, training, midair collisions, newly certified aircraft/engines, and in-flight fire or breakup
- General aviation accidents, some of which are delegated to the Federal Aviation Administration (FAA) for fact finding (it is important to note, however, that probable-cause determinations are never delegated)

In addition, based on the agency's mandate under Annex 13 to the Chicago Convention and related international agreements, the NTSB participates to a greater or lesser degree in the investigation of commercial aviation accidents throughout the world. The NTSB enjoys a worldwide reputation. The major share of the NTSB's air safety recommendations are directed to the FAA. These recommendations have resulted in a wide range of safety improvements in areas such as pilot training, aircraft maintenance and design, air traffic control procedures, and survival equipment requirements. The NTSB is also empowered to conduct special studies of transportation problems. A special study allows the NTSB to break away from the mold of the single accident investigation to examine a safety problem from a broader perspective. In the past, for example, the NTSB has conducted special studies in weather, crashworthiness, in-flight collisions, and commuter airlines. In over 40 years of operation, the NTSB has issued nearly 13,000 safety recommendations.

NTSB ORGANIZATION

As provided on its Web site, there are three levels to the NTSB organization chart which is provided in Fig. 3-1. Current updates of this organization are*

- Top Level: Five board members, each nominated by the U.S. President and confirmed by the Senate to serve 5-year terms. A Chairman and a Vice Chairman are designated from these five board members. NTSB board members establish policy on transportation safety issues and on NTSB goals, objectives, and operations. Board members review and approve major accident reports, safety recommendations, and decide appeals of FAA certificate actions. Individual NTSB board members often serve as spokesman for major accident investigation ("Go Teams"), preside over public hearings, make major speeches, and testify before Congressional Committees.

- Second Level: The following offices report to the Chairman of the NTSB:
  - The Office of the Chief Financial Officer
  - The Office of the General Counsel
  - The Office of the Managing Director
  - The Office of Communications
  - The Office of Equal Employment Opportunity (EEO)

The Office of the Managing Director supports the NTSB mission by providing overall leadership for the management of the Board. The office coordinates the activities of the entire NTSB staff, supervises eight offices at the next level, and develops plans to achieve NTSB program objectives.

- Third Level: The following offices report to the NTSB Managing Director:
  - The Office of Research and Engineering
  - The Office of Railroad, Pipeline, and Hazardous Materials Investigations
  - The Office of Aviation Safety (further described later in this chapter)
  - The Office of Highway Safety
  - The Office of Marine Safety
  - The Office of the Chief Information Officer
  - The Office of Administration
  - The Office of Administrative Law Judges

The Office of Aviation Safety has the responsibility for investigating aviation accidents and incidents, and for proposing probable causes for the Board's

*Available at www.NTSB.gov.

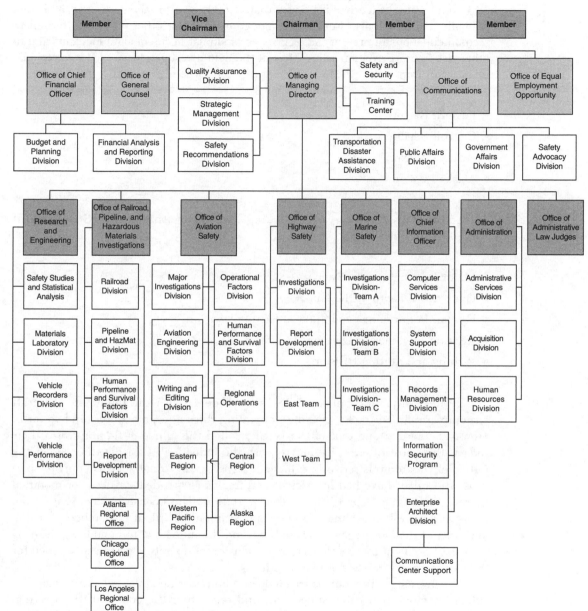

FIGURE 3-1 National Transportation Safety Board organizational chart, Nov 2010. (*Source: www.ntsb.gov*)

approval. With only 80 accident investigators among a total staff of about 120 people, the office handles more than 1,600 aviation accidents and incidents annually. The office also works, in conjunction with the other offices, to formulate recommendations to prevent the recurrence of similar accidents and incidents and to otherwise improve aviation safety. Most of the field investigations are led by a regional investigator from one of the four offices located in these regions across the country:

- Eastern Region
- Central Region
- Western Region
- Alaska Region

Additionally, there are five working divisions within the NTSB Office of Aviation Safety as follows:

- Major Investigation Division
- Operational Factors Division
- Human Performance and Survival Factors Division
- Aviation Engineering Division
- Writing and Editing Division

By necessity, much of the NTSB's work deals with the inanimate—aircraft structures, railroad tracks, pipelines, operating rules and procedures, and so forth. However, there is one unit of the NTSB, called the *Office of Administrative Law Judges,* that most often deals directly with the individual. Basically, the role of the Office of the Administrative Law Judge is to act as an initial appeals court for persons who might have had licenses or certificates suspended, revoked, or modified by the FAA or the Department of Transportation. The license holders range from pilots and aircraft mechanics to merchant seamen and flight dispatchers. But the authority of the law judge also extends beyond the individual to include the hearing appeals that might involve the loss or suspension of operating certificates issued for individual aircraft models or to airline firms.

The law judges function as trial judges, administering oaths, receiving evidence, ruling on motions, issuing subpoenas, and regulating the course of the hearing. Many hearings are held outside the Washington, D.C., area. Pursuant to authority in the Equal Access to Justice Act of 1980, the judges also review applications from airmen who prevail over the FAA in appeals brought under Section 609 of the Federal Aviation Act of 1958. The review of the applications for attorney fees and expenses for the most part is a determination of whether an award will be granted based on the written record of the earlier proceeding. However, the law judge assigned to the application might set the matter for informal conference or an evidentiary hearing when necessary for full and fair resolution of the issues arising from the application.

The law judge's initial decisions and orders are appealable to the full NTSB. Either party to the proceeding, the airman or the FAA, may appeal the judge's decision to the NTSB. After the NTSB has issued its opinion and order, either party may petition the NTSB for reconsideration. If a petition for reconsideration is not filed, then the NTSB's order becomes final if not appealed to the U.S. Court of Appeals. Only the airman or seaman can take an appeal to the U.S. Court of Appeals. The FAA and the U.S. Coast Guard, in the case of seamen, do not have the right of appeal to the court. On review, the court has the power to affirm, modify, or set aside the full NTSB's opinion and order, in whole or in part, and if need is found, to order further proceedings by the NTSB.

ACCIDENT INVESTIGATION PROCESS

When a major commercial aviation accident occurs, an NTSB go-team, led by an *investigator-in-charge (IIC)*, is dispatched from the agency's Washington, D.C., headquarters to the accident site, usually within a couple of hours of notification of the event. The IIC, a senior air safety investigator with the NTSB's Office of Aviation Safety (OAS), organizes, conducts, and manages the field phase of the investigation, regardless of whether a board member is also present on the scene. This activity includes investigating the factual circumstances of the crash (on site and afterward), preparing final reports for submission to the board members, initiating safety recommendations to prevent future accidents, and participating in foreign accident investigations. The IIC has the responsibility and authority to supervise and coordinate all resources and activities of the field investigators. The NTSB go-team may form as many as 10 investigative groups. Specialist "working group" teams may be formed around subject matter areas, such as power plants, systems, structures, operations, air traffic control, human factors, weather, and survivability. Cockpit voice recorder and flight data recorder groups are formed at the NTSB laboratory in Washington. All NTSB staff assigned to a particular investigation are under the direction of the IIC.

PARTY PROCESS

Increasingly, the NTSB has no choice but to conduct its investigations in the glare of intense media attention and public scrutiny. As commercial air travel has become routine for millions of passengers, major accidents have come to be viewed as nothing short of national catastrophes. At the same time, an NTSB statement of cause may be nothing short of catastrophic for the airline, aircraft manufacturer, or other entity that may be deemed responsible for a mishap. A very real, albeit unintended, consequence of the NTSB's safety investigation is the assignment of fault or blame for the accident by both the courts and the media. Hundreds of millions of dollars in liability payments, as well as the international competitiveness of some of the most influential U.S. corporations, rest on the NTSB's conclusions about the cause of a major accident. This was not the system that was intended by those who supported the creation of an independent investigative authority more than 40 years ago, but it is the environment in which the investigative work of the agency is performed today.

The NTSB relies on teamwork to resolve accidents, naming "parties" to participate in the investigation that include manufacturers, operators, and, by law, the FAA. The *party system* enables the NTSB to leverage its limited resources and personnel by bringing into an investigation the technical expertise of the companies, entities (such as the pilots' union), and individuals who were involved in the accident or who might be able to provide specialized knowledge to assist in determining the probable cause. Except for the FAA, party status is a privilege, not a right. The IIC has the discretion to designate the parties that are allowed to participate in an investigation, and each party representative must work under the direction of the IIC or senior NTSB investigators at all times. No members of the news media, lawyers, or insurance personnel are permitted to participate in any phase of the investigation. Claimants or litigants (victims or family members) are also specifically prohibited from serving as party members.

The specialists that any party assigns to an investigation must be employees of the party and must possess expertise to assist the NTSB in its investigation. Providing the safety board with technical assistance gives parties many opportunities to learn what happened and to formulate theories as to the cause of the accident. Party representatives are not permitted to relay information to corporate headquarters without the consent of the IIC, and then only when necessary for accident prevention purposes. Information is not to be used for litigation preparation or for public relations. Sanctions for failing to abide by the NTSB party rules and procedures include the dismissal of individuals or even the party from the investigation team. Party representatives must sign a party pledge, a written statement agreeing to abide by the NTSB rules governing the party process.

The first 2 days following an accident are critical because the evidence is fresh and undisturbed. After people start going through the wreckage, the clues begin to disappear. An airspeed indicator's needle might be moved, or a fuel line might drain. Subtle clues are lost that could reveal possible causes of the accident. Consequently, crash sites are protected from the untrained until the go-team arrives on the scene.

THE GO-TEAM

On 24-hour alert, *go-team* personnel possess a wide range of accident investigation skills. For aviation accidents, a go-team roster could include one of the five members of the NTSB, an air traffic control specialist, a meteorologist, a human-performance expert, an expert trained in witness interrogation, an engine specialist, as well as experts in hydraulics, electrical systems, and maintenance records. Some go-team members are completely intermodal in that their area of expertise is applicable to each mode. Human-factors experts fall into this category, as do the NTSB's metallurgists, meteorologists, and hazardous-materials experts.

Go-team duty is rotated. Immediately after one team has been dispatched, a new list is posted. Like firefighters, go-team members spend many hours doing office work and working on special studies until the inevitable call comes. The FAA usually gets the first word of an accident, then the director of the NTSB's regional or field office. This office notifies the go-team, the board member on duty, the NTSB chair, and the public affairs division. The team is normally on its way within 2 hours. Until it arrives, an investigator from the nearest NTSB field office secures the crash site with the help of local authorities. Representatives from the aircraft manufacturer, the airline, the engine manufacturer, and the FAA also arrive. If the accident is major, a member of the NTSB accompanies the team. The investigator-in-charge calls a meeting and assigns each of these individuals to a section of the go-team.

ACCIDENT SITE

The length of time a go-team remains on the accident site varies with need, but generally a team completes its work in 10 to 14 days. However, accident investigations often can require off-site engineering studies or laboratory tests that might extend the fact-finding stage. In cases of crew fatalities, a local coroner usually performs autopsies on the flightcrew to determine at the outset whether pilot incapacitation might have been a factor. An autopsy can also reveal who was sitting where in the cockpit and who was flying the aircraft.

After the preliminary steps are completed, the detailed work begins. The go-team is organized into groups of experts, each of which focuses on specific aspects of the investigation. Each group, headed by a group chairperson, concentrates on a specific portion of the investigation. Coordination is effected among group chairpersons to ensure investigative coverage in areas where more than one group may have a responsibility. Using their combined knowledge of flying in general and of this aircraft in particular, they compare what they know with what they find in the wreckage. Simple cameras are an important tool of the trade. Before the team members touch any of the wreckage, they take pictures from various angles and distances and make verbal notes into tape recorders.

Operational factors experts in three disciplines (air traffic control, operations, and weather) support major investigations with intensive work in their specialties. Air traffic control (ATC) specialists examine ATC facilities, procedures, and flight handling, including ground-to-air voice transmissions, and develop flight histories from Air Route Traffic Control Center (ARTCC) and terminal facility radar records. Other specialists examine factors involved in the flight operations of the carrier and the airport and in the flight training and experience of the flightcrew. Weather specialists examine meteorological and environmental conditions that may have caused or contributed to an accident.

Human-performance specialists examine the background and performance of persons associated with the circumstances surrounding an accident, including the person's knowledge, experience, training, physical abilities, decisions, actions, and work habits. Also examined are company policies and procedures, management relationships, equipment design and ergonomics, and the work environment.

Aviation engineering experts in four areas provide strong technical investigative skills. Power plant specialists examine the airworthiness of aircraft engines, while structures experts examine the integrity of aircraft structures and flight controls as well as the adequacy of design and certification. Systems specialists examine the airworthiness of aircraft flight controls and electrical, hydraulic, and avionic systems. And maintenance specialists examine the service history and maintenance of aircraft systems, structures, and power plants.

Survival-factors experts investigate factors that affect the survival of persons involved in accidents, including the causes of injuries and fatalities. These investigators also examine cabin safety and emergency procedures, crashworthiness, equipment design, emergency responsiveness, and airport certification.

LABORATORY

While the investigators work on site, the NTSB's materials laboratory in Washington, D.C., performs detailed analyses on items found at the site. One of the finest of its kind in the world, the laboratory is designed to support investigators in the field. For example, the laboratory has the capability to "read out" aircraft cockpit voice recorders (CVRs) and decipher flight data recorders (FDRs), which provide investigators with such key factors as airspeed, altitude, vertical acceleration, and elapsed time. These two *black boxes* provide investigators with a profile of an aircraft during the often crucial last minutes of flight.

Metallurgy is another of the laboratory's critical skills. NTSB metallurgists perform postaccident analysis of wreckage parts. The laboratory is capable of determining whether failures resulted from inadequate design strength, excessive loading, or deterioration in static strength through fatigue or corrosion.

The investigation of the American Airlines DC-10 that lost its left engine after takeoff from Chicago's O'Hare Airport in May 1979 probably could not have been concluded without the help of the materials lab. Preliminary investigations led metallurgists to focus on the aft bulkhead of the left engine pylon—the vertical member of the wing from which the engine is suspended. They found the overstressed area where the engine broke off. As suspected, a trail of fatigue marks also was found leading up to the overstressed area. But the real mystery turned up when the metallurgists followed the fatigue marks to their point of origin, only to discover another overstressed area, and nothing else. The first overstress had caused the fatigue, and the fatigue had caused the final break. But what had caused the initial overstress?

The metallurgists and specialists reviewed the aircraft's maintenance records and found that when removing the engines, a maintenance crew had used a forklift to help lower the entire engine-pylon assembly. Although the crew didn't realize it at the time, the method was causing hidden damage at the points where the engine and pylon were fastened to the wing. As a result of the findings, the engine removal procedure was changed.

ACCIDENT REPORT PREPARATION

Following completion of the on-scene phase of the investigation (which may last for several days or weeks), each NTSB group chair (the senior investigator overseeing a specific area of the investigation) completes a factual report on his or her area of responsibility. The reports are likely to include proposed safety recommendations to correct deficiencies and prevent future similar accidents. All factual material is placed in the public docket that is open and available for public review. Thereafter, the investigators involved in the case begin an often lengthy period of further fact gathering, usually involving one or more public hearings, and final analysis of the factual information collected.

There is no time limit on NTSB investigative activity. Safety board procedures have a target date for completion of the final accident report within 1 year of the date of the accident, but major commercial aviation accident investigations have taken as little as 4 months and as much as more than 4 years.

A key milestone in the report preparation process is the group chairs' preparation of analytical reports in their respective areas of expertise. The parties may contribute to the analytical reports through their continued contact with the NTSB group chairs and the IIC, but parties are not allowed to review, edit, or comment on the analytical reports themselves. The parties also contribute to the safety board's analytical process through written submissions, which are sometimes extensive and become part of the public docket.

PUBLIC HEARING

Following an accident, the NTSB might decide to hold a public hearing to collect added information and to discuss at a public forum the issues involved in an accident. Every effort is made to hold the hearing promptly and close to the accident site.

A hearing involves NTSB investigators, other parties to the investigation, and expert witnesses called to testify. At each hearing, a *board of inquiry* is established that is made up of senior safety board staff, chaired by the presiding NTSB member. The Board of Inquiry is assisted by a technical panel. Some of the NTSB investigators who have participated in the investigation serve on the technical panel. Depending on the topics to be addressed at the hearing, the panel often includes specialists in the areas of aircraft performance, power plants, systems, structures, operations, air traffic control, weather, survival factors, and human factors. Those involved in reading out the cockpit voice recorder and flight data recorder and in reviewing witness and maintenance records also might participate in the hearing.

Parties to the hearing are designated by the NTSB member who is the presiding officer of the hearing. They include those persons, government agencies, companies, and associations whose participation in the hearing is deemed necessary in the public interest and whose special knowledge will contribute to the development of pertinent evidence. Typically, they include the FAA, operator, airframe manufacturer, engine manufacturer, pilots' union, and any other organization that can assist the safety board in completing its record of the investigation. Except for the FAA, party status is a privilege, not a right. Parties are asked to appoint a single spokesperson for the hearing.

Expert witnesses are called to testify under oath about selected topics to assist the safety board in its investigation. The testimony is intended to expand the public record and to demonstrate to the public that a complete, open, and objective investigation is being conducted. The witnesses who are called to testify are selected because of their ability to provide the best available information on the issues related to the accident.

News media, family members, lawyers, and insurance personnel are not parties to the investigation and are not permitted to participate in the public hearings.

Following the hearing, investigators will gather additional needed information and conduct further tests identified as necessary during the hearing. After the investigation is complete and all parties have had an opportunity to review the factual record, from both the hearing and other investigative activities, a technical review meeting of all parties is convened. That meeting is held to ensure that no errors exist in the investigation and that there is agreement that all that is necessary has been done.

On rare occasions, the hearing may be reopened when significant new additional information becomes available or follow-up investigation reveals additional issues that call for an airing in a public forum such as a hearing.

FINAL ACCIDENT REPORT

With the completion of the fact-finding phase, the accident investigation process enters its final stage—analysis of the factual findings. The analysis is conducted at the NTSB's Washington, D.C., headquarters. The final accident report includes a list of factual findings concerning the accident, analysis of those findings, recommendations to prevent a repetition of the accident, and a probable-cause statement.

The IIC and the NTSB senior staff create a final draft report, called the *notation draft,* for presentation to the board members. This draft includes safety recommendations and a finding of probable cause. Following a period for review of the draft report, a public meeting of the board members is held in Washington. The NTSB staff will present and comment on the draft report; party representatives are permitted to attend but may not make any kind of presentation or comment. At this meeting, the board members may vote to adopt this draft, in its entirety, as the final accident report; may require further investigation or revisions; or may adopt the final accident report with changes that are discussed during the meeting.

Safety recommendations resulting from major investigations generally are included in the final accident report; however, in the interest of safety, they may be issued at any time during the course of an investigation if the NTSB deems it necessary.

Technically, NTSB investigations are never closed. Parties to the investigation may petition the board to reconsider and modify the findings and/or probable-cause statement if the findings are believed to be erroneous or if the party discovers new evidence. Petitions from nonparties will not be considered.

SAFETY RECOMMENDATIONS

The *safety recommendation* made to the FAA is the NTSB's end product. Nothing takes a higher priority, and nothing is more carefully evaluated. In effect, the recommendation is vital to the NTSB's basic role of accident prevention because it is the lever used to bring changes and improvements in safety to the nation's transportation system. Close to 80 percent of the recommendations made to the FAA are acted upon favorably. With human lives involved, timeliness also is an essential part of the recommendation process. As a result, the NTSB issues a safety recommendation as soon as a problem is identified without necessarily waiting until an investigation is completed and the probable cause of an accident determined. In its mandate to the NTSB, Congress clearly emphasized the importance of the safety recommendation, saying the NTSB shall "advocate meaningful responses to reduce the likelihood of recurrence of transportation accidents." Each recommendation issued by the NTSB designates the person, or the party, expected to take action, describes the action that the NTSB expects, and clearly states the safety need to be satisfied.

Recommendations are based on findings of the investigation and may address deficiencies that do not pertain directly to what is ultimately determined to be the cause of the accident. For example, in the course of its investigation of the crash landing of a DC-10 in Sioux City, Iowa, in 1989, the NTSB issued recommendations on four separate occasions before issuing its final report. In the TWA Flight 800 investigation in 1996, once it was determined that an explosion in the center fuel tank caused the breakup of the aircraft, the NTSB issued urgent safety recommendations aimed at eliminating explosive fuel/air vapors in airliner fuel tanks.

Occasionally, a single crash investigation can have major ramifications on the entire commercial aviation industry. Such was the case of Colgan Air Flight 3407, a commuter airline accident which occurred on February 12, 2009 near Buffalo, New York. From an NTSB point of view, this accident investigation was extraordinary in its sweeping scope, number of recommendations, and speed of delivery to the public. The issues presented and explored during the public hearing and investigation were the following:

- Effect of icing on aircraft performance
- Cold weather operations
- The "sterile cockpit" (inappropriate discussions between the pilots)

- Flight crew experience and training
- Fatigue management
- Stall recovery

The final NTSB report was issued with unprecedented speed less than 1 year after the accident and the Board makes 28 safety recommendations in this significant document. These recommendations cover a wide range of safety issues that were factors in this accident, especially pilot training and fatigue. As stated in Chapter 1, this single accident has had a profound effect on commercial aviation safety which will be addressed in detail in appropriate sections throughout this book.

INVESTIGATING A GENERAL-AVIATION ACCIDENT

The investigation of general-aviation accidents is a simpler process requiring fewer staff members per accident. Inasmuch as the NTSB investigates many general-aviation accidents per year, abbreviated investigations are generally necessary, given the agency's limited staff and budgetary resources. Most general-aviation accident investigations are conducted by one of the NTSB's regional or field offices. In a *field investigation,* at least one investigator goes to the crash site; a *limited investigation* is carried out by correspondence or telephone. Some, but by no means all, general-aviation accidents generate safety recommendations approved by the NTSB members.

FAMILY ASSISTANCE AND THE TRANSPORTATION DISASTER ASSISTANCE OFFICE

Following the enactment of the *Aviation Disaster Family Assistance Act of 1996,* the President designated the NTSB as the lead federal agency for the coordination of federal government assets at the scene of a major aviation accident and as the liaison between the airline and the families. The role of the NTSB includes integrating the resources of the federal government and other organizations to support the efforts of state and local governments and the airlines to aid aviation disaster victims and their families. The NTSB's Transportation Disaster Assistance Office assists in making federal resources available to local authorities and the airlines, for example, to aid in rescue and salvage operations and to coordinate the provision of family counseling, victim identification, and forensic services. The safety board has sought to maintain a distinct separation between family assistance activities and the NTSB's technical investigative staff.

FAA RESPONSIBILITIES DURING AN INVESTIGATION

Accident investigation is largely the responsibility of each FAA Flight Standards District Office (FSDO), which maintains a preaccident plan that is tailored to that office's specific requirements (e.g., geographic location, climate, staffing, and resources). The FAA works very closely with the NTSB, and the formal agreement between agencies can be found in *FAA Order 8020.11C* dated 2/2/2010, entitled, *Aircraft Accident and Incident Notification, Investigation, and Reporting*. FAA accident investigation responsibilities include the following:

Ensuring that
- All facts and circumstances leading to the accident are recorded and evaluated.
- Actions are taken to prevent similar accidents in the future.

Determining whether:
- Performance of FAA facilities or functions was a factor.
- Performance of non-FAA owned and operated air traffic control (ATC) facilities or a navigational aid was a factor.
- Airworthiness of FAA-certified aircraft was a factor.
- Competency of FAA-certified airmen, air agencies, commercial operators, or air carriers was involved.
- Federal Aviation Regulations were adequate.
- Airport certification safety standards or operations were involved.
- Airport security standards or operations were involved.
- Airman medical qualifications were involved.
- There was a violation of Federal Aviation Regulations.

The FAA conducts investigations and submits factual reports of the investigations to the NTSB on accidents delegated to the FAA by the NTSB. This delegation of certain NTSB accident investigation responsibilities is exercised under Section 304(a)(1) of the Independent Safety Board Act of 1974.

The FAA's principal investigator at an accident is called the *investigator-in-charge*. This individual directs and controls all FAA participation in the accident until the investigation is complete. Included is the authority to procure and use the services of all needed FAA personnel, facilities, equipment, and records.

The FAA investigator-in-charge is under the control and direction of the NTSB investigator-in-charge in an NTSB-conducted investigation. When accident investigations are delegated to the FAA by the NTSB, the FAA investigator-in-charge becomes an authorized representative of the NTSB. All the investigative authority prescribed in the applicable NTSB regulations falls to this person. All other FAA personal report to the investigator-in-charge and are responsible to that person for all reports they have prepared or received during the investigation.

NTSB ACCIDENT DATABASES

The NTSB aviation accident database and synopses, which go back as far as 1962, are available online at http://www.ntsb.gov. These databases contain information on civil aviation accidents and selected incidents that occur within the shores of the

United States, within its territories, and in international waters. Within a few days of an accident a preliminary report is normally available online. This is followed by a factual preliminary report that gets replaced with a final report outlining probable cause, when the investigation is completed. Complete information may not be available for cases under revision or where the NTSB did not have primary responsibility for investigating an accident. A summary of information available on this site includes

- Interactive search capability for the NTSB database that is updated daily. Search information is possible by
  Day, month, and year since 1962
  City and state in which the accident occurred
  Severity—incident, accident, fatal, nonfatal
  Aircraft category (airplane, helicopter, etc.)
  Type of assembly—amateur or commercial
  Aircraft make, model, and registration
  Operation (Part 121, Part 135, etc.)
  Name of airline
  NTSB accident number
  Word strings
- Monthly listing of accidents sorted by date that is updated daily.
- List of investigations nearing completion.
- Complete downloadable data sets for each year beginning from 1982 and updated monthly in Microsoft Access 2000 MDB format.
- Complete description of Government Information Locator System (GILS) accident databases, including definition of *accident* and *incident*.
- Complete FAA incident database information about incidents, including those not investigated by NTSB.
- Lists of data, information products, and other sources of information about aviation accidents, including publications, dockets, and press releases.

NTSB MOST WANTED AVIATION SAFETY IMPROVEMENTS

Since the NTSB lacks regulatory authority over the FAA and aviation industry, one of its most important functions is to strongly recommend and advocate appropriate FAA actions that will improve aviation safety. The Board satisfies this mandate by publishing and frequently updating its "Most Wanted List" of aviation safety improvements. A current NTSB "Most Wanted List" can be found on its Web site at www.ntsb.gov

As of this writing, there are seven aviation issues on this list, and NTSB "color codes" FAA action in three categories:

Red = Unacceptable response

Yellow = Acceptable response, proceeding slowly

Green = Acceptable response, proceeding in a timely manner

Below is the current NTSB most wanted list, and the reader is encouraged to check the most current NTSB Web site for updates and emerging information on the latest safety issues.

NTSB SEVEN MOST WANTED LIST ITEMS (JUNE 2011)

1. Improve Oversight of Pilot Proficiency

    Status: Red

    Action needed by FAA:
    - Evaluate prior flight check failures for pilot applicants before hiring
    - Provide training and additional oversight that considers full performance histories for flight crewmembers demonstrating performance deficiencies

2. Require (Cockpit) Image Recorders

    Status: Red

    Action needed by FAA:
    - Install crash-protected image recorders in cockpits to give investigators more information to solve complex accidents

3. Improve the Safety of (Helicopter) Emergency Medical Services (EMS) Flights

    Status: Red

    Action needed by FAA:
    - Conduct all flights with medical personnel on board in accordance with stricter commuter aircraft regulations
    - Develop and implement flight risk evaluation programs for EMS operators
    - Require formalized dispatch and flight-following procedures including up-to-date weather information
    - Install terrain awareness and warning systems (TAWS) on aircraft used for EMS operations

4. Improve Runway Safety (Runway Incursions)

    Status: Yellow

    Action needed by FAA:
    - Give immediate warnings of probable collisions/incursions directly to flight crews in the cockpit

- Require specific air traffic control (ATC) clearance for each runway crossing
- Require operators to install cockpit moving map displays or an automatic system that alerts pilots when a takeoff is attempted on a taxiway or a runway other than the one intended
- Require a landing distance assessment with an adequate safety margin for every landing

5. Reduce Dangers to Aircraft Flying in Icing Conditions

   Status: Red

   Action needed by FAA:
   - Use current research on freezing rain and large water droplets to revise the way aircraft are designed and approved for flight in icing conditions
   - Apply revised icing requirements to currently certified aircraft
   - Require that airplanes with pneumatic deice boots activate the boots as soon as the airplane enters icing conditions

6. Improve Crew Resource Management (CRM)

   Status: Yellow

   Action needed by FAA:
   - Require commuter and on-demand air taxi flight crews to receive crew resource management training

7. Reduce Accidents and Incidents Caused by Human Fatigue in the Aviation Industry

   Status: Red

   Action needed by FAA:
   - Set working hour limits for flight crews, aviation mechanics, and air traffic controllers based on fatigue research, circadian rhythms, and sleep and rest requirements
   - Develop a fatigue awareness and countermeasures training program for controllers and those who schedule them for duty
   - Develop guidance for operators to establish fatigue management systems, including a methodology that will continually assess the effectiveness of these systems.

KEY TERMS

ICAO Annex 13—Aircraft Accident and Incident Investigation

Accredited Representative

National Transportation Safety Board (NTSB)

Independent Safety Board Act of 1974

NTSB Board Members

Office of the Managing Director

Office of Aviation Safety

Office of Administrative Law Judges

Investigator-in-charge

Go-team

Party system

Probable cause

Black boxes

Board of Inquiry

Final accident report

Safety recommendations

Notation draft

Field investigation

Aviation Disaster Family Assistance Act of 1996

Transportation Disaster Assistance Office

FAA Order 8020.11C, Aircraft Accident and Incident Notification, Investigation, and Reporting

NTSB Most Wanted List

REVIEW QUESTIONS

1. Discuss the role of ICAO and NTSB in international aviation accident investigations. What is an accredited representative?

2. What are the primary responsibilities of the National Transportation Safety Board (NTSB)? How did passage of the Independent Safety Board Act of 1974 affect the NTSB? Describe the types of accidents investigated by the NTSB. Describe the organizational structure of the NTSB.

3. Explain the role of the investigator-in-charge (IIC) and the go-team. What is the so-called party system that enables the NTSB to leverage its limited resources? Identify the steps taken in a major accident investigation. What types of activities are performed at the NTSB's laboratory in Washington, D.C.? When are safety recommendations made?

4. What is the purpose of a public hearing? Are hearings ever reopened? What information is included in the final accident report? Distinguish between a field investigation and a limited investigation of a general-aviation accident.

5. What is the role of the NTSB under the Aviation Disaster Family Assistance Act of 1996? What is the Transportation Disaster Assistance Office?

6. Describe the responsibilities of the FAA during a major accident investigation. Describe some of the functions of the NTSB besides accident investigation.

7. What is the purpose of the NTSB Most Wanted List?

REFERENCES

ICAO—Chicago Convention Article 26/Annex 13.

NTSB Strategic Plan FY2010 - FY 2015.

Testimony of Honorable Deborah A.P. Hersman, Chairman NTSB, before the Aviation Subcommittee, Committee on Commerce, Science and Transportation, U.S. Senate—Hearing on Aviation Safety: One year after the crash of (Colgan) Flight 3407, February 25, 2010.

WEB REFERENCES

http://www.icao.int

http://www.ntsb.gov

RECORDING AND REPORTING OF SAFETY DATA

LEARNING OBJECTIVES

After completing this chapter, you should be able to

- Recognize the importance of studying accidents and incidents for the purpose of developing insights, information, and recommendations leading to accident prevention.
- Explain and give examples of an accident, incident, and injury as defined by the NTSB and ICAO.
- Explain the importance of establishing a culture of safety reporting.
- Distinguish between mandatory and voluntary reporting systems.
- Identify and describe some of the accident and incident reporting systems that are maintained by organizations within the FAA.
- Describe the Aviation Accident Data System managed by the NTSB.
- Discuss the role of NASA in managing the Aviation Safety Reporting System.
- Describe the information obtained from flight data recorders and cockpit voice recorders.
- Identify several reports provided by ICAO based on ADREP data.
- Explain OSHA's definition of an injury and illness.
- Highlight some of OSHA's recording requirements.
- Recognize what constitutes an EPA environmental spill and what are its reporting requirements.

ACCIDENTS

Before any discussion on accident reporting, investigation, and prevention can occur, there should be an understanding of the definitions used to classify accidents and incidents. The ICAO Annex 13 definition of aircraft accident is very similar to that used by the National Transportation Safety Board (NTSB); therefore for discussion purposes, this text will focus on the NTSB's definition. Some NTSB definitions used in measuring aviation safety in commercial passenger transportation are as follow:

- *Accident*—an occurrence associated with the operation of an aircraft that takes place between the time any person boards the aircraft with the intention of flight and the time all such persons have disembarked, and in which any person (occupant or nonoccupant) suffers a fatal or serious injury or the aircraft receives substantial damage.
- *Fatal injury*—any injury that results in death within 30 days of the accident.

- *Serious injury*—any injury that requires hospitalization for more than 48 hours, results in a bone fracture, or involves internal organs or burns.

- *Substantial damage*—damage or failure that adversely affects the structural strength, performance, or flight characteristics of the aircraft and that would normally require major repair or replacement of the affected component.

- *Incident*—an occurrence other than an accident associated with the operation of an aircraft that affects or could affect the safety of operations.

- *Major accident*—an accident in which a Part 121 aircraft was destroyed, or there were multiple fatalities, or there was 1 fatality and a Part 121 aircraft was substantially damaged.

- *Serious accident*—an accident in which there was 1 fatality without substantial damage to a Part 121 aircraft, or there was at least 1 serious injury and a Part 121 aircraft was substantially damaged.

- *Injury*—a nonfatal accident with at least 1 serious injury without substantial damage to a Part 121 aircraft.

- *Damage*—an accident in which no person was killed or seriously injured, but in which any aircraft was substantially damaged.

The accident and its investigation remain the most conspicuous source of insights and information leading to accident prevention. Accidents provide compelling and incontrovertible evidence of the severity of hazards. The often catastrophic and very expensive nature of accidents provides the incentive for allocating resources to accident prevention to an extent otherwise unlikely.

In an accident investigation, it is essential that a clear and accurate analysis of the relevant factors be developed without delay. Further, the focus of the investigation should be directed toward effective preventive action. This focus applies particularly to government authorities and operators. With the investigation directed away from "pursuit of the guilty party" and toward effective preventive action, cooperation is fostered among those involved in the accident, facilitating the discovery of the true causes of the accident. It is emphasized that the short-term expediency of finding someone to blame for an accident is detrimental to the long-term goal of preventing accidents.

Since, by definition, an accident involves at least serious injury or substantial aircraft damage, there is a likelihood that a legal process will result from an accident. As the official authority on the accident, the investigator is often seen as a ready source of information with which to establish culpability in the courts. Consequently, witnesses and other persons involved in an accident may be inclined to withhold information from the investigator, thereby preventing a full understanding of what occurred, particularly with respect to the human-factor elements involved.

An accident investigation includes an analysis of the evidence to determine all the causes that induced the accident—a process leading to the formulation of safety recommendations. Safety recommendations regarding serious hazards should be made as soon as the hazards have been positively identified, rather than waiting until the investigation is completed. These safety recommendations should be included in the final report on the investigation. This publicity of safety recommendations fulfills several functions:

- It helps ensure that the recommendations are reasonable and realistic in the circumstances.
- It enables other countries, organizations, and individuals to see what action was recommended. Although the recommendation was not specifically addressed to them, it may enable them to take actions that avoid similar hazards.
- It can provide pressure for a prompt and reasonable response.

Recommendations must cover all hazards revealed during the investigation, not just those directly concerned with the causes. In this way, accident investigation forms the basis of an effective accident prevention program.

INCIDENTS

Incidents are events that can be defined loosely as *near-accidents*. Both ICAO and NTSB define an incident as an occurrence, other than an accident, associated with the operations of an aircraft which affects or could affect the safety of operation. Causal factors leading to accidents also lead to incidents, and all accidents begin as near-accidents. The various combinations of possibly unsafe acts and conditions that occur each day usually end as incidents rather than accidents, and the larger number of incidents offers wider opportunities for safety trend analyses and for suggesting potential accident prevention measures. However, for an aviation incident to be widely known, it must be reported by at least one of the people involved. Yet, the definition of an incident is often subject to the interpretation of the observer, and what appears to be an incident to one person might not appear so to another. NTSB regulations at 49 CFR Section 830.5 require immediate notification (by the most expeditious means available) of aircraft accidents and any of the following *serious* incidents:

- Flight control system malfunction or failure
- Inability of required flight crewmember to perform normal flight duties as the result of injury or illness
- Failure of any internal turbine engine component that results in the escape of debris other than out the exhaust path
- In-flight fire
- Aircraft collision in flight

- Damage to property, other than the aircraft, estimated to exceed $25,000 for repair

- For large multi-engine aircraft: In-flight failure of electrical or hydraulic systems, loss of power on two or more engines, or emergency evacuation

- Release of a propeller blade, excluding release from ground contact

- A complete loss of information, excluding flickering, from more than 50 percent of an aircraft's (glass) cockpit displays

- Certain airborne collision and avoidance system advisories

- Damage to helicopter tail or main rotor blades that requires major repair or replacement

- Certain airplane landings on a taxiway, incorrect runway, or a runway incursion that requires another aircraft to take immediate corrective action to avoid a collision

ICAO Annex 13 states that the difference between an accident and a serious incident lies only in the result.

Under the new ICAO Safety Management System procedures, the reporting, investigation, and analysis of incidents can be a highly effective means of accident prevention. The most important characteristics of incidents are as follows:

- They are similar to accidents, except that they lack the terminal event that causes the injury or damage in an accident. Incidents can, therefore, reveal the same hazards as accidents, without the associated injury or damage.

- They are far more numerous than accidents (estimates range from 10 to 100 times more numerous). Thus, they are a plentiful source of hazard information.

- The people involved in incidents are available to provide additional information on the hazards that caused them.

It stands to reason that the introduction of a comprehensive incident reporting and investigation system requires money and labor hours. However, experience has shown that such systems are cost-effective, as incident investigation offers true "before the accident" prevention. Further details of accident and incident prevention will be explored in Chapter 13.

ESTABLISHING A CULTURE OF SAFETY REPORTING

Timely and accurate reporting of safety occurrences is critical to modern aviation safety management programs. ICAO has long recognized that establishment of a proper safety culture is key to effective safety reporting. According to the ICAO

Safety Management Manual (ICAO Doc. 9859, 2nd edition), certain attributes and traits are necessary for an organization to perform well in this area. The five basic attributes of an effective safety reporting organization are as follows:

- A demonstrated willingness to report errors and experiences
- Information is received from knowledgeable aviation professionals
- Flexibility of reporting is available for information to travel directly to the appropriate decision maker in unusual circumstances
- A learning organization with the ability to draw conclusions and implement major reforms
- An accountable organization where people are encouraged and rewarded for providing essential safety related information.

Figure 4-1 emphasizes the five basic traits of effective safety reporting and additional information is provided at www.icao.int.

Experience indicates that successful incident reporting systems employ most of the following characteristics:

- *Trust.* Persons reporting incidents must be able to trust the recipient organization and be confident that any information they provide will not be used against them. Without such confidence, people are reluctant to report their mistakes,

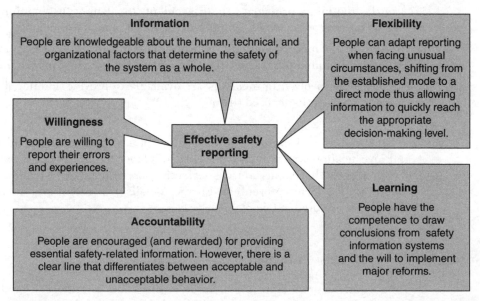

FIGURE 4-1   Effective safety reporting—five basic traits. (*Source: www.icao.int ICAO SMM 9859 2nd ed.*)

and they may also be reluctant to report other hazards they are aware of. For an incident reporting system to be successful, it needs to be perceived as being *non-punitive* with regard to unintentional errors or mistakes.

- *Confidentiality.* Non-punitive systems are based on confidential reporting. The person reporting an incident must be sure that his identity and other information that may be used to identify those involved will not be disclosed.

- *Independence.* Ideally, an incident reporting system should be run by an organization divorced from the federal agency that is also responsible for the enforcement of aviation regulations. Accordingly, some countries, including the United States, use a *third party* for the management of so-called voluntary reporting systems. The third party receives, processes, and analyzes the submitted incident reports and feeds the results back to the federal agency and the aviation community. With so-called mandatory reporting systems, it may not be possible to employ a third party. Nevertheless, it is desirable that the federal agency give a clear understanding that any information received will be used for accident prevention purposes only. This principle also applies to an airline or any other aircraft operator that uses incident reporting as part of its accident prevention program.

- *Ease of reporting.* The task of submitting incident reports should be as easy as possible for the reporter. Reporting forms should be readily available so that anyone wishing to file a report can do so easily. They should be simple to compile, with adequate space for a descriptive narrative, and they should also encourage suggestions on how to improve the situation or prevent a recurrence. Classifying information such as type of operation, light conditions, type of flight plan, weather, and so forth can be presented in a "check" format. The forms should ideally be self-addressed and postage-free.

- *Acknowledgment.* The reporting of incidents requires considerable time and effort by the user and should be appropriately acknowledged. Whenever possible, feedback on the actions taken in response to a report should be provided to the reporting person.

- *Motivation and promotion.* The information received from an incident reporting system should be made available to the aviation community as soon as possible, as this may help to motivate people to report further incidents. Such promotion activities may take the form of monthly regular e-mail or internet bulletins, newsletters or periodic summaries. Ideally, all such methods would be used with a view to achieving maximum exposure.

INCIDENT REPORTING SYSTEMS

Effective incident reporting systems can be organized in different ways. However, there are two main types that characterize the majority of systems used by aviation authorities: mandatory and voluntary.

MANDATORY INCIDENT REPORTING SYSTEMS

In a mandatory reporting system, people are required to report certain types of incidents, which necessitates detailed regulations outlining who shall report and what shall be reported. Otherwise, the mandatory system could not be enforced. To achieve this goal and avoid unnecessary duplication, those items requiring an incident report must be segregated from the day-to-day problems, defects, and so forth for which adequate control systems and procedures should already exist. In effect, this means establishing a base level, in terms of hazards, below which an incident report is not necessary. Unless this is done, the mandatory system may be flooded with reports, possibly obscuring important items. It is important to concentrate what are usually limited resources where they will be most effective.

The number of variables in aircraft operations is so great that it is very difficult to provide a complete list of items or conditions that should be reported. For example, loss of a single hydraulic system on an aircraft with only one such system is critical; on a type with three or four systems, it may not be. A relatively minor problem in one set of circumstances can, when the circumstances change, result in a hazardous situation. The rule should be: *If in doubt, report.*

Because mandatory systems deal mainly with specific and concrete matters, they tend to collect more information on technical failures than on the human-factor aspects. To help overcome this problem, some countries with a well-developed mandatory reporting system also have a voluntary incident reporting system aimed specifically at acquiring more information on the human-factor aspects.

VOLUNTARY INCIDENT REPORTING SYSTEMS

In a *voluntary reporting system,* pilots, controllers, and others involved in aviation are invited (rather than required) to report hazards, discrepancies, or deficiencies in which they were involved or which they observed. Experience in a number of countries, including the United States, has shown that a voluntary system requires a trusted third party (i.e., NASA) to manage the system. The reason is simply that people are reluctant to report their mistakes to the carrier that employs them or the government agency that licenses them.

In voluntary systems, *confidentiality* is usually achieved by deidentification, or not recording any identifying information. Because of this confidentiality, voluntary systems tend to be more successful than mandatory systems in collecting human factor–related information without fear of retribution or embarrassment.

The next sections discuss the salient features of accident/incident recording and reporting systems that are related to aviation, general industry, and the environment.

AVIATION RECORDING AND REPORTING SYSTEMS

The U.S. federal government collects vast amounts of aviation data to support its responsibility for overseeing aviation safety. The four major organizations that collect and analyze aviation safety and safety-related data are the FAA, the NTSB, The Bureau of Transportation Statistics (BTS) of the Research and Innovative Technology Administration. and the National Aeronautics and Space Administration (NASA).

FEDERAL AVIATION ADMINISTRATION REPORTING SYSTEMS

The FAA collects a wide range of aviation information and automated database systems. Most of the safety-related data are collected and managed under the FAA Associate Administrator for Aviation Safety (AVS) and its subordinate offices. In today's era of internet databases and Web sites, the reader is urged to check the most recent Web site pages for updated information at www.faa.gov.

THE AVIATION SAFETY INFORMATION ANALYSIS AND SHARING SYSTEM (ASIAS) (www.asias.faa.gov)

As indicated on its Web site, the FAA promotes open exchange of safety information in order to continually improve aviation safety. The FAA developed the Aviation Safety Information Analysis and Sharing system to enable users to perform integrated searches across multiple databases. The ASIAS portal to these databases is an evolving data warehouse of safety data which contains weblinks and crosslinks to a variety of sources organized as follows:

- *Lessons learned from transport airplane accidents*—This link features a large library of videos and animations of major aircraft accidents and illustrates the complex interrelationship of accident causes. It is an excellent resource for students beginning the study of commercial aviation safety.
- *Data and information link*—This tab provides access to the multiple databases and tools alphabetically or by subject.
- *Subjects and studies link*—These tabs provide access to high level findings, studies, and reports. Among the more important of these databases are the following:
  - FAA Accident/Incident Data System (AIDS)
  - Aviation Safety Reporting System (ASRS)
  - Bureau of Transportation Statistics (BTS) on Airline Traffic
  - Near Midair Collision System (NMACS)
  - NTSB Aviation Accident and Incident Data System (NTSB)
  - NTSB Safety Recommendations to the FAA with FAA responses
  - Runway Safety Office, Runway Incursions
  - World Aircraft Accident Summary (WAAS) produced by the British Civil Aviation Authority
- Other database services available on FAA.gov
  - *Service Difficulty Reporting System (SDRS)*. These reports are filed by air carriers, repair stations, manufacturers, FAA inspectors, and others concerning specific types of mechanical problems, aircraft failures, or malfunction.
  - *Air Traffic Activity Data System (ATADS)*. This system contains the official National Airspace System air traffic operations data available for public release.

- *Operational errors.* An operational error is a violation of ATC separation standards that define minimum safe distances between aircraft, between aircraft and other physical structures, and between aircraft and otherwise restricted airspace. Separation is one of the fundamental principles of aviation safety. FAA air traffic facilities have a software program that detects possible operational errors and sends alert messages to FAA supervisory personnel. The FAA conducts annual reviews of reported data and summarizes it in the Air Traffic Operational Error and Deviation database.

- *Pilot deviations.* A pilot deviation is defined as an action of a pilot that results in a violation of a FAR regulation while in flight. Such deviations could result in the loss of separation between one airplane and another, or a safety of flight violation. The most frequent types of pilot deviations are altitude, course clearance, airspace and airspeed violations, and flying VFR into Instrument Meteorological Conditions (IMC). When ATC observes a pilot deviation, it asks the pilot to telephone the controlling agency, and files a report to the local Flight Standards District Office (FSDO).

INDUSTRY VOLUNTARY SAFETY REPORTING INITIATIVES

As noted in Chapter 1, the serious nature of the circumstances surrounding the crash of Colgan Air Flight 3407 near Buffalo, New York, has prompted new safety legislation signed into law August 1, 2010 entitled the *Airline Safety and Federal Aviation Administration Extension Act of 2010.* Section 213 of this statute strongly endorses four existing voluntary safety programs:

- Aviation Safety Action Program (ASAP)
- Flight Operational Quality Assurance (FOQA)
- Advanced Qualification Program (AQP)
- Line Operations Safety Audit (LOSA)

Section 214 of this law provides for an ASAP and FOQA implementation plan for all FAR Part 121 air carriers not later than one year after enactment of the Act (August 1, 2011), and requires a detailed report if an air carrier is not using AQP and LOSA as part of their safety system. Thus, these complementary "voluntary" safety programs will likely be required as FAA mandated programs in the near future.

AVIATION SAFETY ACTION PROGRAM (ASAP—FAA ADVISORY CIRCULAR 120-66B).   The goal of ASAP is to prevent accidents and incidents by identifying unsafe practices and correcting them. Voluntary reporting of safety issues by employees is encouraged. Under ASAP, safety issues are resolved through corrective action rather than through punishment or discipline. The program's strategy is to create a nonpunitive environment through enforcement-related incentives for employees to report safety issues, even though these issues may involve an alleged violation of aviation regulations. In ASAP an airline enters into a safety partnership with the FAA and could include a third party such as the employee's labor union. Information contained in ASAP reports may be considered sensitive by some airlines so the FAA has tried to overcome this problem by proposing regulations that would protect ASAP reports and other sensitive safety and security information from being disseminated to the public.

FLIGHT OPERATIONAL QUALITY ASSURANCE (FOQA—FAA ADVISORY CIRCULAR 120-82).   Flight operational quality assurance programs involve collecting and analyzing data recorded during flight to improve the safety of flight operations (for more on this see Automatic Recording Systems later). In excess of 200 flight parameters can be accessed from digital flight data recorders downloaded frequently through quick-access recorders. While FOQA programs are quite popular with international airlines, only a few U.S. airlines use FOQA because of data confidentiality issues. In 2001 the FAA took steps to address the confidentiality issue by codifying carrier data and ensuring confidential enforcement protection for any operator who operates aircraft under an FOQA program approved by the FAA. The regulations would ban disclosure of sensitive safety information and security information that was voluntarily submitted to an approved FAA program.

ADVANCED QUALIFICATION PROGRAM (AQP—FAA ADVISORY CIRCULAR 120-54A).   The overall goals of AQP are to increase aviation safety by using innovative training and qualification concepts, and to be responsive to changes in aircraft technology, operations, and training methodologies. Some distinguishing features of AQP:

- It is proficiency based using a full cockpit crew environment.
- It is aircraft specific by make, model, and series of aircraft using full flight simulators or other simulator flight training equipment.
- It emphasizes the concept of crew resource management (CRM) training, which is mandatory.

The ultimate goal of AQP is to achieve the highest possible standards of individual and crew performance to levels above the present pilot training standards in FAR Parts 121 and 135.

LINE OPERATIONS SAFETY AUDIT (LOSA—FAA ADVISORY CIRCULAR 120-90).   As indicated in the advisory circular, a LOSA is a voluntary formal process that uses highly trained observers to collect safety-related date on regularly scheduled airline flights. The observer is often a trusted airline captain who rides the jump seat in the cockpit to obtain data about flight crew behavior and crew strategy for managing threats and errors under conditions of operational complexity. The concept of Threat and Error Management (TEM) provides that such threats and errors are part of everyday operations that flightcrews must manage to maintain flight safety.

ICAO also promotes the importance of LOSA to the international aviation community in its Line Operations Safety Audit Manual (ICAO document 9803).

The initial research on LOSA was developed during a joint effort between FAA, Continental Airlines and the University of Texas at Austin. This research became known as the University of Texas Human Factors Research Project. Typically, a LOSA project can take between 6 and 12 months to conduct, and the FAA recommends a LOSA be conducted every 3 years. Additional information will be provided in Chapter 7, entitled Human Factors in Aviation Safety.

Air Transportation Oversight System (ATOS). The Air Transportation Oversight System was implemented in 1998 as a new approach to FAA certification and surveillance oversight, using system safety principles and systematic processes to ensure that air carriers are in compliance with FAA regulations and have safety built into their operating systems. Unlike the traditional oversight methods, ATOS incorporates the structured application of new inspection tasks, analytical processes, and data collection techniques into the oversight of individual air carriers. This approach enables Flight Standards inspectors to be more effective in the oversight of air carriers by focusing on the most critical safety aspects of an air carrier's operation. As currently applied, ATOS provides a systematic process for conducting surveillance, identifying and dealing with risks, and providing data and analysis to guide the oversight of each carrier. As provided on its Web site, ATOS implements FAA policy by providing safety controls of aviation business organization that fall under FAA regulations. Three major functions define this oversight system (source: www.faa.gov).

- *Design assessment.* Is the ATOS function that ensures an air carrier's operating systems comply with FAA regulations and safety standards.
- *Performance assessments.* Confirm that an air carrier's operating systems produce intended results, including mitigation or control of hazards and associated risks.
- *Risk management.* This process deals with hazards and associated risks, and is used to manage FAA resources according to risk-based priorities.

By collecting and analyzing data on the many airline systems, FAA inspectors are better able to target areas for improvement. ATOS is a significant shift in the way in which the FAA oversees airlines and how its inspectors operate. It should lead to a more collaborative partnership toward system evaluation and may open doors for additional information sharing between air carriers and the FAA. With the cooperation of the air carriers, ATOS will foster a more proactive relationship with the FAA on safety-related issues as emerging trends and concerns are identified. This will aid in achieving the AFS mission, which is to provide the public with the safest aircraft operations in the world. The relationship between FAA oversight and the concept of Safety Management Systems (SMS) will be further explored in Chapter 13 of this book.

AFS is in the process of developing a centralized analysis and information management system to accumulate, analyze, and disseminate safety data and information within Flight Standards and to assist airlines in the interpretation of data. Information will be disseminated from a computer-based decision support tool called the *Safety Performance Analysis System (SPAS)*. It will include selected air carrier data and data summaries from ATOS and other sources including Aviation Safety Reporting System (ASRS) reports and Office of System Safety information. This information will assist inspectors and air carriers in decision making with respect to targeting surveillance resources and taking corrective actions to mitigate safety risks. Planned enhancements will continue to be validated for effectiveness and added to the system as appropriate.

NTSB ACCIDENT/INCIDENT REPORTING SYSTEM

Since its inception in 1967, the NTSB has kept records of civil aircraft accidents and its Aviation Incident/Accident Database contains information on every known civil aviation accident in the United States. Accidents involving only military or public-use aircraft are not usually investigated by the National Transportation Safety Board. While this incident/accident reporting system was primarily designed for administrative purposes, the system does have analytical capabilities. The NTSB Aviation Accident/Incident database is the official repository of aviation accident data and causal factors. The NTSB publishes annual reviews of aircraft accident data and occasional special studies, which are supported by statistical analyses accomplished with the data system. As discussed earlier, the NTSB is responsible for investigating all aircraft accidents and certain incidents, determining their probable causes, and making recommendations to the FAA. The NTSB uses Form 6120.19A to generate a preliminary report that is completed within 5 working days. It then develops the more informative factual report using Form 6120.4F. The final report may take months and even years to complete. Preliminary reports contain a limited amount of information (date, location, aircraft operator, type of aircraft, etc.). The NTSB Aviation Accident/Incident database is presented in the following categories: location information, aircraft information, operator information, narrative, sequence of events, findings, injury summary, weather/environmental information, and pilot information.

DEPARTMENT OF TRANSPORTATION REPORTING SYSTEM

The *Bureau of Transportation Statistics (BTS)*, a unit of the U.S. Department of Transportation, compiles and analyzes data on the nation's transportation system. By law, the airlines are required to provide financial, activity, and other data to BTS. They are used to develop and provide information to the flying public on the condition and performance of the nation's air transportation system. Some of the measures cataloged include the number of departures, flight hours, and miles performed in domestic commercial service during the most recent 5 calendar years, by airline.

These are also the same measures tracked by government and industry to calculate accident or incident rates for the air transportation system.

On February 20, 2005, BTS became a part of the Research and Innovative Technology Administration (RITA). RITA is composed of BTS, the former Research Office of the Research and Special Programs Administration (RSPA), Volpe National Transportation Systems Center (formerly with RSPA), Transportation Safety Institute (formerly with RSPA), and Office of Intermodalism (formerly with the Office of the Secretary). BTS is headed by a Director, appointed by the Secretary of Transportation, and the Director reports to the RITA Administrator.

NATIONAL AERONAUTICS AND SPACE ADMINISTRATION REPORTING SYSTEM

The National Aeronautics and Space Administration provides and supports aviation research and development and administers the confidential, voluntary, and non-punitive *Aviation Safety Reporting System (ASRS)*. The Aviation Safety Reporting System is designed to encourage reports by pilots and air traffic controllers concerning errors and operational problems in the aviation system by guaranteeing anonymity and immunity from prosecution for all reporters. More than 880,000 incident reports have been submitted to date without any reporter's identity ever being revealed. System data can provide an alternate federal insight into the nature and trends of aviation incidents.

NASA AVIATION SAFETY REPORTING SYSTEM (NASA REPORT). The Aviation Safety Reporting System is an extremely successful joint effort by FAA, and NASA, to provide a voluntary reporting system where pilots, controllers, flight attendants, mechanics and others can submit accounts of safety-related aviation incidents. The system is funded mainly by the FAA, administered by NASA, and maintained by Battelle Laboratories. Reports are sent to the Aviation Safety Reporting System office at NASA Ames Research Center, where the data are analyzed and entered into a computer by employees of Battelle.

Since its inception in 1976, the NASA report process has been very successful as a voluntary, nonpunitive means to report and correct aviation safety discrepancies and this process has been widely copied in the strong immunity policy in FAA Advisory Circular 00-46D and FAR Section 91.25 which provide that FAA will not use NASA ASRP reports in any enforcement action absent a crime, intentional violation or serious accident excluded from the program. The only significant limitation is that the aviation professional must file the report within 10 calendar days of the incident to avoid FAA fines and penalties. This 10-day deadline is strictly construed by FAA, but the electronic report submission available at Web site http://asrs.arc.nasa.gov/ facilitates such reports. A copy of page one of the NASA Report format is provided in Fig. 4-2 for your information.

The Aviation Safety Reporting System report form was designed to gather the maximum amount of information without discouraging the reporter. Structured information blocks and key words are provided, not only to guide the reporter, but also to aid subsequent data retrieval and research. Narrative descriptions are

DO NOT REPORT AIRCRAFT ACCIDENTS AND CRIMINAL ACTIVITIES ON THIS FORM.
ACCIDENTS AND CRIMINAL ACTIVITIES ARE NOT INCLUDED IN THE ASRS PROGRAM AND SHOULD NOT BE SUBMITTED TO NASA.
ALL IDENTITIES CONTAINED IN THIS REPORT WILL BE REMOVED TO ASSURE COMPLETE REPORTER ANONYMITY.

(SPACE BELOW RESERVED FOR ASRS DATE/TIME STAMP)

IDENTIFICATION STRIP: *Please fill in all blanks to ensure return of strip.*
NO RECORD WILL BE KEPT OF YOUR IDENTITY. This section will be returned to you.

TELEPHONE NUMBERS where we may reach you for further details of this occurrence:

HOME　Area _____　No. _____　Hours _____

WORK　Area _____　No. _____　Hours _____

NAME _____

ADDRESS/PO BOX _____

_____

CITY _____ STATE _____ ZIP _____

TYPE OF EVENT/SITUATION

_____

_____

DATE OF OCCURRENCE _____
(MM/DD/YYYY)

LOCAL TIME (24 hr. clock) _____
(HH:MM)

PLEASE FILL IN APPROPRIATE SPACES AND CHECK ALL ITEMS WHICH APPLY TO THIS EVENT OR SITUATION.

REPORTER	FLYING TIME (in hours)	CERTIFICATES & RATINGS	ATC EXPERIENCE
☐ Captain ☐ Single Pilot ☐ First Officer ☐ Instructor ☐ pilot flying ☐ Trainee ☐ pilot not flying ☐ relief pilot ☐ Dispatcher: _____ yrs ☐ check airman ☐ Other: _____	Total Time _____ hrs Last 90 Days _____ hrs Time in Type _____ hrs	☐ Student ☐ Flight Instructor ☐ Sport/Rec ☐ Multiengine ☐ Private ☐ Instrument ☐ Commercial ☐ Flight Engineer ☐ ATP ☐ Other: _____	☐ FPL ☐ Developmental radar _____ yrs non-radar _____ yrs supervisory _____ yrs military _____ yrs

AIRSPACE	CONDITIONS/WEATHER ELEMENTS	LIGHT/VISIBILITY	ATC/ADVISORY SVC.
☐ Class A ☐ Class E ☐ Class B ☐ Class G ☐ Class C ☐ Special Use ☐ Class D ☐ TFR	☐ VMC ☐ fog ☐ snow ☐ IMC ☐ hail ☐ thunderstorm 　　　 ☐ haze/smoke ☐ turbulence ☐ Mixed ☐ icing ☐ windshear ☐ Marginal ☐ rain ☐ other: _____	☐ dawn ☐ night ☐ daylight ☐ dusk Ceiling _____ feet Visibility _____ miles RVR _____ feet	☐ Ramp ☐ Center ☐ Ground ☐ FSS ☐ Tower ☐ UNICOM ☐ TRACON ☐ CTAF ATC Facility Name: _____

	AIRCRAFT 1		AIRCRAFT 2	
Your Aircraft Type (Make/Model) (e.g. B737) NOT "N #", Flt #, etc.: _____		Operating FAR Part: _____	Other Aircraft: _____	Operating FAR Part: _____
Operator	☐ air carrier ☐ fractional ☐ military ☐ air taxi ☐ FBO ☐ personal ☐ corporate ☐ government ☐ other: _____		☐ air carrier ☐ fractional ☐ military ☐ air taxi ☐ FBO ☐ personal ☐ corporate ☐ government ☐ other: _____	
Mission	☐ passenger ☐ cargo/freight ☐ ferry ☐ personal ☐ training ☐ other: _____		☐ passenger ☐ cargo/freight ☐ ferry ☐ personal ☐ training ☐ other: _____	
Flight Plan	☐ VFR ☐ SVFR ☐ none ☐ IFR ☐ DVFR		☐ VFR ☐ SVFR ☐ none ☐ IFR ☐ DVFR	
Flight Phase	☐ taxi ☐ climb ☐ final approach ☐ parked ☐ cruise ☐ missed/GAR ☐ takeoff ☐ descent ☐ landing ☐ initial climb ☐ initial approach ☐ other: _____		☐ taxi ☐ climb ☐ final approach ☐ parked ☐ cruise ☐ missed/GAR ☐ takeoff ☐ descent ☐ landing ☐ initial climb ☐ initial approach ☐ other: _____	
Route in Use	☐ airway (ID): _____ ☐ STAR (ID): _____ ☐ visual approach ☐ direct ☐ oceanic ☐ none ☐ SID (ID): _____ ☐ vectors ☐ other: _____		☐ airway (ID): _____ ☐ STAR (ID): _____ ☐ visual approach ☐ direct ☐ oceanic ☐ none ☐ SID (ID): _____ ☐ vectors ☐ other: _____	

If more than two aircraft were involved, please describe the additional aircraft in the "Describe Event/Situation" section.

LOCATION	CONFLICTS
Altitude: _____ (single value) ☐ MSL ☐ AGL Distance: _____ and/or Radial (bearing): _____ from: ☐ Airport _____ ☐ ATC Fac _____ ☐ Intersection _____ ☐ NAVAID _____	Estimated miss distance in feet: horiz _____ vert _____ Was evasive action taken?　　　　　　　　　☐ Yes ☐ No Was TCAS a factor?　　　　　　☐ TA ☐ RA ☐ No Did terrain warning system activate? [Reset] ☐ Yes ☐ No

FIGURE 4-2　NASA Aviation safety reporting system. (*Source: http://asrs.arc.nasa.gov/*)

encouraged. Space is provided for the reporter's name, address, and telephone number, which permits NASA to acknowledge the report's receipt by return mail and also allows the Battelle analyst to contact the reporter for follow-up data. Information that identifies the reporter is deleted before data are entered into the computer.

Under the guidance of NASA, Battelle receives the incident reports, processes and analyzes the data, and publishes reports of the findings. Human factors in aviation safety, a continuing concern at the NASA Ames Research Center, were a major consideration in the development of the Aviation Safety Reporting System. The data analysts, primarily experts in aircraft operations and air traffic control, provide insight into the nature of the human error or other underlying factors in the incidents. Although the reports are encoded in detail, the complete narrative text of each report is retained for later reevaluation. As the result of this process, over 5000 safety alert messages have been issued, correcting aviation safety discrepancies, and a wealth of human factors information is readily available from the ASRS database.

AUTOMATIC RECORDING SYSTEMS

Many modern air transport aircraft have automatic recording devices installed. The *flight data recorder* (FDR), which monitors selected parameters of the flight, and the *cockpit voice recorder* (CVR), which records voices and cockpit sounds, are installed to assist with the investigation of accidents and, in some cases, incidents. An engineering recorder may also be installed to monitor aircraft systems. The data from this recorder can be used to detect impending failures and to verify the adequacy of component life and overhaul schedules.

Automatic recorders that are installed in air traffic control and communication systems primarily for accident investigation purposes may also be used as a check on correct operating procedures.

FDRs and CVRs were initially installed to assist accident investigators in determining accident causes, particularly for catastrophic accidents to large aircraft. In some countries, professional groups such as pilots and air traffic controllers accepted the philosophy that the installation of these recorders would be of significant benefit to the aviation industry in helping to determine accident causes. Accordingly, they agreed to their use, provided that guarantees by operational and administrative agencies were negotiated and honored, preventing disciplinary action from being taken on the basis of information determined from the recorders unless willful negligence or dereliction of duty could be proved (see FOQA earlier).

Many countries routinely use flight recorder information for accident prevention. They regard this information as an invaluable source of safety insights and information on the operation of their aircraft. Standard flight profiles are usually programmed into a computer along with acceptable deviations. Recorded data are then compared with these standard profiles. Significant deviations are then examined to see if hazards could be present. If so, corrective action can then be taken.

This method need not require the identification of individuals, since it is often the number and type of deviations that reveal hazards.

Some carriers that routinely examine FDR records for indications of hazards or deviations from standard operating procedures have the findings reviewed by a committee consisting of retired captains or flightcrews. This group has the respect of both management and pilots and thus avoids direct employer/employee contact. The fear of job loss or punishment is avoided, and the accident prevention insights are more readily obtained.

ICAO INTERNATIONAL REPORTING SYSTEMS

The international exchange of accident and incident data provides a broad range of experience on which to base safety guidance. Such information can be of particular value to smaller countries or carriers that are not in a position to maintain an accident or incident reporting system, or whose database is too limited to permit the identification of potential hazards. Safety data interchange is, therefore, strongly encouraged among civil aviation authorities and safety organizations.

Annex 13 to the Chicago Convention entitled "Aircraft Accident Investigation" provides the basic standards which require member States to report to ICAO pertinent information on aircraft accidents and incidents (occurrences). Furthermore, ICAO standards provide that the "State of occurrence" shall forward notification of an accident or serious incident with minimum delay and by the most suitable and quickest means available to the following:

- State of registry of the aircraft
- State of the operator
- State of design
- State of manufacture

ICAO has developed a detailed process for such notification called the ICAO Accident/Incident Data Reporting System (ADREP). The version of the ADREP system in current use is ADREP 2000. Complete information on this system is available of www.icao.int and in ICAO Document 9156, its Accident/Incident Reporting Manual. Additional information is available in ICAO Document 9859, Safety Management Manual (2nd edition).

The ADREP system operates using a software program called the European Coordination Center for Aviation Incident Reporting System (ECCAIRS). This software system is provided to ICAO member States free of charge and has greatly improved ADREP reporting and classification of occurrences. Additionally, the ADREP/ECCAIRS system allows member States to easily exchange accident analysis tools. Many of the larger countries and safety organizations distribute material, such as films, videos, magazines, and summaries of accidents which are available to the general public.

ICAO ADREP SYSTEM. The ICAO ADREP system is a databank of worldwide accident and incident information for large commercial aircraft. Thus, ICAO can provide countries with accident prevention information based on wide international experience. The ICAO Accident/Incident Reporting Manual (ADREP) (Document 9156) contains detailed information on this system. ICAO provides the following information based on ADREP data:

- *ADREP Bi-monthly Summary* is a computer-generated publication containing the ADREP preliminary reports and data reports received by ICAO during a 2-month period. The last issue of each year contains a comprehensive index of the accident and incident reports reported to ICAO during that year.

- *ADREP Annual Statistics* is an ICAO circular containing annual statistics from the databank. These statistics may be useful for safety studies and accident prevention programs.

- *ADREP Information Requests* are computer printouts provided by ICAO in response to specific requests from countries. Guidance for the formulation of ADREP requests is contained in the ADREP manual.

The ADREP computer programs are available to countries wishing to utilize their computer systems for accident or incident recording. These programs, in addition to promoting standardized coding, offer significant financial savings to smaller ICAO member countries.

OTHER ICAO SAFETY INFORMATION. The ICAO also publishes the following aircraft accident and incident information:

- *Integrated Safety Trend Analysis and Reporting System* (iSTARS). Similar to the FAA ASIAS portal, ICAO recently released a *Beta* test version of its new iSTARS system. The iSTARS is a safety risk-based decision making tool which has the capability to facilitate identification of hazards and resolution of unacceptable safety risks through analysis of multiple safety related factors. ICAO member States are able to become registered users of iSTARS and contribute comments, maps, charts, documentation, or updates to ICAO accident data. As the system develops, this new portal of interactive safety information will greatly facilitate the sharing of international safety information.

- *Aircraft Accident Digest* is a publication that contains, narrative-type accident or incident final reports, selected for their contribution to accident prevention or the use of new or effective investigative techniques.

- *List of Final Reports Available from Countries* is a listing of narrative-type aircraft accident final reports available on request from the reporting countries. The list is updated every 6 months on the basis of information supplied by countries.

No single measurement or statistic provides a complete picture of commercial aviation safety. Although accident and fatality statistics are the best measures of long-term past risk in commercial aviation, they are of limited value over short periods and are not suitable monitors of short-term effects of policy decisions. For example, the consequences of rulings requiring collision avoidance systems on commercial transports and transponders on many general-aviation aircraft might not be apparent in the accident data for years to come. The safety benefits of the next generation of navigation and air traffic control communication systems will also be a major area of analysis and study for the foreseeable future.

Nonaccident safety data, while not substitutes for accident and fatality data, are valuable supplements. If properly collected and maintained, nonaccident data can help identify and estimate the magnitude of safety problems and permit the monitoring of safety programs. The usefulness of such data will be discussed further in later sections of the book.

GENERAL INDUSTRY (OSHA) RECORDING AND REPORTING SYSTEMS

APPLICABILITY

OSHA regulations listed in 29 CFR 1904 require every employer covered by the OSHA Act with 11 or more employees to record and report all employee occupational (work-related) deaths, injuries, and illnesses on OSHA Forms 300 and 301. Employers with 10 or fewer employees are required to keep injury and illness records only if OSHA or the Bureau of Labor Statistics (BLS) specifically notifies them in writing that they must keep these records. However, all employers covered by the OSHA Act must report all deaths and events that cause in-patient hospitalization of 3 or more employees orally to OSHA (in person or via phone to the OSHA area or central office) within 8 hours of being aware of the events. Injuries and illnesses that require only first aid are exempt from being recorded. *Injuries* include, but are not limited to, cuts, fractures, sprains, or amputations. Illnesses include both acute and chronic illnesses such as skin disease, respiratory disorder, or poisoning.

Injury and illness records are used for several purposes. OSHA collects data through the OSHA Data Initiative (ODI) to help direct its programs and measure its own performance. Inspectors also use the data during inspections to help direct their efforts to the hazards that are hurting workers. Records are used by employers and employees to implement safety and health programs at individual workplaces. Analysis of the data is a widely recognized method for discovering workplace safety and health problems and for tracking progress in solving those problems. Records also provide the base data for the BLS Annual Survey of Occupational Injuries and Illnesses, the nation's primary source of occupational injury and illness data. Each year, the BLS uses a stratified random data collection survey process to collect injury and operational data (e.g., hours worked) from companies across the United States. It then analyzes, categorizes, and publishes the data by industry groupings called *standard industrial classification* (SIC) codes. Companies can use these data to compare themselves to their peer industries. More on SIC codes and injury statistics will be covered in Chapter 5 on injury statistics.

OSHA FORM 300—LOG OF WORK-RELATED INJURIES AND ILLNESSES

This form requires the employer to record all occupationally related injuries and illnesses other than first aid and requires the following information:

- Employee's name, employee's job title, date, and the place (department) where the injury/illness occurred
- A very short description of the injury/illness including the body part that was affected and what caused it
- The number of lost or restricted workdays that the injury/illness caused
- The type or classification of the injury/illness

OSHA FORM 301—INJURY AND ILLNESS INCIDENT REPORT

This incident report Form 301 contains more detailed information on the injuries/illnesses than that listed in Form 300. Every injury or illness listed in Form 300 must have an accompanying Form 301. Information on Form 301 describes in greater detail how the injury or illness occurred, what time of day it happened, what the employee was doing when he or she was injured, and the extent of medical treatment the injury or illness required. Actual copies of OSHA Forms 300 and 301 can be downloaded from the OSHA Web site (www.osha.gov). Note that the FAA uses the Mishap Report Form 3900-6 (or succeeding form) for its own internal facilities and operations.

OSHA FORM 300A—SUMMARY OF WORK-RELATED INJURIES AND ILLNESSES

OSHA Form 300A, the annual summary of illnesses and injuries form and the OSHA Form 301 incident report must be retained on file by an employer for 5 years following the end of the calendar year that these records cover (29 CFR 1904.33). An employer is required to post a summary (Form 300A) of the total number of injuries and illnesses for the (calendar) year in a conspicuous place (preferably on the employee bulletin boards) at each establishment. Totals for the previous year must be displayed continuously from February 1 through April 30 of the year, following the year covered by the form.

OPEN GOVERNMENT INITIATIVE (TRANSPARENCY IN GOVERNMENT)

OSHA is considering implementation of a new, improved and modernized record-keeping system which will be made public to encourage innovative ideas in improving occupational, safety and health. Current information on this new process is available at www.osha.gov.

ENVIRONMENTAL (EPA) RECORDING AND REPORTING

Under federal environmental laws, any facility or vessel (e.g., oil spills from ships) must report to government authorities about any hazardous substance that is released into the environment in quantities that exceed a threshold amount. These threshold amounts are referred to as *reportable quantities* (*RQs*) and must be reported in a timely manner. In addition to release reporting required under CWA, CERCLA, (Superfund), EPCRA (SARA Title III), RCRA, and TSCA, state statutes may require additional release reporting. The environment can be air, water, soil, or a combination of the three media. A release includes any contact of a hazardous substance with the environment through either intentional or unintentional discharge and by any method (e.g., spilling, leaking, dumping, leaching, and abandonment of barrels or other closed containers containing hazardous substances). Notification of releases is usually required immediately or within 24 hours of knowledge of the release. EPA regulations cover a variety of reporting requirement that include

- Medium (environments) into which the release is applicable
- Listing of hazardous substances together with their quantities
- Persons responsible for reporting
- Time line for reporting
- The agency to contact in case of a release
- Fines for not reporting

In some cases there is also a requirement for follow-up reporting which must include the original notification, response actions taken, acute or chronic health risks associated with the release, and advice on medical care to exposed victims.

All these reporting release requirements help ensure accountability for clean-up, provide data for regulatory and public policy purposes, and enable federal, state, and local authorities to effectively prepare for and respond to emergencies that could harm humans and the environment. The EPA does not necessarily take enforcement action against releasers, but is quite strict against those violators who fail to report releases.

CERCLA Reporting Requirements

Under this act, if CERCLA-defined hazardous substances greater than or equal to their RQs are released into the environment, then the person in charge of the vessel or facility from where the release occurred must immediately notify the *National Response Center* (*NRC*). The NRC is staffed by U.S. Coast Guard personnel who are on watch 24 hours a day, 365 days a year. The NRC can be reached at (800) 424-8802 or on the web at www.nrc.uscg.mil/. The NRC acts as the single federal agency point of contact for all hazardous release reporting.

CERCLA-defined hazardous substances include substances listed in 40 CFR Part 302.4 and under certain sections of the CAA, CWA, RCRA, and TSCA. As and when substances get regulated under the CAA, CWA, RCRA, or TSCA, they automatically become CERCLA hazardous substances. When a substance is delisted from the CAA, CWA, RCRA, or TSCA lists, it may still be regulated under CERCLA if the EPA administrator independently determines the substance to be hazardous.

EPCRA REPORTING REQUIREMENTS

EPCRA (SARA Title III) was enacted in 1986, with the primary objective of informing communities of chemical hazards in their areas. All rules pertaining to this section are found in the EPA regulation of 40 CFR Subchapter J, Superfund, Emergency Planning, and Community Right-to-Know Programs (Parts 300 through 399). EPCRA reporting requirements are the same as CERCLA requirements, but in addition, EPCRA regulations list *extremely hazardous substances* (*EHSs*) and requires their reporting in case of a release in excess of the RQs. EPCRA requires facilities to notify *state emergency response commissions* (*SERCs*) and *local emergency planning committees* (*LEPCs*). SERCs and LEPCs are dedicated emergency response groups that exist at the state and local levels and were established under EPCRA. The governor of each state must designate a SERC. Many SERCs include state department agencies involved with the environment, natural resources, emergency services, public health, occupational safety, and transportation. The SERC designates local emergency planning districts and appoints LEPCs for each district. The LEPC has representation from elected state and local officials, police, fire, civil defense, public health professionals, environmental, hospital, emergency planning, and transportation officials.

If an EPCRA release also qualifies as a CERCLA release, then in addition to the SERC and the LEPC, the NRC must be notified. Since different criteria were used to compile the list of chemicals, there are chemicals common to both CERCLA and EPCRA while there are some chemicals that appear on one list, but not the other.

EPCRA's emergency planning requirements (40 CFR Parts 302 to 303 and 355.30) requires SERC and LEPC notification if an EHS equal to or in excess of its *threshold planning quantity* (*TPQ*) is stored or used at a facility. EHSs and their TPQs are identified in 40 CFR Part 355. For example, the FAA triggers this reporting requirement as it has sulfuric acid (used in battery electrolyte) present at the Air Route Traffic Control Centers (ARTCCs) and the Air Traffic Control Towers (ATCTs) in excess of its 1,000-pound TPQ. In addition, the FAA's Aeronautical Center and Technical Center has chlorine (used in water and wastewater treatment) in excess of 100 pounds, calcium hypochlorite (used in water and wastewater treatment) in excess of 10 pounds, and methyl chloroform (used as a solvent cleaner) in excess of 100 pounds.

EPCRA Section 313 requires EPA and the states to collect data on chemical releases and make the data available to the public through the *Toxics Release Inventory* (*TRI*) program. And 40 CFR Part 370 (Hazardous Chemical Reporting: Community Right-to-Know) establishes reporting requirements of hazardous chemicals to enhance community awareness and to facilitate the development of state and local emergency response plans. And 40 CFR Part 372 (Toxic Chemical Release Reporting: Community Right-to-Know) requires timely reporting of chemical releases to the EPA. There are several different types of reports that cover diverse topics ranging from oil spills and chemical accidents to releases of dangerous substances into the water or air. The EPA compiles and organizes all this reported information annually, and it makes this information available to the public through printed documents and online databases such as the Toxics Release Inventory. The goal of TRI is to educate the public by disseminating this information so that communities may be empowered to hold companies and local governments accountable for how toxic chemicals are managed. Currently, there are about 650 chemicals included in the TRI. Extended information on TRI including chemical lists, toxicity, and other reporting and program information can be found at http://www.epa.gov.

RCRA SUBTITLE I (UNDERGROUND STORAGE TANKS)

The sections within this part (40 CFR Parts 280 and 281) set technical standards and corrective action requirements for owners and operators of *underground storage tanks* (*USTs*). Underground fuel tanks are of primary concern in fueling handling and storage at airports and fuel supplier work sites.

Reporting requirements. Facilities must submit the following information:
- Notification for all UST systems which includes certification of installation for new UST systems
- Reports of all releases including suspected releases, spills and overfills, and confirmed releases
- Corrective actions planned or taken including initial abatement measures, initial site characterization, free product removal, and investigation of soil and groundwater cleanup
- A notification before permanent closure or change in service

Recordkeeping requirements. Facilities must maintain the following information:
- A corrosion expert's analysis of site corrosion potential if corrosion protection equipment is not used
- Documentation of operation of corrosion protection equipment
- Documentation of UST system repairs
- Recent compliance with release detection requirements
- Results of the site investigation conducted at permanent closure

OTHER REPORTING REQUIREMENTS

Other regulations (in addition to CERCLA and EPCRA) may also require reporting of releases of hazardous materials. Some of these are as follows:

- *The Hazardous Materials* (HAZMAT) *Transportation Act* (HMTA) requires reporting of hazardous material releases during shipping and handling. These regulations are administered by the Department of Transportation (DOT).
- *The Toxic Substances Control Act* (TSCA) requires reporting of releases of polychlorinated biphenyls (PCBs), and also addresses asbestos, radon and lead-based paint.
- *The Clean Water Act* (CWA) requires reporting of hazardous substances listed under 40 CFR 116.4 and oil releases from vessels and facilities into navigable waters.
- *RCRA listed* hazardous wastes (F, K, P, U) under 40 CFR Part 261, Subpart D, and *characteristic* wastes (I, C, R, TC) under 40 CFR Part 261, Subpart C, are reportable if the release equals or exceeds the designated RQ.

CONTROL OF AIR POLLUTION FROM AIRCRAFT AND AIRCRAFT ENGINES

The sections within this part (40 CFR Part 87) deal with setting limits on exhaust emissions of smoke from aircraft engines, ban venting of fuel emissions into the atmosphere from certain types of aircraft engines, and require the elimination of intentional discharge to the atmosphere of fuel drained from fuel nozzle manifolds after engines are shut down. The EPA is also reviewing new regulations regarding lead emissions from piston engine aircraft using AVGAS. Current information is available on www.epa.gov.

KEY TERMS

Accident

Fatal injury

Serious injury

Substantial damage

Incident

NTSB accident classifications: major, serious, injury, damage

Near-midair collision

Mandatory reporting system

Voluntary reporting system

Aviation Safety Information Analysis and Sharing System (ASIAS)

FAA Accident/Incident Data System

Service Difficulty Reporting System

Operational Error and Pilot Deviation Reporting System

Aviation Safety Reporting System

Aviation Safety Action Program (ASAP)

Flight Operational Quality Assurance (FOQA)

Advanced Qualification Program (AQP)

Line Operations Safety Audit (LOSA)

Air Transportation Oversight System (ATOS)

Flight data recorder (FDR)

Cockpit voice recorder (CVR)

ICAO Accident/Incident Reporting System (ADREP)

ICAO Integrated Safety Trend Analysis and Reporting Systems (iSTARS)

OSHA Forms 300, 300A, and 301

OSHA injury and illness

Standard industrial classification (SIC) codes

CWA, CERCLA, EPCRA (SARA Title III), RCRA, RCRA Subtitle I, TSCA, and HMTA reporting

Underground storage tanks (USTs)

Pollution from aircraft and aircraft engines—40 CFR Part 87

REVIEW QUESTIONS

1. What is the difference between accidents and incidents? Identify several types of incidents.

2. List and define NTSB's classification of accidents.

3. Why do accident investigations still provide the best insight and information leading to accident prevention? What is the purpose of publicizing safety recommendations following an accident?

4. Explain the importance of establishing a culture of safety reporting. What is the difference between a mandatory and voluntary reporting system? What characteristic of voluntary systems is particularly important?

5. What is included in the FAA Accident Incident Data System? What is the purpose of the Service Difficulty Reporting System?

6. How does the Aviation Safety Information Analysis and Sharing System work?

7. Discuss the salient features of ASAP, FOQA, AQP and LOSA.

8. Describe the database system managed by the NTSB. Why was NASA chosen to manage the Aviation Safety Reporting System? What is its function?

9. What is the purpose of flight data recorders and cockpit voice recorders? How might they be used for accident prevention purposes? What is the ICAO ADREP? Give several examples of information provided by ICAO and based on ADREP data.

10. Discuss OSHA's recording requirements for injuries and illnesses and discuss the reasons for collecting these data.

11. Discuss the role that BLS plays in recording and analyzing occupational injury/illness data.

12. Discuss some of the salient features of EPA's hazardous substance reporting requirements, giving reasons for the same.

13. What does 40 CFR Part 87 address?

REFERENCES

International Civil Aviation Organization. 2009. Safety Management Manual (SMM), 2d ed. Montreal, Canada: www.icao.int
____ 2006. Accident/Incident Reporting Manual (ADREP Manual) Montreal, Canada (ICAO Doc 9156).

WEB REFERENCES

http://www.icao.int/
www.ntsb.gov/
http://asias.faa.gov/
http://asrs.arc.nasa.gov/
www.epa.gov/
www.osha.gov/
www.nrc.uscg.mil/

REVIEW OF SAFETY STATISTICS

LEARNING OBJECTIVES

After completing this chapter, you should be able to

- Explain why the accident rate is a better measure of safety than accident counts.

- Discuss some of the issues to be aware of when analyzing and comparing commercial aviation accident statistics.

- Discuss the results of Boeing's *Summary of Commercial Jet Transport Aircraft Accidents.*

- Summarize the trend in aviation accident statistics as reported by United States and international commercial aviation during the 1980s, 1990s, 2000s, and 2010s.

- Discuss future concepts in ICAO accident prevention.

- Discuss OSHA injury and illness rate statistics.

INTRODUCTION

In passenger transportation, *safety factors* (also referred to as *causal factors*) are events that are associated with or influence fatality rates. A *safety indicator* is a measurable safety factor. The probability of death (or injury) as a result of traveling on a given mode, when quantified, is the primary benchmark of passenger transportation safety. Vehicle accident rates also are commonly used as safety indicators since most passenger fatalities occur as a result of vehicle accidents.

If one were to look at risk as the probability of death, then the risk due to traveling can be established from fatality rates. However, commercial aviation fatality rates are poor indicators of short-term risk changes, as a single large jet can result in the deaths of a few hundred people. Hence, a single accident can greatly influence fatality rates. Therefore, the trending of fatality rates requires data from extended time periods (5 years or more). As an alternative, accident rates can be used instead of fatality rates as indicators of safety levels. The number of fatalities, even for a specific type of accident, varies greatly with each crash. However, while the number of accidents may have a narrower range of annual variance than the number of fatalities, it has similar analysis issues. While accidents exceed fatalities in numbers, they are still quite small. Also, accidents can vary significantly from one year to the next, and hence accident rates are also poor indicators of short-term estimates of risk trends.

Before one can use accident statistics for safety analysis, it is necessary to understand the denominators of *exposure data* used in transportation accident or fatality rates. Exposure data are information that indicates the amount of opportunity for an event to occur. Cycles, distance, and time for passengers and vehicles are the principal exposure types. They are used in the denominator of rates, such as fatalities per passenger departure or electrical system failures per aircraft-hour. The choice of type of exposure data will affect how rates are compared between and within the transportation modes. Passenger-miles (the number of passengers multiplied by the miles traveled) are most appropriate when one is comparing air transportation with other modes of transportation, as they allow generalized broad-based system comparisons. Using trips between the same city pairs to compare risks for different modes of travel would be very appealing. However, the number of city pairs in the United States is extremely large, and some modes of transportation do not have passenger data in the format required for this type of comparison. In addition, risk per passenger-mile is not uniform over a trip and can vary by the type of route taken or the time of day.

It is clear that the probability of an accident is significantly higher during takeoff or landing than in any other phase of flight. Hence, the bulk of commercial aviation passenger-miles occur under conditions when the overall average risk is much lower. Vehicle-miles (as opposed to passenger numbers) are preferred in the denominator for comparison of exposure and accident rates, as the number of passengers per vehicle usually vary. Finally, vehicle-miles traveled do not reflect the risks associated with the size of the vehicle and/or its speed.

Aviation accidents can vary greatly in severity, ranging from a flight attendant back injury from pushing carts during in-flight service or handling overhead luggage, to several hundred deaths due to an aircraft crash. To render a more equitable perspective on things, some databases categorize accidents as fatal and nonfatal. However, even here, a single death on a ramp during an aircraft pushback would still be classified as an aviation accident. To address this issue, the NTSB classifies accidents as *major, serious, injury,* or *damage* (see Chapter 4 for definitions).

Another classification (popular with the insurance and aircraft manufacturing industries) of accidents includes (aircraft) *hull loss* and *nonhull loss*. A hull loss is airplane damage that is beyond economic repair.

Accident counts by themselves cannot be reliably used to measure relative safety among organizations and their products. All things being equal, an airline that has a larger fleet of aircraft could be expected (statistically speaking) to have a larger number of accidents than an airline with far fewer planes. Similarly, aircraft models that are flown more often would be expected to be involved in more accidents than less frequently used models. For this reason the *accident rate*, which is the number of accidents divided by some common base variable (e.g., flight hours, departures, and miles flown), is a more valid indicator of relative safety than just accident counts. In this context, time (flight hours) is a popular measure of exposure in many types of risk analysis. Since flight-hour data are needed for economic, operational, and maintenance reasons, airlines keep accurate records of such data which are readily available for safety rate comparisons. Also, rate data are used by the NTSB to compare safety performances among airlines and other sectors of the air transportation industry.

It goes without saying that takeoffs and landings are the riskiest phases of flight where most accidents occur. So, if takeoffs and/or landings are used as a base for rate calculations, then the airline that predominately flies smaller aircraft more frequently on shorter routes would be favored more than the airline that predominately flies larger aircraft on long routes (and would consequently have fewer takeoffs and landings). The bias reverses in the above example if hours flown are used as the base for accident rate calculation. Then again, since most of the risk involved with air transportation is associated with takeoffs and landings, a 1000-mile trip is similar to a 100-mile trip when both are compared for safety. Overall, there is a stronger correlation between accidents and departures than there is between accidents and flight hours or between accidents and passenger-miles. Therefore, aircraft departures (or landings) are a valid exposure parameter for air transportation.

The air transportation industry is extremely dynamic with technological advances, operational strategies (point-to-point versus hub and spoke), and mergers, consolidations, and bankruptcies rapidly changing the aviation landscape. The aviation industry of today is very different from what it was 30 years ago. Hence, one should exercise care when comparing safety statistics over long periods. While there is no clear answer as to what time domain is appropriate for analyzing and comparing aviation safety data, the most recent 5- to 10-year span appears to be the accepted norm among federal regulators.

Aviation accidents are rare occurrences, and the risk of death or serious injury by air travel is minuscule. The domestic airlines as a group average just under four catastrophic accidents per year (even less if one were to exclude sabotage and terrorism). With such small numbers, making statistical inferences from the data to evaluate safety performance of the industry is very difficult. Regulatory agencies and the aviation industry maintain a wide variety of safety-related information. However, one must take into consideration the accuracy and completeness of these data when developing safety trends and recommending control measures.

Safety factors other than fatalities or accidents that include the nature and causes of accidents (covered in Chapter 6) should be studied and used for preemptive strikes against safety exposures and to provide timely feedback to management and policymakers.

In summary, no single measurement provides the complete safety picture. Passenger-exposure data are used when passenger risk is to be described. Passenger-miles are used when the influence of vehicle size or speed on the data is nonexistent. Departure-exposure data account for nonuniform risk over a trip. Time is the most widely used exposure measure, and time-based data are often readily available.

AVIATION ACCIDENT STATISTICS

Chapter 4 reviewed various data reporting systems, both voluntary and mandatory, that are used in aviation. While ensuring compliance or consistency in safety reporting is difficult in general, voluntary systems have added complexity in that they are inconsistent and less reliable. There are several reliable sources of accident data. One of the most easily accessible accident databases is maintained by Boeing, which publishes an annual *Statistical Summary of Commercial Jet Airplane Accidents*. Another excellent source document is the Aviation Safety Network of the Flight Safety Foundation found at http://aviation-safety.net/. Finally, the definitive and mandatory source of aircraft accident data in the United States is the NTSB database.

MANUFACTURERS' INVOLVEMENT WITH SAFETY DATA

Safety professionals with the airframe and engine manufacturers can be found at accident sites participating in the investigation. They collect and analyze data. They recommend improvements in the way aircraft are designed and built and in the way they are operated and maintained. The mandate is to learn to prevent future mishaps.

While manufacturers may undertake similar functions, the organizational framework for carrying out those functions can differ markedly between them. For example, accident investigation and safety data analysis are separate units at Boeing, while they are combined at other manufacturers and often linked to customer support in a single organization.

The closest thing to a constant among manufacturers' safety departments is their responsibility for accident investigation. When a company's product is involved in an accident, that company has a duty to help find the cause. In the parlance of the NTSB, the company becomes a party to the investigation. That means the manufacturer's representatives work alongside NTSB investigators in examining evidence at the accident site. The parties conduct subsequent tests and suggest findings, but the final analysis and published report are the NTSB's exclusively.

Safety departments scrutinize incidents as well as accidents. Committees are formed to represent various departments using new safety management systems (SMS) procedures.

Based on the events reviewed, the committees recommend design changes or revisions in maintenance or operating procedures. Accident reports produce a lot of data. Still more data come from incident reports and other files compiled by airlines, manufacturers, and government agencies. The FAA, for example, records Service Difficulty Reports. Manufacturers' safety departments have come to view these data as a resource to be developed and cultivated. The data come from a variety of sources. Manufacturers' service representatives around the world report regularly on anomalous events both large and small.

Airlines often report directly to manufacturers. Other data sources include civil authorities and international organizations such as the International Air Transport Association, the U.K. CAA, and U.S. FAA, as well as insurance underwriters and publications. Even with all those collection efforts, safety staff don't claim that a record of every event unfailingly finds its way to the appropriate database. When little or no damage is involved, airlines sometimes make no external reports.

While manufacturers of large airframes maintain what probably are the most extensive databases, others also are collecting such information. Makers of engines and other components are also in the business of collecting statistical data.

The purpose of gathering the data is trend analysis. Computers may be able to discern patterns that a human analyst would miss. Boeing has taken a proactive stance in the area of safety. Manufacturers can alert carriers to problems they didn't know they had. A team assembled by Boeing uses accident records and other data to identify current safety issues. For example, Boeing was a leading proponent in stressing the importance of installing GPWS and the need for proper training in its use.

Similarly, to prevent approach and landing accidents, Boeing has urged that every runway used by commercial transports be equipped with an instrument landing system (ILS). And several training measures have been advocated for pilots and controllers to alleviate the dangers associated with nonstabilized approaches.

Boeing research also has addressed "crew-caused accidents." Its data and other sources identify flightcrew error as the primary cause in close to 70 percent of commercial jet hull-loss accidents (see next section). However, accidents rarely have single causes. Manufacturers' prevention strategies address all the factors, not just the primary one. Removing even one link in the chain of events can prevent many accidents.

Airlines maintain databases and conduct trend analyses of their own. These analyses can benefit from work done by the manufacturers. For example, American studied a series of 757 tail strikes. After assessing its own records, the carrier checked with Boeing about the experiences of other 757 operators. Once the problem was identified, American and Boeing worked to solve it by revising training methods.

Metro operators receive a monthly trend monitoring report with statistics on factors such as dispatch reliability, in-flight engine shutdowns, and unscheduled component removals. The reports are based on operating data from more than 25 operators of close to 250 Metros.

The large airframe manufacturers also publish annual reports on their compilations of industrywide accident data, although they differ in the way the data are analyzed and presented. For day-to-day problem solving, airlines have more contact with a manufacturer's accident and incident investigators than its data analysts.

In the discussions that follow, databases from Boeing, the Flight Safety Foundation, and the NTSB will be explored to analyze aviation accident statistics.

BOEING'S ACCIDENT STATISTICAL SUMMARY

According to the Boeing statistical summary, in the 50-year history (worldwide) of scheduled commercial jet operations (1959 to 2009) there have been 1704 accidents resulting in 28,296 onboard fatalities. Of the 1704 accidents, 592 accidents caused fatal injuries of which 479 were hull-loss accidents, 25 were attributed to substantial damage accidents, and 88 were a result of less than substantial damage accidents. If one were to consider accidents by type of operation, 79 percent (1344) of the 1704 accidents were passenger operations, 14 percent (244) were cargo operations, and the remaining 116 accidents occurred during testing, training, demonstration, or ferrying. United States and Canadian operations collectively had about 31 percent (530) of the 1704 worldwide accidents, contributing to about 22 percent (6153) of the 28296 global onboard fatalities. Over this period, there have been in excess of 563 million cumulative departures and more than 993 million cumulative flight hours. The industry is now operating 20025 jet aircraft, with little over 21 million departures each year.

If we look at the plot of all accidents for the worldwide commercial jet fleet for the period 1959 to 2001, we can see that the rates for all accidents and those involving hull losses has been fairly stable for the past 35 years (Fig. 5-1). If we apply control limits to these data, generally 3 standard deviations on either side of the number of hull-loss accidents each year, we can clearly see that the jet airplane transportation process has been under statistical control for a number of years. However, even if this low accident rate remains constant over the next few years, we can expect to see an increase in the actual number of hull-loss accidents each year as the fleet increases in number of departures.

Hull losses and fatal accidents were also analyzed according to the phase of flight in which they occurred (Fig. 5-2). The combined final approach-and-landing phases accounted for 34 percent of the hull-loss and fatal accidents, followed by the combined phases from loading through initial climb (38 percent). Cruise, which accounts for about 57 percent of flight time in a 1.5-hour flight, occasioned only 10 percent of hull-loss accidents.

In past years, the summary also considered primary cause factors for commercial operation hull-loss accidents for the period 1992 to 2001 (Fig. 5-3). For accidents with known causes, flightcrews were considered the primary cause in most accidents—66 percent over the 10-year period from 1992 to 2001 (Fig. 5-3).

Finally, fatalities by accident categories were covered for the period 2000 to 2009 (Fig. 5-4). Loss of control in flight accounted for 1759 onboard fatalities followed by controlled flight into terrain (CFIT) which had 961 fatalities during the recent 9-year period.

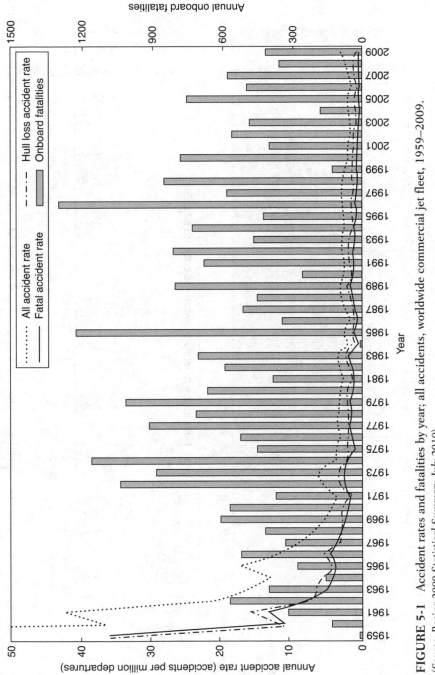

FIGURE 5-1 Accident rates and fatalities by year; all accidents, worldwide commercial jet fleet, 1959–2009.

(*Source: Boeing 2009 Statistical Summary, July 2010*)

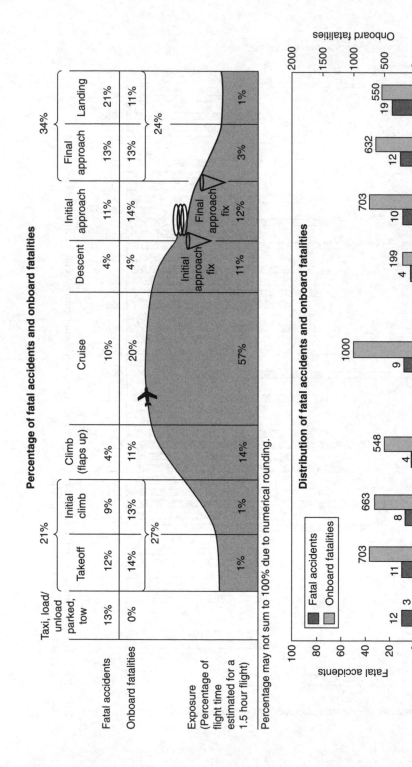

FIGURE 5-2 Accidents and onboard fatalities by phase of flight; hull-loss and/or fatal accidents, worldwide commercial jet fleet, 1992–2009. (*Source: Boeing 2009 Statistical Summary, July 2010*)

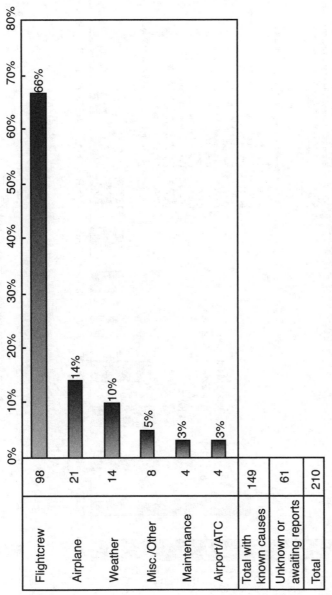

Flightcrew	98	66%
Airplane	21	14%
Weather	14	10%
Misc./Other	8	5%
Maintenance	4	3%
Airport/ATC	4	3%
Total with known causes	149	
Unknown or awaiting reports	61	
Total	210	

FIGURE 5-3 Accidents by primary cause,* hull loss, worldwide commercial jet fleet, 1992–2001. (*As determined by the investigating authority: *Source: Boeing 2001 Statistical Summary, June 2002.*)

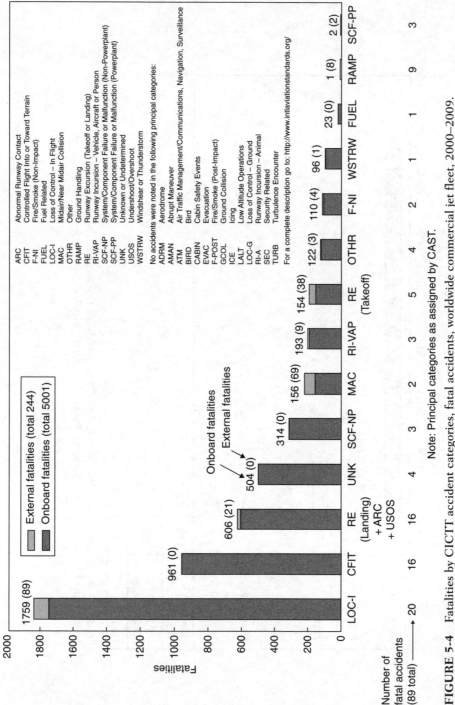

FIGURE 5-4 Fatalities by CICTT accident categories, fatal accidents, worldwide commercial jet fleet, 2000–2009.
(*Source: Boeing 2009 Statistical Summary, July 2010*)

Boeing's accident data exclude turboprop aircraft as well as those with maximum gross weight of 60,000 lb (27,216 kg) or less; Soviet Union and Commonwealth of Independent States accidents; and accidents resulting from sabotage, hijacking, suicide, and military action.

Several important lessons can be learned from the Boeing data. The most obvious of these is the issue of human error. The human factor is clearly and consistently the most frequent cause of incidents and accidents in the airline industry. It must be kept in mind that Fig. 5-3 only shows the tip of the human-error problem; flight crewmembers are not the only source of error in the aviation system. Hidden in each of the other categories of primary cause factors is a significant, although sometimes hard-to-determine, human-error component. In the maintenance category, for example, recent unpublished data suggest that more than one-half of these involve human factors, and it seems likely that similar rates of human-error involvement can be found in the other categories as well. This is why estimates of the total contribution of human error to aviation incidents and accidents can range as high as 80 to 90 percent.

Based largely on data such as these, it has become increasingly apparent that the management and control of human error are the largest single challenge facing airline safety management. This recognition is not new. In 1974, the Flight Safety Foundation held its 28th Annual International Aviation Safety Seminar in Williamsburg, Va. The theme was human factors in flight operations, and it is believed that this seminar was the first major airline industry safety conference devoted to human error. In 1975, IATA sponsored its 20th Technical Conference: Safety in Flight Operations. The conference proceedings almost exclusively focused on human factors and human-error issues, which were termed "the last frontier of aviation safety." These early industry efforts were at the vanguard of present-day approaches to the problem of management and control of human error in aviation operations.

The University of Texas, Human Factors Research project conducted aviation safety research funded by the FAA and industry for over 25 years. More recently, a current safety initiative of the Flight Safety Foundation is a project entitled Operators Guide to Human Factors in Aviation which contains an extensive compendium of Human Factors information on the flight safety website. The Flight Safety Foundation, in conjunction with AIRBUS, established an annual Human Factors in Aviation Safety Award which is presented at the FSF European Aviation Safety Seminar. Likewise, IATA and ICAO are very active in Human Factors training on an international basis. The subject of Human Factors is thoroughly explored later in Chapter 7 of this text.

UNITED STATES AND INTERNATIONAL ACCIDENT STATISTICS

For the United States, the primary source of safety data is the NTSB. Tables 5-1 through 5-5 present a year-by-year summary of accidents, fatalities, and accident rates for U.S. carriers operating in scheduled and nonscheduled service under FAR Part 121 (major air carriers) and Part 135 (commuter air carriers and on-demand air taxis), respectively. Note that these data reflect the NTSB's official definition of an accident, of which fatal accidents are a subset.

TABLE 5-1 Accidents, Fatalities, and Rates, 1990 through 2009, for U.S. Air Carriers Operating Under 14 CFR 121, Scheduled and Nonscheduled Service (Airlines)

Year	Accidents All	Accidents Fatal	Fatalities Total	Fatalities Aboard	Flight Hours	Miles Flown	Departures	Accidents per 100,000 Flight Hours All	Fatal	Accidents per 1,000,000 Miles Flown All	Fatal	Accidents per 100,000 Departures All	Fatal
1990	24	6	39	12	12,150,116	4,947,832,000	8,092,306	0.198	0.049	0.0049	0.0012	0.297	0.074
1991	26	4	62	49	11,780,610	4,824,824,000	7,814,875	0.221	0.034	0.0054	0.0008	0.333	0.051
1992	18	4	33	31	12,359,715	5,039,435,000	7,880,707	0.146	0.032	0.0036	0.0008	0.228	0.051
1993	23	1	1	0	12,706,206	5,249,469,000	8,073,173	0.181	0.008	0.0044	0.0002	0.285	0.012
1994*	23	4	239	237	13,124,315	5,478,118,000	8,238,306	0.168	0.030	0.0040	0.0007	0.267	0.049
1995	36	3	168	162	13,505,257	5,654,069,000	8,457,465	0.267	0.022	0.0064	0.0005	0.426	0.035
1996	37	5	380	350	13,746,112	5,873,108,000	8,228,810	0.269	0.036	0.0063	0.0009	0.450	0.061
1997	49	4	8	6	15,838,109	6,696,638,000	10,318,383	0.309	0.025	0.0073	0.0006	0.475	0.039
1998	50	1	1	0	16,816,555	6,736,543,000	10,979,762	0.297	0.006	0.0074	0.0001	0.455	0.009
1999	51	2	12	11	17,555,208	7,101,314,000	11,308,762	0.291	0.011	0.0072	0.0003	0.451	0.018
2000	56	3	92	92	18,299,257	7,524,027,000	11,468,229	0.306	0.016	0.0074	0.0004	0.488	0.026
2001*	46	6	531	525	17,814,191	7,294,191,000	10,954,832	0.236	0.011	0.0058	0.0003	0.383	0.018
2002	41	0	0	0	17,290,198	7,192,501,000	10,508,473	0.237	—	0.0057	—	0.390	—
2003	54	2	22	21	17,467,700	7,280,383,000	10,433,164	0.309	0.011	0.0074	0.0003	0.518	0.019
2004	30	2	14	14	18,882,503	7,930,159,000	11,023,128	0.159	0.011	0.0038	0.0003	0.272	0.018
2005	40	3	22	20	19,390,029	8,165,643,000	11,130,407	0.206	0.015	0.0049	0.0004	0.359	0.027
2006	33	2	50	49	19,263,209	8,139,357,000	10,820,915	0.171	0.010	0.0041	0.0002	0.305	0.018
2007	28	1	1	1	19,637,322	8,315,905,000	10,928,432	0.143	0.005	0.0034	0.0001	0.256	0.009
2008	28	2	3	1	19,097,962	8,068,288,000	10,437,197	0.147	0.010	0.0034	0.0002	0.268	0.019
2009	30	2	52	51	18,001,000	7,557,161,000	10,349,200	0.167	0.011	0.0040	0.0003	0.290	0.019

NOTES:
- 2009 data are preliminary.
- Flight hours, miles, and departures are compiled by the Federal Aviation Administration.
- Since March 20, 1997, aircraft with 10 or more seats used in scheduled passenger service have been operated under 14 CFR 121.
- Years followed by the symbol * are those in which an illegal act was responsible for an occurrence in this category. These acts, such as suicide, sabotage and terrorism are included in the totals for accidents and fatalities but are excluded for the purpose of accident rate computation. Table 12 contains a list of illegal act occurrences involving US air carriers for the period covered by this table. Other than the persons aboard aircraft who were killed, fatalities resulting from the September 11, 2001 terrorist act are excluded from this table.

SOURCE: http://www.ntsb.gov

TABLE 5-2 Accidents, Fatalities, and Rates, 1990 through 2009, for U.S. Air Carriers Operating Under 14 CFR 121, Scheduled Service (Airlines)

Year	Accidents All	Accidents Fatal	Fatalities Total	Fatalities Aboard	Flight Hours	Miles Flown	Departures	Accidents per 100,000 Flight Hours All	Fatal	Accidents per 1,000,000 Miles Flown All	Fatal	Accidents per 100,000 Departures All	Fatal
1990	19	4	11	9	11,524,726	4,689,287,000	7,795,761	0.165	0.035	0.0041	0.0009	0.244	0.051
1991	21	3	60	47	11,139,166	4,558,537,000	7,503,873	0.189	0.027	0.0046	0.0007	0.280	0.040
1992	15	3	29	27	11,732,026	4,767,344,000	7,515,373	0.128	0.026	0.0031	0.0006	0.200	0.040
1993	22	1	1	0	11,981,347	4,936,067,000	7,721,870	0.184	0.008	0.0045	0.0002	0.285	0.013
1994*	18	4	239	237	12,292,356	5,112,633,000	7,824,802	0.138	0.033	0.0033	0.0008	0.217	0.051
1995	30	1	160	160	12,776,679	5,328,969,000	8,105,570	0.235	0.008	0.0056	0.0002	0.370	0.012
1996	31	3	342	342	12,971,676	5,449,997,000	7,851,298	0.239	0.023	0.0057	0.0006	0.395	0.038
1997	43	3	3	2	15,061,662	6,339,432,000	9,925,058	0.285	0.020	0.0068	0.0005	0.433	0.030
1998	41	1	1	0	15,921,447	6,343,690,000	10,535,196	0.258	0.006	0.0065	0.0002	0.389	0.009
1999	40	2	12	11	16,693,365	6,689,327,000	10,860,692	0.240	0.012	0.0060	0.0003	0.368	0.018
2000	49	2	89	89	17,478,519	7,152,260,000	11,053,826	0.280	0.011	0.0069	0.0003	0.443	0.018
2001*	41	6	531	525	17,157,858	6,994,939,000	10,632,880	0.216	0.012	0.0053	0.0003	0.348	0.019
2002	34	0	0	0	16,718,781	6,927,954,000	10,276,107	0.203	—	0.0049	—	0.331	—
2003	51	2	22	21	16,887,756	7,015,935,000	10,227,924	0.302	0.012	0.0073	0.0003	0.499	0.020
2004	23	1	13	13	18,184,016	7,604,248,000	10,782,989	0.126	0.005	0.0030	0.0001	0.213	0.009
2005	34	3	22	20	18,712,191	7,843,717,000	10,910,460	0.182	0.016	0.0043	0.0004	0.312	0.027
2006	26	2	50	49	18,647,896	7,851,864,000	10,627,481	0.139	0.011	0.0033	0.0003	0.245	0.019
2007	26	0	0	0	19,014,677	8,024,313,000	10,734,170	0.137	—	0.0032	—	0.242	—
2008	20	0	0	0	18,551,362	7,821,849,000	10,271,639	0.108	—	0.0026	—	0.195	—
2009	26	1	50	49	17,470,000	7,316,000,000	10,199,000	0.149	0.006	0.0036	0.0001	0.255	0.010

NOTES:
- 2009 data are preliminary.
- Flight hours, miles, and departures are compiled by the Federal Aviation Administration.
- Since March 20, 1997, aircraft with 10 or more seats used in scheduled passenger service have been operated under 14 CFR 121.
- Years followed by the symbol * are those in which an illegal act was responsible for an occurrence in this category. These acts, such as suicide and sabotage are included in the totals for accidents and fatalities but are excluded for the purpose of accident rate computation. Table 12 contains a list of illegal act occurrences involving US air carriers for the period covered by this table. Other than the persons aboard aircraft who were killed, fatalities resulting from the September 11, 2001 terrorist acts are excluded from this table.

SOURCE: http://www.ntsb.gov

TABLE 5-3 Accidents, Fatalities, and Rates, 1990 through 2009, for U.S. Air Carriers Operating Under 14 CFR 121, Nonscheduled Service (Airlines)

| YEAR | ACCIDENTS | | FATALITIES | | FLIGHT HOURS | MILES FLOWN | DEPARTURES | ACCIDENTS PER 100,000 FLIGHT HOURS | | ACCIDENTS PER 1,000,000 MILES FLOWN | | ACCIDENTS PER 100,000 DEPARTURES | |
	ALL	FATAL	TOTAL	ABOARD				ALL	FATAL	ALL	FATAL	ALL	FATAL
1990	5	2	28	3	625,390	258,545,000	296,545	0.800	0.320	0.0193	0.0077	1.686	0.674
1991	5	1	2	2	641,444	266,287,000	311,002	0.779	0.156	0.0188	0.0038	1.608	0.322
1992	3	1	4	4	627,689	272,091,000	365,334	0.478	0.159	0.0110	0.0037	0.821	0.274
1993	1	0	0	0	724,859	313,402,000	351,303	0.138	—	0.0032	—	0.285	—
1994	5	0	0	0	831,959	365,485,000	413,504	0.601	—	0.0137	—	1.209	—
1995	6	2	8	2	728,578	325,100,000	351,895	0.824	0.275	0.0185	0.0062	1.705	0.568
1996	6	2	38	8	774,436	423,111,000	377,512	0.775	0.258	0.0142	0.0047	1.589	0.530
1997	6	1	5	4	776,447	357,206,000	393,325	0.773	0.129	0.0168	0.0028	1.525	0.254
1998	9	0	0	0	895,108	392,853,000	444,566	1.005	—	0.0229	—	2.024	—
1999	11	0	0	0	861,843	411,987,000	448,070	1.276	—	0.0267	—	2.455	—
2000	7	1	3	3	820,738	371,767,000	414,403	0.853	0.122	0.0188	0.0027	1.689	0.241
2001	5	0	0	0	656,333	299,252,000	321,952	0.762	—	0.0167	—	1.553	—
2002	7	0	0	0	571,417	264,547,000	232,366	1.225	—	0.0265	—	3.012	—
2003	3	0	0	0	579,944	264,448,000	205,240	0.517	—	0.0113	—	1.462	—
2004	7	1	1	1	698,487	325,911,000	240,139	1.002	0.143	0.0215	0.0031	2.915	0.416
2005	6	0	0	0	677,838	321,926,000	219,947	0.885	—	0.0186	—	2.728	—
2006	7	0	0	0	615,313	287,493,000	193,434	1.138	—	0.0243	—	3.619	—
2007	2	1	1	1	622,645	291,592,000	194,262	0.321	0.161	0.0069	0.0034	1.030	0.515
2008	8	2	3	1	546,600	246,439,000	165,558	1.464	0.366	0.0325	0.0081	4.832	1.208
2009	4	1	2	2	531,000	241,161,000	150,200	0.753	0.188	0.0166	0.0041	2.663	0.666

NOTES:
- 2009 data are preliminary.
- Flight hours, miles, and departures are compiled by the Federal Aviation Administration

SOURCE: http://www.ntsb.gov

TABLE 5-4 Accidents, Fatalities, and Rates, 1990 through 2009, for U.S. Air Carriers Operating Under 14 CFR 135, Scheduled Service

Year	Accidents		Fatalities		Flight Hours	Miles Flown	Departures	Accidents per 100,000 Flight Hours		Accidents per 1,000,000 Miles Flown		Accidents per 100,000 Departures	
	All	Fatal	Total	Aboard				All	Fatal	All	Fatal	All	Fatal
1990	15	3	6	4	2,341,760	450,133,000	3,160,089	0.641	0.128	0.0333	0.0067	0.475	0.095
1991	23	8	99	77	2,291,581	433,900,000	2,820,440	1.004	0.349	0.0530	0.0184	0.815	0.284
1992*	23	7	21	21	2,335,349	507,985,000	3,114,932	0.942	0.300	0.0433	0.0138	0.706	0.225
1993	16	4	24	23	2,638,347	554,549,000	3,601,902	0.606	0.152	0.0289	0.0072	0.444	0.111
1994	10	3	25	25	2,784,129	594,134,000	3,581,189	0.359	0.108	0.0168	0.0050	0.279	0.084
1995	12	2	9	9	2,627,866	550,377,000	3,220,262	0.457	0.076	0.0218	0.0036	0.373	0.062
1996	11	1	14	12	2,756,755	590,727,000	3,515,040	0.399	0.036	0.0186	0.0017	0.313	0.028
1997	16	5	46	46	982,764	246,029,000	1,394,096	1.628	0.509	0.0650	0.0203	1.148	0.359
1998	8	0	0	0	353,670	50,773,000	707,071	2.262	—	0.1576	—	1.131	—
1999	13	5	12	12	342,731	52,403,000	672,278	3.793	1.459	0.2481	0.0954	1.934	0.744
2000	12	1	5	5	369,535	44,943,000	603,659	3.247	0.271	0.2670	0.0223	1.988	0.166
2001	7	2	13	13	300,432	43,099,000	558,052	2.330	0.666	0.1624	0.0464	1.254	0.358
2002	7	0	0	0	273,559	41,633,000	513,452	2.559	—	0.1681	—	1.363	—
2003	2	1	2	2	319,206	47,404,000	572,260	0.627	0.313	0.0422	0.0211	0.349	0.175
2004	4	0	0	0	302,218	46,809,000	538,077	1.324	—	0.0855	—	0.743	—
2005	6	0	0	0	299,775	45,721,000	527,267	2.002	—	0.1312	—	1.138	—
2006	3	1	2	2	301,495	46,503,000	568,464	0.995	0.332	0.0645	0.0215	0.528	0.176
2007	3	0	0	0	291,701	46,049,000	592,577	1.028	—	0.0651	—	0.506	—
2008	7	0	0	0	293,499	46,428,000	576,203	2.385	—	0.1508	—	1.215	—
2009	2	0	0	0	292,000	46,251,000	566,000	0.685	—	0.0432	—	0.353	—

NOTES:
- 2009 data are preliminary.
- Flight hours, miles, and departures are compiled by the Federal Aviation Administration.
- Since March 20, 1997, aircraft with 10 or more seats used in scheduled passenger service have been operated under 14 CFR 121.
- Years followed by the symbol * are those in which an illegal act was responsible for an occurrence in this category. These acts, such as suicide, sabotage and terrorism are included in the totals for accidents and fatalities but are excluded for the purpose of accident rate computation. Table 12 contains a list of illegal act occurrences involving US air carriers for the period covered by this table.

SOURCE: http://www.ntsb.gov

TABLE 5-5   Accidents, Fatalities, and Rates, 1990 through 2009, for U.S. Air Carriers Operating Under 14 CFR 135, On-Demand Operations

	ACCIDENTS		FATALITIES			ACCIDENTS PER 100,000 DEPARTURES	
YEAR	ALL	FATAL	TOTAL	ABOARD	FLIGHT HOURS	ALL	FATAL
1990	107	29	51	49	2,249,000	4.76	1.29
1991	88	28	78	74	2,241,000	3.93	1.25
1992	76	24	68	65	2,844,000	2.67	0.84
1993	69	19	42	42	2,324,000	2.97	0.82
1994	85	26	63	62	2,465,000	3.45	1.05
1995	75	24	52	52	2,486,000	3.02	0.97
1996	90	29	63	63	3,220,000	2.80	0.90
1997	82	15	39	39	3,098,000	2.65	0.48
1998	77	17	45	41	3,802,000	2.03	0.45
1999	74	12	38	38	3,204,000	2.31	0.37
2000	80	22	71	68	3,930,000	2.04	0.56
2001	72	18	60	59	2,997,000	2.40	0.60
2002	60	18	35	35	2,911,000	2.06	0.62
2003	73	18	42	40	2,927,000	2.49	0.61
2004	66	23	64	63	3,238,000	2.04	0.71
2005	65	11	18	16	3,815,000	1.70	0.29
2006	52	10	16	16	3,742,000	1.39	0.27
2007	62	14	43	43	4,033,000	1.54	0.35
2008	58	20	69	69	3,205,000	1.81	0.62
2009	47	2	17	14	2,875,000	1.63	0.07

NOTES:
- 2009 data are preliminary.
- Flight hours are estimated by the Federal Aviation Administration (FAA). Miles flown and departure information for on-demand Part 135 operations is not available.

SOURCE: http://www.ntsb.gov

Because of the NTSB's very broad definition of an accident (any incident that involves, e.g., a broken bone is classified as an accident), the United States experiences an average of close to 36 reportable accidents involving scheduled and nonscheduled air service each year. However, serious accidents, those involving fatalities, are much rarer. The average number of fatal accidents from 1990 to 2009 has been about per year for scheduled and nonscheduled (Part 121) operations. Over this same period scheduled and nonscheduled (Part 121) departures increased from 8.1 million to 10.4 million having reached a high point of 11.5 million departures in 2000 before the terrorist incidents of September 11, 2001. Clearly, the trend in fatal accident rates in the United States is in the desired direction; however, the worldwide trend in developing third world aviation countries is not so optimistic.

THE 1980s.   In the NTSB's first 15 years, airline safety improved steadily. In 1967, the airline fatal accident rate was 0.006 for every 1 million aircraft miles flown. By 1980, this rate was down to 0.001, a reduction of 83 percent. And on January 1, 1982, U.S. airlines had completed 26 months without a catastrophic crash of a pure-jet transport. The period spanned 1980 and 1981; never before had there been even one calendar year without such an accident. The previous record had been 15 months, from September 1971 to December 1972. The airlines flew more than half a billion passengers on more than 10 million flights in the 26-month period—more than half a trillion passenger miles. The aerial transportation involved would have taken every man, woman, and child in the country on a flight of more than 2000 miles.

Just 13 days into 1982, the air carriers' remarkable record came to a shattering end when Air Florida Flight 90, a Boeing 737, taking off from Washington National Airport, crashed onto a bridge in a snowstorm. Seventy passengers, four crewmembers, and four persons on the bridge were killed. There were five fatal accidents in 1982 that produced a fatal accident rate of 0.057 per 100,000 aircraft hours. This rate was down 7 percent from the 0.061 recorded in 1981, when there were four bizarre single-fatality accidents. The fatality total, however, was 235, the third highest in a decade, compared with 4 in 1981 and none in 1980.

In 1983, the commuter airlines achieved sharply lower accident rates for the second successive year. The commuters had only 16 total accidents, 2 of them fatal, compared to 26 total accidents in 1982, of which 5 were fatal. Fatalities dropped from 14 to 11. The rate of 0.687 total accident per 100,000 departures, the rate most often used to measure commuter safety, was 46 percent lower than that in 1982.

The scheduled airlines had 4 fatal accidents in 1983 that produced 15 fatalities. There were 22 total accidents in 1983 compared with 16 in 1982 for a 1983 rate of 0.318 total accident per 100,000 aircraft hours.

Only 13 accidents, fatal and nonfatal, occurred in scheduled airline service in 1984. A record low, this produced a total accident rate of 0.229 per 100,000 departures. By comparison, the same rate was almost 3 times higher, 0.659, in 1975.

By the end of 1984, there had not been a catastrophic crash of a U.S. turbojet airliner since the July 9, 1982, takeoff crash of a Pan American B-727 at New Orleans. That period of nearly 30 months marked the second time since 1979 that the airlines had operated for more than 2 years without a catastrophic crash of a jet airliner.

The year 1985 began in tragedy. On January 1, Eastern Airlines Flight 980, a B-727 smashed into the side of Mount Illimani, Bolivia, 22,000 feet above sea level during a descent for landing into La Paz. All 29 passengers and crew were killed. Although an NTSB-led expedition reached the accident site, deep snow, and severe cold made recovering any of the bodies or the cockpit voice and flight data recorders impossible.

The Bolivian accident was the first of 13 fatal U.S. commuter and major airline accidents in 1985, in which 563 people died. In addition to the Bolivian accident, major accidents involving substantial loss of life included these two serious crashes:

- Delta Airlines Flight 191, a L-1011 TriStar aircraft which slammed into the ground as it encountered a wind shear on final approach into the Dallas–Fort Worth airport on August 2. *Fatalities:* 135. This accident prompted major changes in FAA training and improvements in Doppler radar weather technology.

- A military-chartered DC-8, Arrow Air Flight 1285, which crashed on takeoff from Gander, Canada, due to icing, on December 12. *Fatalities: 256.*

Although the loss of life in commercial passenger flights was higher in 1985 than in any previous year in the 1980s, the overall record for U.S. civil aviation, as judged by accident rates and fatal accidents, was mixed.

For the scheduled (Part 121) airlines, the fatal accident rate of 0.066 per 100,000 departures in 1985 was more than 3.5 times that of 1984. For unscheduled (Part 121) airline service (the charters), the fatal accident rate of 1.261 per 100,000 departures was more than twice that of 1982, the last year before 1985 where there had been a fatal charter airline accident. In 1985, the total number of fatalities for both the scheduled and nonscheduled (chartered, Part 121) airlines was 526, making it the second-worst year in history for accidents involving U.S. air carriers. Only in 1977, when two charter B-747s collided on the runway at Tenerife Airport in the Canary Islands, were fatalities higher.

After 21 accidents and 5 fatalities in 1986, major mishaps of large, scheduled (Part 121) U.S. airlines hit a 13-year high in 1987 with 32 accidents, resulting in 231 deaths. This number was the highest since 42 accidents occurred in 1974 when departures were almost one-half of those in 1987. The 32 accidents for scheduled airlines resulted in an accident rate of 0.425 per 100,000 departures, up from the 0.289 (a 47 percent increase) recorded in 1986. However, the fatal accident rate of 0.041 represented an almost 200 percent increase from the year before. Among the major fatal accidents in 1987 was:

- Northwest Airlines Flight 255, an MD-82, crashed after liftoff from Detroit Metropolitan Wayne County International Airport due to the flight crew's failure to use the checklist to ensure the flaps and slats were extended for takeoff. *Fatalities:* 156, including 2 on the ground.

In the commuter or regional airlines, there also was a sharp turnabout from 1986. There were 33 accidents in 1987, of which 10 were fatal—the highest since 38 accidents were recorded in 1980. Fatalities hit an 8-year high of 59. There were 1.174 accidents per 100,000 departures, compared with 0.50 in 1986. The fatal accident rate of 0.356 rose from 0.071 in 1986 to the highest level since 1979.

Accident rates for U.S. scheduled air carriers declined in 1988 from the year before. According to the NTSB's statistics, there were 29 major air carrier accidents in 1988, which is 3 less than in 1987. Of the 29 accidents, 3 were fatal, resulting in 285 fatalities. The accident rate dropped from 0.425 per 100,000 departures in 1987 to 0.381 in 1988. The fatal accident rate fell from 0.041 per 100,000 departures to 0.027.

Commuter airline accidents decreased from 33 in 1987 to 18 in 1988. There were only 2 fatal accidents, compared to the 10 that occurred the previous year, with fatalities down from 59 to 21. The accident rate was 0.619 per 100,000 departures, compared with 1.174 in 1987, for the second-lowest rate in this decade.

The NTSB investigated two unusual major aviation accidents in 1988:

- Aloha Airlines Flight 243, which suffered an explosive decompression caused by metal fatigue, but was able to land safely in Maui with only one fatality. The U.S.

Congress passed the Aviation Safety Research Act of 1988 in the wake of the disaster which provided for enhanced research in the critical areas of aircraft maintenance and structural technology.

- Pan Am Flight 103, the explosion of a B-747 in the sky over Lockerbie, Scotland on December 21, 1988, claiming all 259 persons on board and an estimated 11 on the ground. When British investigators announced a week later that the aircraft was brought down by a bomb, the tragedy became the most deadly act of sabotage ever perpetrated against a U.S. airliner.

Although the total accident rates for U.S. air carriers and commuters declined in 1989, the fatal accident rates rose. The actual number of fatal accidents for the air carriers was the highest since 1968.

There were 28 accidents involving major U.S. scheduled and charter airlines in 1989, down from 30 in 1988. Of those 28, 11 involved fatalities, which was the highest since 15 fatal accidents in 1968. The fatal accident rate of 0.144 per 100,000 departures was the highest of the decade, up from 0.026 in 1988.

The major U.S. scheduled airlines suffered 24 accidents in 1989, down from 29 in the previous year. Of those 24, eight involved fatalities, the most since 1973. Of the 131 fatalities registered in 1989, 111 of them occurred in the crash of Flight 232, a United Airlines DC-10 in Sioux City, Iowa, on July 19, 1989, due to a fatigue crack in the engine, resulting in explosion and catastrophic loss of hydraulic pressure. United Flight 232's captain credited Crew Resource Management (CRM) techniques for saving the lives of the 185 survivors of this crash.

Commuter air carriers had 19 accidents in 1989, which is one more than in the previous year. The five fatal accidents, up from 2 in 1988, resulted in 31 fatalities, 10 more than in 1988. The 0.674 accident rate per 100,000 departures was slightly higher than the 0.619 rate in 1988, but the fatal accident rate rose from 0.069 to 0.177.

THE 1990s.   The decade of the 1990s started off favorably with a decline in all accident rates for the scheduled, nonscheduled, and commuter air carriers in 1990.

The major U.S. scheduled airlines experienced 25 accidents in 1991, up slightly from the previous year. Of those 25, four involved fatalities. Of the total 62 fatalities in 1991, 59 involved two major accidents:

- On February 1, 1991, U.S. Air Flight 1493, a B-737, collided with a Skywest Metroliner on the runway while landing at Los Angeles International Airport (LAX). Both planes were destroyed in the accident, which killed 34 persons. The cause of the accident was a situational awareness human error made by an air traffic controller at this extremely busy airport. This type of accident is known as a classic "runway incursion" which has been on the list of the NTSB's most wanted safety improvements for many years.

- On March 3, 1991, all 25 persons aboard United Airlines Flight 585, a B-737, were killed when the plane crashed after an uncontrolled rudder flight problem during final approach to Colorado Springs. The aircraft was approximately 1000 feet above the ground when the upset occurred. The Boeing 737 had an enviable safety record until that time, and the cause of this accident was undetermined by the NTSB for a number of years. Later similar 737 rudder problems near Pittsburgh and Richmond provided answers to solve the perplexing mystery of this accident.

Overall, commuter air carriers were involved in 23 accidents in 1991, resulting in 99 fatalities, the largest number in two decades. Fatal accidents per 100,000 departures rose from 0.095 in 1990 to 0.284, an increase of 200 percent.

In 1992, there were seven fatal commuter accidents, compared to eight in 1991. The number of fatalities aboard commuter air carriers dropped to 21 from 77 a year earlier. The accident rate for commuter airlines had not yet become a major concern to the flying public.

Scheduled air carriers recorded four fatal accidents and 33 fatalities in 1992, compared to the same number of accidents and 62 deaths in the year before. The 33 fatalities represented the lowest number of deaths since 1986, when five persons in that category died.

In 1993 the major scheduled airlines experienced only one fatal accident, that involving a ground crewmember being struck by a propeller. The fatal accident rate of 0.013 per 100,000 departures was the lowest since 1980, when there were no fatal accidents among the scheduled airlines.

Paradoxically, the scheduled air carriers experienced more accidents in 1993 (22) than in the previous year (16), resulting in a higher total accident rate, 0.285 versus 0.213.

The fatal accident rate for commuter airlines dropped from 0.225 to 0.111 per 100,000 departures, but fatalities rose from 21 in 1992 to 24 in 1993. The total accident rate dropped from 0.706 to 0.444.

In 1994 the scheduled air carriers had 19 accidents, four of them fatal, for a total of 239 deaths, versus 22 accidents and one fatality in 1993.

- On July 2, USAir Flight 1016, a DC-9 approaching Charlotte/Douglas International Airport, with thunderstorm activity in the area, crashed when the crew encountered a windshear and attempted to abort the landing. There were 37 fatalities; 20 people on board survived the accident. This brought to an end a 27-month period in which the major U.S. scheduled airlines did not suffer a passenger fatality.

- On September 8, USAir Boeing Flight 427, another 737-300 aircraft crashed while on approach to the Greater Pittsburgh International Airport. All 132 people on board were killed, and the plane was destroyed by impact, making it one of the worst aviation accidents in U.S. history. This accident was extremely similar to the United Flight 585 rudder control problem which occurred in Colorado Springs. Again, NTSB investigators were baffled at the cause of the uncontrolled rudder problem until another 737 survived a similar situation near Richmond in 1996. A thermal shock test finally confirmed that the 737 rudder could be subject to a hard over rudder reversal under certain conditions and it was ultimately reengineered.

- Finally, on October 31, a Simmons Airlines ATR-72, operating as American Eagle flight 4184, crashed south of Roselawn, Ind. The flight, en route from Indianapolis to Chicago's O'Hare Airport, had been placed in a holding pattern for about 32 minutes because of traffic delays. The weather conditions during the period of holding were characterized by icing, a temperature near freezing and visible moisture. All 64 passengers and four crewmembers were killed in the accident.

The 1994 fatal accident rate per 1 million miles flown for scheduled service rose to 0.0008, compared with 0.0002 in 1993. Per 100,000 aircraft departures, the fatal rate was up to 0.051 from 0.013.

Nonscheduled air carriers experienced their fifth consecutive year without a fatality. There were four accidents, compared with one in 1993. The 1994 total accident rate per 1 million miles flown was 0.0109 versus 0.0032 in 1993; 0.481 per 100,000 aircraft hours versus 0.138; and 0.967 per 100,000 departures versus 0.285.

Commuter air carrier fatalities rose from 24 persons to 25 in 10 accidents in 1994, of which three were fatal. There were 16 accidents in 1993, including four fatal. The 1994 fatal accident rate per 1 million aircraft miles flown declined to 0.005 from 0.007 in 1993, while the rate per 100,000 departures fell to 0.084 from 0.111 the year before. It was the fourth consecutive annual decline in accident rates.

The 1995 fatal accident rate per 1 million miles flown for the scheduled airlines declined to 0.0004 from 0.0008 the year before. Based on 100,000 departures, the fatal rate was 0.025, down from 0.051 in 1994.

- The fatal mountain crash of American Airlines Flight 965, a B-757, near Cali, Colombia in December, 1995, was the result of a navigation error entered in the aircraft's Flight Management System (FMS). This accident is a classic "Controlled Flight into Terrain" (CFIT) situation which caused 160 deaths.

The Part 121 charter airlines, after 5 years without fatalities, had two deaths in 1995 for a fatal accident rate per 1 million miles flown of 0.0031, and 0.284 in terms of 100,000 departures.

Commuter or regional airline fatalities dropped to nine persons from 25 in 1994. The fatal accident rate fell both in terms of million miles flown to 0.0036 from 0.005 in 1994, and from 0.084 to 0.062 in terms of 100,000 departures. It was the fourth consecutive annual decline in fatal accident rates.

Total accidents for the scheduled airlines declined slightly in 1996 from 34 to 32, but the three fatal accidents resulted in the loss of 342 persons.

- On May 11, 1996, ValuJet flight 592, a DC-9, crashed into the Everglades shortly after takeoff from Miami International Airport, en route to Atlanta. All 105 passengers and five crewmembers aboard were killed. This accident was caused by the improper packaging and storage of hazardous materials (oxygen canisters) which resulted in a serious aircraft fire, and the ultimate bankruptcy of the airline.

• On July 17, 1996, TWA flight 800, a Boeing B-747 on a regularly scheduled flight to Paris, France, crashed into the Atlantic Ocean off the coast of Long Island, N.Y., shortly after takeoff from John F. Kennedy International Airport. All 230 people on board the aircraft were killed. The cause of the accident was an explosion from a short circuit electrical charge entering a fuel tank during an extended ground delay. These two disastrous crashes in 1996 resulted in the highest number of fatalities for a single year during the past two decades. The 1996 fatal accident rate per 100,000 departures for scheduled service increased from 0.025 to 0.038, and the rate per 100,000 flight hours rose from 0.016 to 0.023.

The nonscheduled carriers experienced five accidents, including two fatal ones, resulting in 38 fatalities. The total accident rate per 1 million aircraft miles almost doubled, from 0.0062 in 1995 to 0.0118 in 1996. The fatal accident rate per 100,000 departures increased from 0.284 to 0.530.

Commuter airlines had only one fatal accident in 1996. The fatal accident rate per miles, hours, and departures all declined for the fifth consecutive year. There were 90 accidents, including 29 fatal ones, for the air taxis during the same year, resulting in 63 fatalities. The accident rate per 100,000 hours declined to 2.80 from 3.02 in 1995, while the fatal accident rate declined to 0.90 from 0.97 in 1995.

The scheduled carriers experienced 44 accidents in 1997, an increase of 12 over the 32 in 1996. There were three fatal accidents, the same number as in 1996, which resulted in three fatalities. Given the increase in miles and hours flown and departures, the accident rates stayed about the same for the years 1996 and 1997.

Under the Commuter Safety Initiative, Effective March 20, 1997, aircraft with 10 or more seats were required to conduct scheduled passenger operations under 14 CFR 121, which resulted in a significant reduction in miles and hours flown and departures for the commuter air carriers that formerly operated under 14 CFR 135. Total accidents increased from 11 in 1996 to 16 in 1997, while fatal accidents rose during the same period from 1 to 5. As a result, the total accident rate per 1 million miles increased from 0.019 in 1996 to 0.065 in 1997. The rates per 100,000 aircraft hours went from 0.399 to 1.628 and, per 100,000 departures, rose from 0.313 to 1.148 during the same period.

In 1998, the scheduled air carriers experienced only one fatal accident and two fewer total accidents than in 1997. As a result, all accident rates declined. Charter air carriers had a total of seven accidents, an increase of two over 1997. There were no fatal accidents in 1998 or 1999. Commuter airlines also had an exceptionally good year in 1998 with total accidents being halved, from 16 to 8 and no fatal accidents. However, as previously noted in Chapter 1, the fatal crash of Colgan Air Flight 3407 in February of 2009 would result in significant changes in commuter airline operations. The total rate per 100,000 aircraft hours decreased from 2.65 to 2.03, and the fatal rate per 100,000 aircraft decreased from 0.48 to 0.45.

Scheduled commuter carriers recorded 13 total accidents, including five fatal ones with a loss of 12 persons in 1999. On June 1, 1999, American Airlines Flight 1420, an MD-82 landing in a thunderstorm with windshear conditions, overran the end of the runway, went down an embankment, and hit approach light structures at the Little Rock Airport. There were 11 fatalities, including the aircraft captain.

The nonscheduled carriers experienced five total accidents in 1999, 2 less than in 1998. The total accident rate per 1 million aircraft miles fell from 0.0178 to 0.0121, while the rate per 100,000 departures was reduced from 1.575 to 1.116.

Commuter airline accidents increased to 13 in 1999 versus 8 in 1998, and fatal accidents increased from none in 1998 to 5 in 1999. Combined with the decline in total operating statistics caused by more commuters becoming Part 121 operators, accident rates increased by 58 to 71 percent across the board between 1998 and 1999.

THE 2000s.   The start of the new millennium saw an increase in total accidents to 51 from 47 in 1999 for scheduled Part 121 carriers. In 2000, there were three fatal accidents resulting in 92 deaths, as opposed to two fatal accidents resulting in 12 deaths in 1999. This equated to a 50 percent increase in the rate of fatalities per 100,000 departures between the two years. Chartered Part 121 operations continued to have no fatal accidents in 2000. Commuter Part 135 operations had only one fatal accident (as opposed to five in 1999) that resulted in five deaths (as opposed to 12 in 1999). This translated to a 78 percent reduction in the 2000 fatal accident rate of 0.164 per 100,000 departures over the 1999 rate of 0.744.

- On January 31, 2000, Alaska Airlines Flight 261 (McDonnell Douglas MD-83), en route from Puerto Vallarta, Mexico to San Francisco, California, reported control problems with its horizontal stabilizer and a loss of stability before it crashed into the Pacific Ocean near Point Mugu, California. There were 88 fatalities.

- On July 25, 2000, Air France Flight 4590, a Concorde SST, crashed on takeoff from Charles de Gaulle International Airport near Paris after striking a titanium metal strip on the runway from a Continental DC-10. The metal strip punctured a tire of the Concorde and debris ruptured a fuel tank causing a serious fire from which the Concorde could not recover. This type of damage is known as Foreign Object Damage (FOD) when it strikes another aircraft. There were 109 fatalities. This accident marked the beginning of the end of Concorde Supersonic Transport Service, since the aircraft was retired 3 years later due to safety concerns and lack of passengers in the wake of the 9/11 terrorist attacks in New York and Washington.

- On August 23, 2000, Gulf Air Flight 072, an Airbus 320, crashed into the Persian Gulf near Manama, Bahrain due to spatial disorientation. There were 143 fatalities.

- On October 31, 2000, Singapore Airlines Flight 006, a Boeing 747, crashed on takeoff at Taipei, Taiwan. There were 83 fatalities. The crew of this aircraft lost situational awareness in a rain storm and attempted to takeoff on a closed parallel runway, hitting several large pieces of construction equipment.

The year 2001 changed the landscape of aviation for good. On September 11, four U.S. commercial aircraft (Part 121 scheduled) were crashed in separate acts of terrorism that collectively killed 265 individuals on board (and several thousands on the ground). Aviation Security will be covered in detail in Chapter 11. The four aircraft involved in the attacks on September 11, 2001 were as follows:

- American Airlines Flight 11, a B-767 which departed from Boston and struck the north tower of the World Trade Center in New York City.
- United Airlines Flight 175, a B-767 which departed from Boston and struck the south tower of the World Trade Center in New York City.
- American Airlines Flight 77, a B-757 which departed from Dulles and struck the Pentagon in Washington.
- United Airlines Flight 93, a B-757 which departed Newark and crashed in a field near Shanksville, PA.

This tragic day, coupled with two other fatal crashes, resulted in a total of 531 fatalities for Part 121 operations—the worst on record since the 1977 runway collision between two 747s at Tenerife Airport in the Spanish Canary Islands which resulted in the loss of 583 lives. If one were to discount the four terrorist disasters, Part 121 operations had only two fatal accidents (one less than in 2000), but this resulted in 266 deaths—almost a threefold increase over that in 2000. While the fatal rate per 1 million miles flown was not significant between the 2 years, the fatal rate per 100,000 departures actually decreased by 26 percent in 2001. The Part 121 nonscheduled operations continued to have no fatal accidents. Part 135 commuter operations had two fatal accidents (compared to one in 2000), resulting in 13 deaths—eight more than in 2000. This resulted in a 2.5-fold increase in the accident rate per 100,000 departures to 0.402.

- On November 12, 2001, American Airlines Flight 587, an Airbus 300-600 experienced a loss of control upon initial climbout and crashed into a residential area in Queens, NY killing 265. The accident was caused by excessive overuse of the rudder to counter a wake turbulence problem, resulting in a separation of the vertical stabilizer from the aircraft.

The following is a summary of major international and U.S. aircraft accidents which occurred for the remainder of the decade ending in 2010:

- On April 15, 2002, Air China Flight 129, a B-767 from Beijing to Busan South Korea crashed into a hillside during a circling approach in poor weather. There were 129 fatalities in this controlled flight into terrain (CFIT) accident.
- On May 7, 2002, China Northern Airlines Flight 6136, a MD-82 from Beijing to Dalian China crashed after a troubled passenger set the cabin on fire with gasoline to obtain the proceeds from seven insurance policies. There were 112 fatalities.

- On May 25, 2002, China Airlines Flight 611, a B-747 from Taiwan to Hong Kong broke up in flight and crashed due to a previous inadequate structural repair to the hull. There were 225 fatalities.

- On March 6, 2003, Air Algiers Flight 6289, a B-737, had an engine failure on takeoff, veered off the runway and crashed. There were 102 fatalities. Likewise, on July 8, 2003, Sudan Airways Flight 139, another Boeing 737, lost an engine and crashed with 117 fatalities.

- On January 3, 2004, Flash Airlines Flight 604, a B-737 from Egypt to France, crashed in the Red Sea with 148 fatalities. The cause of the accident was distraction of the pilot and spatial disorientation.

- On August 14, 2005, Helios Airways Flight 522, a B-737 traveling from Cyprus to Greece lost pressurization and crashed with 121 fatalities. The apparent cause of the accident was an improper setting of the cabin pressurization switch which incapacitated the crew and passengers due to hypoxia brought about by depressurization of the aircraft. The plane crashed near Athens after fuel starvation of the engines.

- On September 5, 2005, Mandala Flight 091, a B-737, crashed on takeoff in Indonesia. There were 117 fatalities. The cause of the accident was failure to use the checklist resulting in improper setting of the flaps and slats which were not placed in the takeoff configuration.

- On May 3, 2006, Armavia Flight 967, an A-320, crashed into the Black Sea at night on a missed approach in poor weather near Adler/Sochi airport in Russia. There were 113 fatalities. The cause of this accident was controlled flight into terrain (water) while climbing out with weather conditions below established minima. The official investigative report cited the psycho-emotional stress level of the captain, poor cockpit resource management, and air traffic control problems.

- On July 9, 2006, another domestic accident, S7 Airlines Flight 778, an A-310 from Moscow crashed in Siberia with 125 fatalities. The aircraft failed to decelerate on landing due to inadvertent movement of one throttle to the forward thrust position which caused the plane to overrun the runway and crash into a concrete barricade.

- On August 27, 2006, Comair Flight 191, a CRJ-100, crashed on takeoff from Lexington, Kentucky. The aircraft took off on the wrong (short) runway due to a loss of situational awareness and the crew's nonpertinent conversations in violation of the FAA "Sterile Cockpit Rule."

- On September 29, 2006, GTA Flight 1907, a new B-737-800, suffered a midair collision at Flight Level 370 with an Embraer Legacy business jet on a domestic flight over the Brazilian Amazon jungle. There were 154 fatalities. Although the Brazilian authorities initially charged the pilots of the Legacy with negligence, the NTSB report cited a combination of ATC errors that placed both aircraft on the same airway at the same altitude.

- On July 17, 2007, Tam Airlines Flight 3054, an A-320 on another domestic Brazilian flight, overran the runway and crashed on landing at Congonhas

Airport in Sao Paulo. There were 187 fatalities. The investigation indicated that the aircraft thrust reverser was deactivated, causing the plane to run off this short runway.

- On August 20, 2008, Spanair Flight JK 5022, a MD-82, crashed on takeoff in Madrid, Spain. There were 154 fatalities. The investigation indicated that after an interruption, the aircraft tried to take off in the wrong configuration with the flaps and slats retracted.

- Miracle on the Hudson—On January 15, 2009, US Airways Flight 1549 successfully ditched in the Hudson River near New York City after takeoff from LaGuardia Airport. The aircraft was disabled by striking a flock of Canadian Geese and lost thrust in both engines during its initial climb out after takeoff. All 155 occupants safely evacuated the airliner. The NTSB cited the excellent crew resource management, safety equipment, and fast rescue response from ferry boat operators.

- On February 12, 2009, Colgan Air Flight 3407, a Bombardier Dash 8 stalled and crashed on final approach to Buffalo, NY. There were 49 fatalities. The investigation revealed that the crew failed to monitor the airspeed in icing conditions and took an inappropriate response to the stick shaker stall alarm system, as well as failure to comply with the FAA Sterile Cockpit rule.

- On June 1, 2009, Air France Flight 447, an A-330 from Rio de Janeiro to Paris crashed into the Atlantic Ocean. There were 228 fatalities. The aircraft crashed in bad weather after receiving unusual airspeed indications from the plane's Pitot Static System.

- On May 22, 2010, Air India Express Flight 812, a B-737, crashed in Mangalore, India, after overshooting the runway and falling over a cliff on landing. Apparently, the aircraft was attempting to go around after landing long. There were 158 fatalities.

- On July 28, 2010, Air Blue Flight 202, an A-321, crashed into a mountain near Islamabad, Pakistan. There were 152 fatalities. This is a Controlled Flight into Terrain (CFIT) accident where the air traffic controller lost contact with the flight during dense fog and monsoon rain.

LOOKING FORWARD—2011 AND BEYOND

Looking forward in the United States, the Comair Flight 191 accident in Lexington, Kentucky, in 2006, and the Colgan Air Flight 3407 in Buffalo, New York, in 2009 changed the landscape of FAR Part 121 commercial aviation safety for the foreseeable future. The Airline Safety and Pilot Training Improvement Act, signed August 1, 2010, will certainly result in higher standards for pilot training, strong support for safety advocacy programs, and new pilot fatigue rules to improve air passenger safety. The FAA is on a very tight schedule to implement the requirements of this Act, and new FAR regulations will be forthcoming in the next 3 years.

Meanwhile, ICAO has taken a new approach to reducing the number of aviation accidents at the international level. A "High-level Safety Conference" (HLSC) was held at ICAO headquarters in Montreal in late March 2010 comprised of worldwide aviation safety leaders to consider new proactive safety management systems (SMS) concepts and formulate recommendations for improvement of key ICAO safety programs. As the result of HLSC 2010, the SMS proactive approach has been significantly strengthened and extended to ICAO member States through the State Safety Programme (SSP), a comprehensive system for the management of safety within a State.

Additionally, ICAO developed a Global Aviation Safety Plan (GASP) and published 12 Global Safety Initiatives in 2007 to be used as working tools to enhance international aviation safety where most needed. Since certain ICAO regions in the third world had twice the accident rate of modern aviation countries, ICAO recently decided to focus its limited resources to support States and regions whose safety performance was not at an acceptable level. In its 37th annual assembly, which concluded in October 2010, ICAO adopted two specific resolutions, ICAO Global Planning for Safety (Resolution A37-4) and the Universal Safety Oversight Audit Programme (USOAP) Continuous Monitoring Approach (Resolution A37-5). In the latter resolution, the assembly directed an evolution of the USOAP to achieve a continuous monitoring approach (CMA) which will incorporate the analysis of safety risk factors to be applied to assess the oversight capabilities of each member State. This CMA approach is one of the key elements of SMS which advocates continuous improvement as one if its precepts. Additional information on these future developments is provided in an article in the Flight Safety Foundation Magazine Aero Safety World, April 2010 edition entitled "A New Approach." (http://flightsafety.org)

OCCUPATIONAL ACCIDENT STATISTICS—DEPARTMENT OF LABOR, BUREAU OF LABOR STATISTICS (BLS) (www.bls.gov/iif/)

The death toll trend due to occupational injuries has been on the decline over the years because of new technology, stricter safety regulations, and a shift in the economy toward safer service-industry jobs. The Census of Fatal Occupational Injuries program reports that the preliminary count of fatal work injuries in the United States in 2009 was 4,340, down from a revised total of 5,214 in 2008. Overall, fatal work injuries are down 26 percent since 2006, having reached a 20-year peak of 6,632 in 1994.

The Bureau of Labor Statistics still remains the authoritative source for nonfatal occupational (work-related) illnesses and injuries in the United States. As mentioned in Chapter 4, the BLS in cooperation with OSHA and the states uses scientifically developed sample surveys to collect data on nonfatal injuries and illnesses. All deaths are reported directly to OSHA. The surveys exclude the self-employed, farms with fewer than 11 employees, private households, federal government agencies, and employees in state and local governments. The results are published in the *Standard Industrial Classification (SIC) Manual*, which was first developed and published in 1987. Information in the *SIC Manual* is published by industry type. Major groups are further divided into *industry groups* with each industry group having SIC codes within it.

One of the groups of relevance to aviation is Major Group 45, *Transportation by Air*. This Major Group contains Industry Group 451 (*Air Transportation, Scheduled, and Air Courier*), Industry Group 452 (*Air Transportation, Nonscheduled*), and Industry Group 458 (*Airports, Flying Fields, and Airport Terminal*). The other group of relevance to aviation is Major Group 37, *Transportation Equipment*. This Major Group contains Industry Group 372 (*Aircraft and Parts*). SIC codes contain a four-digit coding system to further identify industry segments.

NORTH AMERICAN INDUSTRY CLASSIFICATION SYSTEM (**NAICS**) (www.census.gov)   Developed in cooperation with Canada and Mexico, the North American Industry Classification System (NAICS) is a profound change in government statistical reporting programs. NAICS uses a six-digit coding system to classify all economic activity into twenty service sectors. For example, the category of "Scheduled Air Transportation" is broken down into two codes as follows: 481111, "Scheduled Passenger Air Transportation" and 481112, "Scheduled Freight Air Transportation". The NAICS has been adopted to soon replace the Department of Labor's Standard Industrial Classification (SIC) coding system thus providing a new statistical tool in the area of occupational accident statistics.

The BLS Injuries, Illness, and Facilities program provides annual information on the rate and number of work related injuries, illnesses, and fatal injuries, and how these statistics vary by incident, industry, geography, occupation, and other characteristics. The number of nonfatal occupational injuries and illnesses reported in 2009 declined to 3.3 million cases, compared to 3.7 million cases in 2008. The total recordable case injury and illness incidence rate among private industry employers has declined significantly each year since 2003 when estimates were first published using NAICS data.

In recent years, the number of work-related fatalities, injuries, and illnesses has steadily decreased in the Air Transportation subsector, NAICS category 481. For example, the number of fatalities in this subsector decreased from 38 in 2008 to 28 in 2009, a reduction of over 26 percent. Similarly, the rate of injury and illness cases fell in the category of total recordable cases, cases involving days away from work, and cases involving days of job transfer or restriction. However, when compared with general industry injury and illness trends, the air transportation rates are 8.5 total recordable cases per 100 full-time workers which is nearly twice the national injury average. While general industry injury and illness trends have been at their lowest level in years, the air transportation sector could still use some significant improvement.

Like many collected data, the NAICS/BLS statistics do have some limitations. As in most surveys, the data are subject to sampling error. There are also other sources of error to which the data are subject. Some of these include the inability to obtain information about all cases in the sample, mistakes in coding or recording data, and definition difficulties. These statistics provide an estimate of workplace injuries and illnesses based on logs kept by employers during the year. The employer's understanding of which cases are work-related under OSHA's recordkeeping guidelines may not always be clear. Finally, the number of injuries and illnesses that get reported can be influenced by level of economic activity, working conditions and work practices, worker experience and training, and the number of hours worked.

KEY TERMS

Safety factors

Safety indicator

(Safety) exposure data

Accident rate

Hull loss

ICAO High Level Safety Conference (HLSC)

State Safety Programme (SSP)

Global Aviation Safety Plan (GASP)

Universal Safety Oversight Audit Programme (USOAP)

North American Industry Classification System (NAICS)

NTSB accident definitions: major, serious, injury, damage

Standard Industrial Classification (SIC) Manual

OSHA injuries and illnesses

Bureau of Labor Statistics (BLS)

REVIEW QUESTIONS

1. Why is the accident rate a better measure of safety than accident counts are?

2. Discuss some of the issues to be aware of in analyzing and comparing commercial aviation accident statistics.

3. What are hull-loss accidents and what are their primary cause? Describe the hull-loss accident trend since the early 1970s.

4. What is the most critical phase of flight? What are some lessons that can be learned from the Boeing summary?

5. Describe the general trend in aviation accidents during the 1980s and 1990s and 2000s.

6. Discuss future ICAO safety concepts and programs under the Global Aviation Safety Plan.

7. Explain how OSHA injury and illness rates are calculated.

8. What are SIC and NAICS codes and how are they used?

REFERENCES

Boeing, Commercial Airplane Co. 2010. *Statistical Summary of Commercial Jet Aircraft Accidents: Worldwide Operations* 1959–2009, Seattle, Wash., July 2010 http://www.boeing.com/

National Transportation Safety Board. 1982–2009. Annual reports. Washington, D.C. NTSB. http://www.ntsb.gov/

WEB REFERENCES

Flight Safety Foundation, http://flightsafety.org

International Civil Aviation Organization, http://www.icao.int/

North American Industry Classification System, http://www.census.gov/

OSHA statistics http://www.osha.gov/

U.S. Department of Labor, Bureau of Labor Statistics. www.bls.gov

ACCIDENT CAUSATION MODELS

LEARNING OBJECTIVES

After completing this chapter, you should be able to

- List the reasons for accident modeling.

- Explain Reason's "Swiss Cheese" model of accident causation, giving examples where appropriate.

- Discuss the SHELL Model of accident causation and its implications.

- Give an outline of the avionics equipment in the cockpit of the modern commercial airliner.

- Explain the five causal factors in examining the nature of accidents: man, machine, medium, mission, and management.

INTRODUCTION

Aviation safety concepts have continued to evolve and develop over the years as technology changes. Accident investigation has traditionally focused on preventing accidents by concentrating on simple causation theories to determine "what happened," "who was responsible," and "when did it occur." This is basically a reactive approach to a very complex question.

Modern aviation safety theorists have been led by ICAO and other safety organizations to embrace a new way of thinking to prevent aviation accidents. These experts have focused on the questions:

- "Why did the accident happen?"
- "How can we prevent future accidents?"

In other words, since the turn of the century, aviation accident investigation has matured from assigning blame to the following thinking regarding accident prevention:

Reactive → Proactive → Predictive

FIGURE 6-1   Trend of Modern Aviation Theory. (*Source: www.icao.int*)

Instead of reacting to accidents and punishing the guilty party for failure to act safely, modern safety thinking has become proactive and even predictive of where and when the next accident may occur if current trends continue (Fig. 6-1). The new *ICAO Safety Management Manual* (2nd edition, 2009) (SMM) has an excellent discussion of the evolution of safety thinking over the past 50 years. It points out that from after World War II until the 1970s, safety experts focused on "technical factors" to solve aviation safety problems. As aircraft became more modern, the thinking shifted to "Human Factors" and using crew resource management concepts (for example) to remedy aviation safety woes. Today the concept of "organizational factors" and safety culture of the organization has been embraced as an important factor to consider using the latest safety management system tools (Fig. 6-2).

The purpose of accident investigation is to uncover pervasive, unrecognized causal factors of accidents. This can help prevent similar accidents from occurring in the future. However, since commercial aviation accidents are relatively rare, proactive measures for identifying short-term changes in safety are

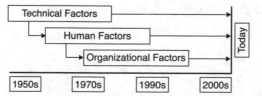

FIGURE 6-2   The evolution of safety thinking.

[*Source: ICAO Safety Management Manual (2nd Edition, 2009); www.icao.int*].

required. The goal of *nonaccident* data analysis and *modeling* is to conduct a pre-emptive strike on the very first accident to prevent it from occurring by addressing the root cause of the accident. While incidents and accidents provide after-the-fact evidence that safety was inadequate, accident *modeling* assists with understanding how accidents happen so that measures (policy decisions or changes in the aviation operating environment) can be taken to prevent potential hazards from materializing.

The main goal of accident investigations is to establish the probable causes of accidents and to recommend control measures. Because most aviation accidents involve a complex maze of diverse events and causes, classifying or categorizing these accidents by type or cause gets quite complicated and involved. Also, accidents that are similar may often require different preventive strategies, although at times a single solution can eliminate or reduce the rate of occurrence of a wide range of accidents. For example, ground-proximity warning devices addressed the wide range of issues involved with controlled flight into terrain accidents for jetliners and helped reduce its occurrence rate.

The NTSB classifies accidents by several methods, such as causes and factors, sequence of events, and phase of operation. While aircraft component failures and encounters with weather are easy to classify, failures due to human errors are harder to trace. Accidents frequently have multiple causes, hence developing causal categories is difficult. In a majority of the cases, each cause is independent of the others, and if one did not exist, the accident might not have occurred. This is known as the "chain of causation" and breaking one of the links in the chain through defense and control measures can be sufficient to prevent the accident.

Accident modeling, as explained in this chapter, helps us understand the nature of accidents. Accident models

- Help explain the relationship between hazards and accidents
- Assist with understanding and explaining reality
- Aid in visualizing things that cannot be directly observed
- Must approximate conditions that exist in reality to be useful

There are several accident models discussed in the literature. Three of these models that have been most frequently associated with aviation are discussed next.

REASON'S "SWISS CHEESE" MODEL

James Reason's accident causation model was published in 1990 to illustrate how human factors at various levels of the organization can lead to accidents. He traces the root cause of accidents to human errors that occur in the management levels of an organization. This model is a good representation of the complex relationship between the individual and the organization. Reason explains that before an active human failure occurs, there are certain latent conditions in the organization which

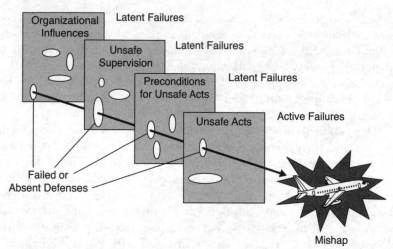

FIGURE 6-3   The Reason Model and Accident Causal Chain. Adapted from Reason (1990).

are the result of management action or inaction. He also states that human error is the active "end result" rather than the root cause of accidents. Some of the important features of Reason's model (Fig. 6-3) are the following:

- Systems are protected by multiple layers of defenses that are designed to prevent hazards or system failures from cascading into accidents.
- Each layer of protection, however, can develop "holes" or flaws though safety deficiencies, resembling Swiss cheese.
- As the number and size of these holes in the defenses increase, the chances of accidents also increase.
- When the holes in each of the layers of defenses line up, an accident occurs.

The model recommends focusing on events beyond the *active failures* of frontline employees to latent preexisting conditions that result from fallible decisions made by high-level decision makers. It is these failures that permit active failures to occur. Management should build defenses by creating an organizational culture in which precursor events are detected and promptly corrected.

Examples of real world problems using Reason's Swiss Cheese model would include the following:

Organizational Influences
- Rapid expansion
- Lack of regulation
- Management "lip service" to safety

Unsafe Supervision
- Risks and hazards neglected
- Poor work scheduling—fatigue
- Insufficient training

Preconditions for Unsafe Acts
- High workload
- Time pressure to perform tasks
- Ignorance of the system

Unsafe Acts
- Aircraft warning system disabled
- Omission of critical checklist item
- Over-reliance on automation

Another way to depict the complex relationship of the organization and human factor is provided in the ICAO SMM. Consistent with Reason's model, various "screens" or defenses can be set up to plug the holes in the "Swiss cheese" and thus prevent the accident. These defenses are controls built into the system by management to protect against the inevitable human error that cannot be completely avoided. Figure 6-4 shows that organizational factors could be either weaknesses (holes in the cheese) or strengths (preventative screens) that could serve as defenses or "safety nets" to prevent an accident situation.

Still another helpful model depiction to illustrate the importance of the organization and the dichotomy between latent conditions and active failures is provided in Fig. 6.5. The top block represents the organizational processes which

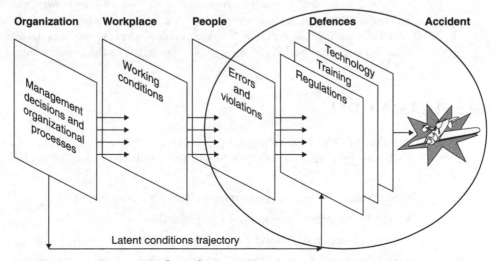

FIGURE 6-4 A concept of accident causation. [*Source: ICAO Safety Management Manual (2nd edition 2009) www.icao.int*].

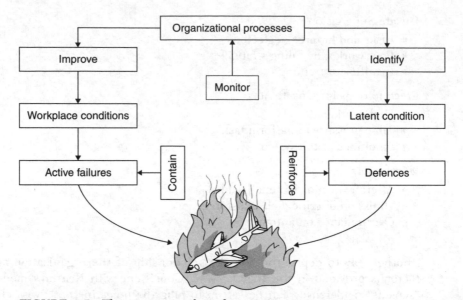

FIGURE 6-5   The organizational accident. [*Source: ICAO Safety Management Manual (2nd edition 2009) www.icao.int*].

are under the control of management such as policy making, communication, resources, etc. These must be continually monitored and enhanced to prevent the organizational accident from occurring. Existing latent conditions can breach aviation system defense screens if not sufficient in the areas of regulations, training, and technology (Fig. 6-4). Mitigation strategies should be developed to reinforce defenses to human errors. After the latent conditions and defenses are bolstered, the organization should shift its focus to improve workplace conditions and thus contain the active "human errors" that always occur in any complex aviation scenario. See Fig. 6-5 for this perspective of the organizational accident.

THE SHELL MODEL

Another widely used conceptual tool in aviation safety is the SHELL Model. As indicated by ICAO, the components of this model (Fig. 6-6) are as follows:

- **S** = Software (such as procedures, checklists, training, etc.)
- **H** = Hardware (machines and equipment)
- **E** = Environment (operating conditions)
- **L** = Liveware (human interface to S, H, and E above)
- **L** = Liveware (again, i.e., human to human interface)

FIGURE 6-6   The SHELL Model.
[*Source: ICAO Safety Management Manual*
(2nd edition 2009) www.icao.int].

To properly understand this model, it is important to note that the L (Liveware or Human Person) is always in the center of the diagram interacting with the other "SHELL" components. Since humans are inconsistent and do not interact perfectly with the other components, we need to consider four important factors affecting human performance (the so-called "4 Ps"):

- Physical factors (strength, height)
- Physiological factors (health, stress, etc.)
- Psychological factors (motivation, judgment, etc.)
- Psycho-social factors (personal issues or tension, etc.)

The SHELL interfaces are in constant interaction with each other and should be matched closely to the human element (Liveware) in the center of the system:

- *Liveware to Software*—The relationship between the human and supporting systems found in the workplace. Not just computer programs, these include user-friendly issues in regulations, manuals, and checklists.
- *Liveware to Hardware*—The relationship between man and machine. Although humans adapt well to poor interfaces, they can easily cause safety hazards if not well designed.
- *Liveware to Liveware*—This is the relationship between the human and other people in the workplace. Communication styles and techniques are important here. Crew Resource Management (CRM) training has made great strides in this area.
- *Liveware to Environment*—The relationship between the human and the internal and external environments. Sleep patterns and fatigue are examples of important considerations here.

(Source: ICAO Safety Management Manual, 2nd edition, 2009, Chapter 2)

THE 5-M MODEL

The *man, machine, medium, mission,* and *management factors* represent another valuable model for examining the nature of accidents (Fig. 6-7). That is, when one seeks causal factors or preventive or remedial action, the diagram of the intertwined circles becomes a meaningful checklist for fact-finding and analysis to ensure that all factors are considered.

The five factors are closely interrelated, although management plays the overall predominant role. Mission is located as the central target or objective to emphasize that effective mission accomplishment is implicit in professional system (aviation) safety work.

MODERN COMMERCIAL AVIONICS EQUIPMENT

In the period from 1970 to 1990, three generations of wide-body airliners appeared. The first generation, introduced around 1970, was the wide-body, long-range airliner. Examples are the Boeing 747, the Lockheed 1011, and the Douglas DC-10. These wide-body airliners were equipped with three to four high-bypass-ratio engines, *inertial navigation systems* (INSs), and an *automatic landing system* (ALS). The cockpits of these airliners were equipped with many electromechanical instruments. These aircraft were flown by a flightcrew of two pilots and one flight engineer.

The second generation, introduced around 1980, was the long-medium-range, wide-body airliner with a new digital avionics system. Examples are the Airbus A-310 and the Boeing 757/767. These new wide-body airliners are equipped with

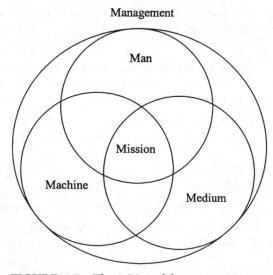

FIGURE 6-7   The 5-M model.

an *electronic flight instrument system* (*EFIS*), a *flight management system* (*FMS*), and either an Engine Indicating Crew Alerting System (EICAS—Boeing) or an Electronic Centralised Aircraft Monitor (ECAM—Airbus). In greater detail, the EFIS, FMS, and EICAS/ECAM provide the following functions:

- The EFIS *primary flight display* (*PFD*) provides a combined presentation of attitude, flight director, *instrument landing system* (*ILS*) deviation, flight mode annunciation, and speed and altitude information on a single *cathode-ray tube* (*CRT*) display, thus reducing the scanning cycle.

- The EFIS multifunction display (MFD) or *navigation display* (*ND*) provides integrated map, horizontal flight path, weather radar, heading, and wind vector information, largely reducing the navigation task (in combination with the FMS) and improving the positional awareness of the pilot.

- The FMS provides integrated navigation and fuel management information, as well as a host of performance and navigation information, largely increasing the pilot's flight management and navigation capabilities.

- The EICAS/ECAM provides the aircraft crew with engine and other system instrumentation and warning annunciations with remedial action required.

The "glass" cockpit is equipped with six color CRT graphics display for the EFIS, as well as two alphanumeric monochrome *control display units* (*CDUs*) for the FMS; apart from the CRTs, a number of electromechanical instruments are still used to enable a safe continuation of flight in case all CRTs should fail. The introduction of the EFIS, FMS, and EICAS/ECAM allowed the elimination of the flight engineer from the flight deck, providing a significant cost reduction for the operations with this type of airliner.

The third-generation airliner, introduced around 1990, is a long-medium-range aircraft with a revolutionary new digital flight control system, no longer using mechanical links between the pilot's control yoke and the hydraulic actuators of the flight control surfaces. This *fly-by-wire* (*FBW*) technology allows for new flight control concepts and envelope protection systems. Examples of these new FBW airliners are the Airbus A-320/330/340 and the Boeing 777. The engines of this new generation of airliners are controlled by *full-authority digital engine control* (*FADEC*) systems. In the cockpit, the CRTs have become larger, and the number of electromechanical instruments has strongly decreased.

Modern scheduled aviation developed into a reliable and economical all-weather transport system. Through the use of ever-improving aerodynamics and engine technology, as well as the increasing use of lightweight composite materials since 1970, the fuel consumption per passenger-mile has been reduced by more than 30 percent. Radio navigation and approach systems, inertial navigation systems, weather radar, and ATC, in combination with ever-improving training and standardized procedures, allow safe flight, also in reduced-visibility conditions. However, in *instrument meteorological conditions* (*IMCs*), the pilot's situational awareness is sometimes poor due to the nonexistence of outside visual attitude, navigation, weather, and terrain information.

Given that today's accident rate is unacceptable, what have we to look forward to, given the constant increase in activity in the same finite blocks of airspace and real estate?

First, we can be encouraged by the progress to date. The accident rate for the newer generation of airplanes, such as the B-757, B-767, the A-310, A-320 and A-330, is considerably better than that for earlier designs. It is reasonable to expect that the current new models, such as the B-777, B-787, and the A-350 and A-380, will be safer yet, as a result of more sophisticated design and applied technology.

In summary, new technology will be available to the flightcrews and controllers during the next decade:

- Better weather detection systems will provide information to airline dispatchers and pilots, allowing more efficient and safe flight around weather systems, both en route and near the airport.

- *Global positioning systems* (GPSs) are widely being used now, but will become the primary source for navigation and surveillance information, replacing ground-based, line-of-sight-limited VOR navigation facilities and radar facilities. GPS will also be the primary means of guidance for precision landings and departures at our nation's airports under the new Next Generation (NextGen) concept.

- Improved air traffic control tools are already being installed in FAA facilities to give the controller more reliable and efficient means to see and communicate with the airplanes under his or her control. Improved collision avoidance will be provided by the new Automatic Dependent Surveillance Broadcast (ADS-B) System.

- Data link will allow clearances, weather, and traffic information to be provided in the cockpit in a fast, error-free, digital form. One of the big advantages of data link will be the elimination of "read-back" errors between the pilot and controller.

- Flight decks will continue to improve, with added redundancy and integrated avionics giving the pilot more options and greater situational awareness.

- Human factors will be a major consideration from the onset of airplane design to ensure that the airplane can be operated and maintained easily within human limits.

Our aviation system has evolved over the past decades to serve a vital role in the economy and our way of life. The system is complex, built on international standards with rigid quality control in all areas from the cockpit to the maintenance hangar to the air traffic control facility.

MAN

While some may see the pilot as the only "man" in the system, others include all persons directly involved with the operation of aircraft—flightcrew, ground crew,

ATC, meteorologists, etc. In its widest sense, the concept should include all human involvement in aviation, such as in design, construction, maintenance, operation, and management. This latter is the meaning intended in this discussion since accident prevention must aim at all hazards, regardless of their origin.

As a result of refinements over the years, the number of accidents caused by the machine has declined, while those caused by man have risen proportionately. Because of this significant shift in the relationship between human and machine causes, a consensus has now emerged that accident prevention activities should be mainly directed toward the human, and the organization using SMS concepts.

People are naturally reluctant to admit to their limitations for a variety of reasons, such as loss of face among peers, self-incrimination, fear of job loss, or considerations of blame and liability. It is not surprising, therefore, that information on the human-factor aspects of accidents or incidents is not readily forthcoming. This is unfortunate since it is often these areas that hold the key to the "why" of a person's actions or inactions.

In the past, the view that the man involved only the pilot led to the inappropriate use of the term *pilot error* as a cause of accidents, often to the exclusion of other human-factor causes. As a consequence, any other hazards revealed by an investigation were often not addressed. Further, since the term tended to describe only *what* happened rather than *why,* it was of little value as a basis for preventive action. Fortunately, the term is now rarely used by investigation authorities.

The pilot is often seen as the last line of defense in preventing an accident. In fact, over the years, the skill and performance of pilots have prevented many accidents when the aircraft or its systems failed or when the environment posed a threat. Such occurrences usually do not receive the same attention and publicity as accidents, sometimes leading to an unbalanced perception of the skill and performance of pilots.

MACHINE

Although the *machine* (aviation technology) has made substantial advances, there are still occasions when hazards are found in the design, manufacture, or maintenance of aircraft. In fact, a number of accidents can be traced to errors in the conceptual, design, and development phases of an aircraft. Modern aircraft design, therefore, attempts to minimize the effect of any one hazard. For instance, good design should seek not only to make system failure unlikely, but also to ensure that should it nevertheless occur, a single failure will not result in an accident. This goal is usually accomplished by so-called fail-safe features and redundancy in critical components or systems. A designer must also attempt to minimize the possibility of a person using or working on the equipment committing errors or mistakes in accordance with the inevitability of Murphy's law: "If something can go wrong, it will." To meet these aims, some form of system safety program is often used during the development of a new aircraft type. Modern design must also take into account the limitations inherent in humans. Therefore, it includes systems that make the human's task easier and that aim to prevent mistakes and errors. The *ground-proximity warning system* (*GPWS*) is an example of such a system. It has significantly reduced the number of accidents in which airworthy aircraft collide with the ground or water while under the control of the pilot.

The level of safety of an aircraft and its equipment is initially set by the airworthiness standards to which it is designed and built. Maintenance is then performed to ensure that an acceptable level of safety is achieved throughout the life of the aircraft. Manufacturing, maintenance, and repair errors can negate design safety features and introduce hazards that may not be immediately apparent.

As the service experience with a particular aircraft type increases, the maintenance program needs to be monitored and its contents developed and updated where necessary to maintain the required levels of safety. Some form of reporting system is required to ensure that component or system malfunctions and defects are assessed and corrected in a timely manner.

The reliability of a component is an expression of the likelihood that it will perform to certain specifications for a defined length of time under prescribed conditions. Various methods can be used to express reliability. A common method for electronic components is the *mean time between failures (MTBF)*, and the reliability of aircraft power plants is usually expressed as the number of shutdowns per 100,000 operating hours.

Failures normally arise in three distinct phases in the life of a component. Initial failures, caused by inadequate design or manufacture, usually occur early in its life. Modifications to the component or its use usually reduce these to a minimum during the main or useful life period. Random failures may occur during this period. Near the end of the life of a component, increased failures occur as the result of its wearing out. Graphic representation of this failure pattern gives rise to the typical "bathtub-shaped" curve (Fig. 6-8).

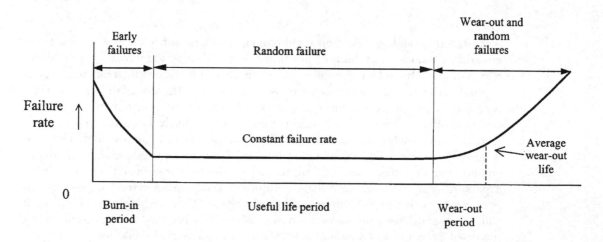

FIGURE 6-8   Aircraft system failure pattern.

MEDIUM

The *medium* (environment) in which aircraft operations take place, equipment is used, and personnel work directly affects safety. From the accident prevention viewpoint, this discussion considers the environment to comprise two parts—the natural environment and the artificial environment.

Weather, topography, and other natural phenomena are thus elements of the natural environment. Their manifestations, in forms such as temperature, wind, rain, ice, lightning, mountains, and volcanic eruptions, are all beyond the control of humans. These manifestations may be hazardous, and since they cannot be eliminated, they must be avoided or allowances must be made for them.

The artificial portion of the environment can be further divided into physical and nonphysical parts. The physical portion includes those artificial objects that form part of the aviation environment. Air traffic control, airports, navigation aids, landing aids, and airfield lighting are examples of the artificial physical environment. The artificial nonphysical environment, sometimes called *system software,* includes those procedural components that determine how a system should or will function. This part of the environment includes national and federal legislation, associated orders and regulations, standard operating procedures, training syllabi, and so forth.

Many hazards continue to exist in the environment because the people responsible do not want to become involved in change, consider that nothing can be done, or are insufficiently motivated to take the necessary actions. Obstructions near runways, malfunctioning or nonexistent airport equipment, errors or omissions on aeronautical charts, faulty procedures, and so forth are examples of artificial environmental hazards that can have a direct effect on aviation safety.

MISSION

Notwithstanding the man-machine-medium concept, some safety experts consider the type of *mission,* or the purpose of the operation, to be equally important. Obviously the risks associated with different types of operation vary considerably. A commuter airline operating out of many small airports during the winter months in the New England area has a completely different mission than an all-cargo carrier flying extensive over-water flights to underdeveloped countries or a major carrier flying from New York to Los Angeles. Each category of operation (mission) has certain intrinsic hazards that have to be accepted. This fact is reflected in the accident rates of the different categories of operation and is the reason why such rates are usually calculated separately.

MANAGEMENT

The responsibility for safety and, thus, accident prevention in any organization ultimately rests with *management,* because only management controls the allocation of resources. For example, airline management selects the type of aircraft to be purchased, the personnel to fly and maintain them, the routes over which they operate, and the training and operating procedures used. Federal authorities promulgate airworthiness standards and personnel licensing criteria and provide air traffic and other services. Manufacturers are responsible for the design and manufacture of aircraft, components, and power plants as well as monitoring of their airworthiness.

The slogan "Safety is everybody's business" means that all persons should be aware of the consequences of their mistakes and strive to avoid them. Unfortunately, not everyone realizes this, even though most people want to do a good job and do it safely. Therefore, management is responsible for fostering this basic motivation so that each employee develops an awareness of safety. To do this, management must provide the proper working environment, adequate training and supervision, and the right facilities and equipment.

Management's involvement and the resources it allocates have a profound effect on the quality of the organization's accident prevention program. Sometimes, because of financial responsibilities, management is reluctant to spend money to improve safety. However, it can usually be shown that not only are accident prevention activities cost-effective, but also they tend to improve the performance of people, reduce waste, and increase the overall efficiency of the organization.

Management's responsibilities for safety go well beyond financial provisions. Encouragement and active support of accident prevention programs must be clearly visible to all staff, if such programs are to be effective. For example, in addition to determining who was responsible for an accident or incident, management's investigation also delve into the underlying factors that induced the human error. Such an investigation may well indicate faults in management's own policies and latent organizational procedures.

Complacency or a false sense of security should not be allowed to develop as a result of long periods without an accident or serious incident. An organization with a good safety record is not necessarily a safe organization. Good fortune rather than good management practices may be responsible for what appears to be a safe operation.

On the whole, management attitudes and behavior have a profound effect on staff. For example, if management is willing to accept a lower standard of maintenance, then the lower standard can easily become the norm. Or, if the company is in serious financial difficulties, staff may be tempted or pressured into lowering their margins of safety by "cutting corners" as a gesture of loyalty to the company or even self-interest in retaining their jobs. Consequently, such practices can and often do lead to the introduction of hazards. Morale within an organization also affects safety. Low morale may develop for many reasons but nearly always leads to loss of pride in one's work, an erosion of self-discipline, and other hazard-creating conditions.

CONCLUSION

In spite of the use of man, machine, medium, mission, and management as broad categories of hazards, a popular theory holds that most accidents or incidents can be traced to a human failure somewhere, not necessarily the person or thing immediately involved in the occurrence. For example, a machine is designed, built, and operated by humans. Thus a failure of the machine is really a failure of the human. Likewise, humans may not avoid or eliminate known environmental hazards, or they may create additional hazards. Thus, these could all be considered failures of humans rather than environmental failures. This interpretation, therefore, accounts for the wide discrepancy in the percentages of accidents attributed to human failure reported by different sources. Typically, these range from around 50% to close to 90%.

The three models discussed in this chapter clearly show that many aviation hazards are brought about by problems at the interface between technology, human, and organizational factors. As humans are always deeply involved, it is vital that we understand the principles of this critical area. In Chapter 7, we cover the subject of human factors in aviation safety.

KEY TERMS

Modeling

Reason's "Swiss cheese" model

Latent and Active failures

SHELL Model

5M model

ICAO Safety Management Manual (SMM)

Man-machine-medium-mission-management factors

Inertial navigation systems (INSs)

Automatic landing system (ALS)

Electronic flight instrument system (EFIS)

Primary Flight Display (PFD)

Multifunction Display (MFD)

Flight management system (FMS)

Engine Indicating Crew Alerting System (EICAS)

Electronic Centralised Aircraft Monitor (ECAM)

Fly-by-wire (FBW)

Instrument meteorological conditions (IMC)

Full-authority digital engine control (FADEC)

Global positioning system (GPS)

Next Generation Air Traffic control (NextGen)

Automatic-dependent surveillance broadcast system (ADS-B)

Ground proximity warning system (GPWS)

Mean time between failures (MTBF)

REVIEW QUESTIONS

1. Why do we model accidents?

2. Explain Reason's "Swiss cheese" model of defensive screens to include the concept of latent and active failures.

3. Discuss the SHELL Model. Why is "Liveware" always in the center?

4. Why can it be said that management plays a predominant role when examining the five-factors model?

5. Why do you think over the years the number of accidents caused by "machine" has declined, while those attributable to man have risen?

6. The medium or environment includes two parts—the natural environment and the artificial environment. Compare and contrast the two.

7. Give several examples of how management can influence the safety program. How can an effective safety program affect efficiency and cost-effectiveness?

8. Outline some of the modern avionics equipment found in the glass cockpit of today's commercial airliner.

9. What is the purpose of the electronic flight information system (EFIS)?

10. Explain the concept of the organizational accident.

REFERENCES

International Civil Aviation Organization. 2009. *Safety Management Manual*, 2nd ed. (Doc 9859 AN/474), www.icao.int

Reason, J. 1990. *Human error,* 1st ed. Cambridge University Press, UK.

Reason, J. 1997. *Managing the Risks of Organizational Accidents*. Aldershot, U.K.: Ashgate.

Salas, E. and Maurino, D. 2010. *Human Factors in Aviation*, 2nd ed. Burlington, MA; Elsevier Academic Press.

Stolzer, A. J., Halford, C. D. and Goglia, J. J. 2008. *Safety Management Systems in Aviation*, Aldershot, U.K.; Ashgate.

Taylor, Laurie. 1997. *Air Travel: How Safe Is It?* 2nd ed. London: Blackwell Science, Ltd.

HUMAN FACTORS
IN AVIATION SAFETY

LEARNING OBJECTIVES

After completing this chapter, you should be able to

- Discuss the significance of human error in major aircraft accidents.
- List and explain the factors that affect human performance, giving examples of each.
- Define human error and explain the HFACS method used to classify human error.
- Explain what is meant by management of human error.
- Discuss the concepts of Crew Resource Management, Line Operations Safety Audits, and Threat and Error Management in modern commercial aviation.

- Explain common control strategies to manage threats and errors such as cockpit standardization, cockpit automation, warning and alerting systems, the flight management system, and air-to-ground communication in human-error management.
- List the significant conclusions of the NTSB Professionalism in Aviation conference.

INTRODUCTION

The people who operate and support the U.S. aviation system are crucial to its safety; the resourcefulness and skills of crewmembers, air traffic controllers, and mechanics help prevent countless mishaps each day. However, despite the excellent safety record, many studies attribute human error as a factor in at least two-thirds of commercial aviation accidents. The leading human factors theorists and modern researchers believe that between 70 percent and 80 percent of all aviation accidents are attributable to human error somewhere in the chain of causation (Wiegmann & Shappel, 2003, 2009). Safety attention at present is, therefore, heavily focused on trying to understand the human decision-making process and how humans react to operational situations and interact with new technology and improvements in aviation safety systems. The way in which human beings are managed affects their attitudes, which affects their performance of critical tasks. Their performance affects the efficiency and, therefore, the economic results of the operation. It is important to understand how people can be managed to yield the highest levels of error-free judgment and performance in critical situations, while at the same time providing them with a satisfactory work environment. A review of accident cockpit voice recordings clearly indicates that distractions must be minimized, and strict compliance with the sterile cockpit rule must be maintained during the critical phases of flight (taxi, takeoff, approach, and landing).

While the emphasis often focuses on the pilots, they are not the lone threat. They are, however, the last link in the chain and are usually in a position to identify and correct errors that result in accidents and incidents.

Basically, the problem is one of poor human decision making. Essentially, three reasons explain why people make poor decisions: They have incomplete information; they use inaccurate or irrelevant information; or they process the information poorly. Psychologists have traditionally explained the limited information processing capabilities of humans by Miller's Law, which states that the number of objects an average human can hold in working memory is 7 (plus or minus 2). This magic number of 7 is improved when a pilot uses both his visual and auditory channels because the information is processed differently in the brain. Modern research has shown that accidents are more likely to occur during high workload, task saturation periods, when there is an overload of one or more of the pilot's processing channels.

In order to reduce the workload during critical task saturation situations, new pilots are taught a task-shedding strategy to focus on the most important task in the cockpit, flying the plane. Stated simply:

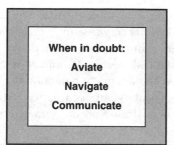

FIGURE 7-1 Aviator's emergency strategy. (*Source:* FAA).

In other words, in an emergency, fly the plane first, then if circumstances permit, navigate; and finally, if the other two tasks are in hand, communicate with air traffic control (Fig. 7.1).

Postaccident investigations usually uncover the details of what happened. With mechanical failures, accident data analysis often leads logically to why the accident occurred. Determining the precise reason for human errors is much more difficult. Without an understanding of human behavior factors in the operation of a system, preventive or corrective actions are impossible.

Understanding human factors is especially important to systems where humans interact regularly with sophisticated machinery and in industries where human-error-induced accidents can have catastrophic consequences. However, human factors are not treated as a technology in commercial aviation. Technical decisions for aircraft design, regulation, production, and operation are based on "hard" sciences, such as aerodynamics, propulsion, and structures. Human capabilities do not lend themselves readily to consistent, precise measurements. And human factors research requires much more time and cooperation than most other aeronautics research. Data on human performance and reliability are regarded by many technical experts as "soft" and receive little attention in some aviation system designs, testing, and certification. Data used in designs are often after the fact. This chapter explores areas of aviation in which human factors are especially important.

HUMAN FACTORS

Human factors is a multidisciplinary science that attempts to optimize the interaction between people, machines, methods, and procedures that interface with one another within an environment in a defined system to achieve a set of systems goals. Human factors encompass fields of study that include, but are not limited to, engineering, psychology, physiology, anthropometry, biomechanics, biology, and certain fields of medicine. Human factors science concentrates on studying the capabilities and limitations of the human in a system with the intent of using this knowledge to design systems that reduce the mismatch between what is required of the human and what the human is capable of doing. If this mismatch is minimized, errors (that could lead to accidents) will be minimized and human performance will be maximized. *Human performance* is a measure of human activity that expresses how well a human has carried out an assigned, well-defined task or a portion of a task (task element), and it is a function of *speed* and *accuracy*. This chapter looks at the accuracy component of human performance. If a task is not performed "accurately" in accordance with its requirements, an error has occurred. An expanded discussion on human error follows in the next section.

Accidents rarely involve a deliberate disregard of procedures. They are generally caused by situations in which a person's capabilities are inadequate or are overwhelmed in an adverse situation. Humans are subject to such a wide range of varying situations and circumstances that not all can be easily foreseen. Careful attention should therefore be given to all the factors that may have influenced the person involved. In other words, consideration must be given not only to the human error (failure to perform as required) but also to why the error occurred.

HUMAN PERFORMANCE

Variables that affect human performance can be grouped into seven categories, i.e., physical factors, physiological factors, psychological factors, psychosocial factors, hardware factors, task factors, and environmental factors. These factors are now briefly reviewed.

1. *Physical Factors* include body dimensions and size (anthropometric measurements), age, strength, aerobic capacity, motor skills, and body senses such as visual, auditory, olfactory, and vestibular.

2. *Physiological Factors* include general health, mental, and medical conditions such as low blood sugar, irregular heart rates, incapacitation, illusions, and history of injury, disability, or disease. Also included in this category are human conditions brought on by lifestyle such as the use of drugs, alcohol, or medication; nutrition; exercise; sports; leisure activities; hobbies; physical stress; and fatigue.

3. *Psychological Factors* include mental and emotional states, mental capacity to process information, and personality types (introverts and extroverts). Some human *personality traits* include the following:

List of Personality Traits
- *Motivation* is a desire of an individual to complete the task at hand. Motivation affects one's ability to focus all the necessary faculties to carry out the task.

- *Memory* allows us to benefit from experience. It is the mental faculty that allows us to prepare and act upon plans. Memory can be improved through the processes of association, visualization, rehearsal, priming, mnemonics, heuristics, and chaining. Memory management organizes remembering skills in a structured procedure while considering time and criticality. It is a step-by-step process to increase the accuracy and completeness of remembering.

- *Complacency* can lead to a reduced awareness of danger. The high degree of automation and reliability present in today's aircraft and the routines involved in their operation are all factors that may cause complacency.

- *Attention* (or its deficit) determines what part of the world exists for you at the moment. Conscious control of attention is needed to balance the environment's pull on attention. An intrapersonal accident prevention approach would describe the hazardous states of attention as distraction, preoccupation, absorption, and attention mismanagement—the inability to cope with tasks requiring flexible attention and focused tracking and steering. The inability to concentrate can lead to lack of (situational) *awareness,* which has been identified as a contributing factor in many accidents and incidents.

- *Attitude* strongly influences the functioning of attention and memory. Attitudes are built from thought patterns. An intrapersonal approach to the attitudes of crewmembers attempts to identify the desirable ranges between such hazardous thought patterns as macho–wimp, impulsive–indecisive, invulnerable–paranoid, resigned–compulsive, and antiauthority–brainwashed.

- *Perceptions* can be faulty. What we perceive is not always what we see or hear. Initial perceptions and perceptions based solely on intended actions are especially susceptible to error. An intrapersonal approach prescribes ways to make self-checking more efficient and reliable.

- *Self-discipline* is an important element of organized activities. Lack of self-discipline encourages negligence and poor performance.

- *Risk taking* is considered by some to be a fundamental trait of human behavior. It is present in all of us to varying extents since an element of risk is present in most normal daily activities. Risk will be present as long as aircraft fly and penalties for failure are high. Accordingly, the taking of risks needs to be carefully weighed against the perceived benefits.

- *Judgment and decision making* are unique capabilities of humans. They enable us to evaluate data from a number of sources in the light of education or past experience and to come to a conclusion. Good judgment is vital for safe aircraft operations. Before a person can respond to a stimulus, he or she must make a judgment. Usually good judgment and sound decision making are the results of training, experience, and correct perceptions. Judgment, however, may be seriously affected by psychological pressures (or stress) or by other human traits, such as personality, emotion, ego, and temperament.
  - *Aeronautical Decision Making (ADM)* The FAA describes ADM as a systematic approach to the mental process used by aircraft pilots to consistently determine the best course of action in response to a given set of circumstances (FAA Advisory Circular 60-22). The FAA Pilot's Handbook of Aeronautical Knowledge recommends the "DECIDE" Model to provide the pilot with a logical way of making decisions. The DECIDE acronym means to Detect, Estimate, Choose a course of action, Identify solutions, Do the necessary actions, and Evaluate the effects of the actions (Fig. 7-2).

The DECIDE Model of Aeronautical Decision Making

1. **Detect.**   The decision maker detects the fact that change has occurred.
2. **Estimate.** The decision maker estimates the need to counter or react to the change.
3. **Choose.**   The decision maker chooses a desirable outcome (in terms of success) for the flight.
4. **Identify.** The decision maker identifies actions which could successfully control the change.
5. **Do.**   The decision maker takes the necessary action.
6. **Evaluate.** The decision maker evaluates the effect(s) of his/her action countering the change.

FIGURE 7-2   The DECIDE model has been recognized worldwide as an effective continuous loop decision-making process. (*Source: www.faa.gov*)

4. *Psychosocial Factors* include mental and emotional states due to death in the family or personal finances, mood swings, and stresses due to relations with family, friends, coworkers, and the work environment. Some of the factors that cause stress are inadequate rest, too much cognitive activity, noise, vibration and glare in the cockpit, anxiety over weather and traffic conditions, anger, frustration, and other emotions. Stress causes fatigue and degrades performance and decision making, and the overall effect of multiple stresses is cumulative. Interactions with coworkers are influenced by two important variables, namely, peer pressure and ego.

- *Peer pressure* can build to dangerous levels in competitive environments with high standards such as aviation in which a person's self-image is based on a high standard of performance relative to his or her peers. Such pressure can be beneficial in someone with the necessary competence and self-discipline, but it may be dangerous in a person with inferior skill, knowledge, or judgment. For example, a young, inexperienced pilot may feel the need to prove himself or herself and may, therefore, attempt tasks beyond his or her capability. Humans have many conflicting "needs," and the need to prove oneself is not limited to the young or inexperienced. Some persons, because of training or background, have a fear that others may consider them lacking in courage or ability. For such persons, the safe course of action may be perceived as involving an unacceptable "loss of face."

- *Ego* relates to a person's sense of individuality or self-esteem. In moderate doses, it has a positive effect on motivation and performance. A strong ego is usually associated with a domineering personality. For pilots in command, this trait may produce good leadership qualities in emergency situations, but it may also result in poor crew or resource management. The domineering personality may discourage advice from others or may disregard established procedures, previous training, or good airmanship. Piloting an aircraft is one situation in which an overriding ego or sense of pride is hazardous. Although usually not specifically identified as such in accident reports, these traits may often be hidden behind such statements as "descended below minima," "failed to divert to an alternate," "attempted operation beyond experience/ability level," "continued flight into known adverse weather," and so forth.

5. *Hardware Factors* include the design of equipment, displays, controls, software, and the interface with humans in the system.

6. *Task Factors* include the nature of the task being performed (vigilance and inspection tasks versus assembly operations), workload (work intensity, multitasking, and/or time constraints), and level of training.

7. *Environmental Factors* include noise, temperature, humidity, partial pressure of oxygen, vibration, and motion/acceleration.

It is important to note that the factors discussed above can act alone or in combination with two or more other factors to further degrade human performance in the occupational setting. These factors can produce synergistic effects on human performance. Some examples include an air traffic controller monitoring air traffic during extremely low air traffic volume while on allergy medication or a quality control inspector monitoring low-defect-rate products while on cold medication.

THE HUMAN FACTORS ANALYSIS AND CLASSIFICATION SYSTEM (HFACS) (www.hfacs.com)

In recent years, the study of aviation human factors has developed, promising new tools and a classification system based upon previous research in this critical area. The father of the "Swiss Cheese" model, Dr. James Reason, used an interesting analogy when discussing the importance of this study when he stated "[Human errors] are like mosquitoes. They can be swatted one by one, but they still keep coming. The best remedies are to create more effective defenses and to drain the swamps in which they breed" (Reason, 2005, p. 769). Starting with Reason's "Swiss Cheese" model and building upon its framework of latent conditions and active failures, Dr. Douglas Wiegmann and Dr. Scott Shappell developed HFACS for the U.S. Navy in response to increasing human errors and a high accident rate. HFACS is a theoretically based tool for investigating, analyzing, and classifying human error associated with aviation accidents and incidents. The HFACS model has been validated through comprehensive research of 1,020 NTSB accident investigations that occurred over a 13-year period. HFACS uses a systems approach whereby human error is not the cause but rather the result of a larger problem in the organization. A review of the "Swiss Cheese" model from Chapter 6 (Figs. 6-3 and 6-4) is helpful to illustrate the defenses (screens or barriers) that should be set up by an organization to stop the human error "chain of causation." The HFACS model uses the same four barrier levels used by Dr. Reason to prevent the accident by controlling:

- Organizational Influences
- Unsafe Supervision
- Preconditions for Unsafe Acts
- The Unsafe Act Itself (The Active Failure)

ORGANIZATIONAL INFLUENCES

Resource Management | Organizational Climate | Organizational Process

UNSAFE SUPERVISIONS

Inadequate Supervision | Planned Inappropriate Operations | Failed to Correct Problem | Supervisory Violations

PRECONDITION FOR UNSAFE ACTS

Environmental Factors | Condition of Operators | Personnel Factors

Physical Environment | Technological Environment | Adverse Mental State | Adverse Physiological State | Physical/ Mental Limitations | Crew Resource Management | Personal Readiness

UNSAFE ACTS

Errors | Violations

Decision Errors | Skill-Based Errors | Perceptual Errors | Routine | Exceptional

FIGURE 7-3   The HFACS framework four barrier levels. (*Adapted from:* Wiegmann and Shappell, 2003, p. 71).

The HFACS Framework Chart, from Shappell et al. (2003) (Fig. 7-3), breaks down these four barrier levels to their component parts. In the next section, we will discuss the active failure level (unsafe act itself) to determine what control strategies may be effective in combating the inevitable human error dilemma.

UNSAFE ACTS—ERRORS OR VIOLATIONS?   Since human error continues to be the largest cause of commercial aviation accidents, the next question becomes, what are these errors and how do we prevent them? The Reason "Swiss Cheese" model and HFACS both describe two general categories of these errors or "unsafe acts." As pointed out by Shappel and Wiegmann in their latest HFACS article entitled *"A Methodology for Assessing Safety Programs Targeting Human Error in Aviation"* (2009), an error is either an honest mistake or a violation, i.e., the willful disregard for the rules and regulations of safety. HFACS further describes three types of honest mistakes (decision, skill-based and perceptual errors) and two types of violations (routine and exceptional). Of course, the primary difference between an error as an honest mistake and a violation is the intent necessary to create a violation. Violations can be routine, habitual, and condoned by the organization, or they can be exceptional or extreme, such as an intentional FAR violation.

The HFACS process can be better understood using a simplified case study from a well-known accident. The case study chosen is the "Attempted Takeoff From Wrong Runway, Comair Flight 5191" accident which occurred in Lexington, Kentucky, on August 27, 2006. (NTSB Accident Report AAR-07/05.) This report explains the accident involving a Bombardier regional commuter jet which crashed during takeoff. The flightcrew was instructed by ATC to take off from the long runway 22, but instead took off on the short runway 26, ran off the end of that runway and crashed, killing 49 people. The NTSB determined the probable cause as follows:

> *The National Transportation Safety Board determines that the probable cause of this accident was the flight crewmembers' failure to use available cues and aids to identify the airplane's location on the airport surface during taxi and their failure to cross-check and verify that the airplane was on the correct runway before takeoff. Contributing to the accident were the flight crew's nonpertinent conversation during taxi, which resulted in a loss of positional awareness, and the Federal Aviation Administration's failure to require that all runway crossings be authorized only by specific air traffic control clearances. (NTSB Report p. 105)*

Working backwards from the probable cause statement, a simplified HFACS analysis would lead the safety analyst to these conclusions:

Organizational Influence
- FAA's failure to require that all runway crossings be authorized only by specific ATC clearances.

Unsafe Supervision
- Captain and ATC failed to correct the problem of lining up and taking off on the wrong runway.

Preconditions for Unsafe Acts
- Personnel Factors—Inadequate Crew Resource Management between the pilots and ATC

Unsafe Acts (Active Failure Level)
- Errors—Perceptual error, copilot's failure to use available location cues and failure to cross check the runway heading

- Violations—Routine, violation of Sterile Cockpit rule, crew's nonpertinent conversation during taxi which resulted in a loss of positional awareness

Figure 7-4 illustrates how HFACS may be used to analyze accident causation.

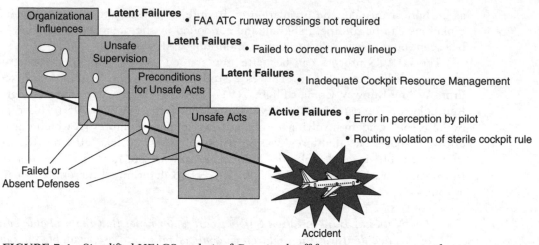

FIGURE 7-4   Simplified HFACS analysis of Comair takeoff from wrong runway accident.
[*Source: Adapted from Wiegmann & Shappell (2003)*].

A review of the HFACS framework in Figs. 7-3 and 7-4 indicates that prevention of the unsafe act at the active failure level is often a complex undertaking indeed. Naturally, the ultimate goal of an organization is to prevent the accident at the earliest level possible, setting the first barrier screen at the "organizational influences" (latent) level, the second barrier at "unsafe supervision," and another barrier at "preconditions" before reaching the "unsafe act" itself. The HFACS team has recently developed a post accident tool to attack the human error problem called the Human Factors Intervention Matrix (HFIX).

Once an accident or incident has occurred, proper utilization of HFACS allows a safety analyst to identify the specific types of human error that occurred at various levels of the organization. Notice once again that the focus of the accident investigation is on "what" happened and specifically "why" it happened in the organization. The next step in this complex process is to review the multi-level framework of factors potentially contributing to the accident. In their recent article (2009), Shappell and Wiegmann propose five intervention strategies to control human error in these categories:

• Organizational/Administrative

• Human/Crew

• Technology/Engineering

• Task/Mission

• Operational/Physical Environment

When plotted against the unsafc acts previously discussed, this relationship becomes the Human Factors Intervention Matrix (HFIX).

The HFACS classification methodology, and the HFIX intervention matrix are modern human factors tools which were designed for sophisticated organizations with high-quality data and knowledgeable human factors personnel. It provides a very useful framework for aviation accident investigators to use to study the organization and its role in accident causation due to latent conditions. Further information on this methodology is available at www.hfacs.com/

MANAGEMENT OF HUMAN ERROR

From the previous discussions of the SHELL model, Reason's Swiss Cheese Model, and HFACS, it is clear that human error is always going to be with us in modern commercial aviation. Therefore, strategies to manage human error are critical to controlling the accident and incident rate. Three strategies have developed over the years which have been built on the sound foundation of modern human factors models and research. These strategies are Crew Resource Management (CRM), Line Operations Safety Audits (LOSA), and Threat Error Management (TEM).

CREW RESOURCE MANAGEMENT (CRM)

After a series of accidents in the 1970s which identified human error as the cause of major air crashed, in 1979 NASA convened a workshop for the aviation industry entitled "Resource Management on the Flightdeck" which began the modern CRM movement in the United States. As pointed out in their article, *"The Evolution of Crew Resource Management Training in Commercial Aviation"* (Helmreich, Merritt, & Wilhelm, 1999), United Airlines led the way with the first comprehensive CRM training program in 1981, followed quickly by Delta Airlines and most U.S. FAR Part 121 carriers. The most vivid example of successful CRM concepts being used to good effect was the landing of United Airlines Flight 232 in Sioux City, Iowa, in 1989, following the total loss of all hydraulics systems aboard the DC-10. The captain of the DC-10 effectively used all available resources including crew, ATC, and ground assets to achieve the miraculous landing of United 232, saving 185 lives in the process.

The CRM movement picked up speed in 1991 when the FAA issued an advisory circular to initiate the Advanced Qualification Program (AQP), a voluntary program to allow airlines to develop innovative human factors training to meet the needs of each specific airline organization. From this, airlines began to institutionalize CRM concepts by adding specific procedures in their checklists, including Line Oriented Flight Training (LOFT) for all flight crews, which considered the cultural perspectives of different regions of the world.

Modern CRM theory accepts that human errors are indeed inevitable; therefore, CRM practices should serve as a set of counter-measures with three distinct lines of defense:

- First—Error avoidance if possible
- Second—Trapping incipient errors before they are committed
- Finally—Mitigating the consequences of those human errors which occur and are not trapped

    (Helmreich et al., 1999)

LINE OPERATIONS SAFETY AUDIT (LOSA)

Closely linked to CRM, LOSA was endorsed by ICAO in 1999 as a primary tool to develop countermeasures to human error. The LOSA process uses an airline observer (usually a pilot) to collect data about flight crew behavior by riding on the "jump seat" during routine flights and observing crew strategy for managing threats, errors and undesirable aircraft states. These observations are conducted under a guarantee of confidentiality with no organizational action taken against the crew for its performance during the flight. Today, LOSA is endorsed by the FAA (Advisory Circular 120-90) and ICAO (LOSA Manual 9803) as an industry best practice to defend against human error in the cockpit.

THREAT AND ERROR MANAGEMENT (TEM)

A natural follow-on to LOSA is the concept of Threat and Error Management which focuses on the normal working environment and the humans working in that environment (Merritt & Klinect, 2006). First, we will define the terms threat, error, and undesired aircraft state; then discuss some control strategies to manage those situations which unfortunately become "undesirable aircraft states."

- **Threats**—Pilots must manage threats that come at them in the normal operating environment. Threats are defined as events or errors that:
  ○ Occur outside the influence of the flight crew (i.e., not caused by the crew);
  ○ Increase the operational complexity of a flight; and
  ○ Require crew attention and management if safety margins are to be maintained.
- **Errors**—Human error comes from the crew and may be innocent or intentional. Errors are defined as flight crew actions or inactions that:
  ○ Lead to a deviation from crew or organizational intentions or expectations;
  ○ Reduce safety margins; and
  ○ Increase the probability of adverse operational events on the ground or during flight.

- **Undesired Aircraft State**—an error not well managed which may lead to an event which compromises safety. An undesired aircraft state (UAS) is defined as a position, speed, altitude, or configuration of an aircraft that:
  - ○ Results from flight crew error, actions, or inaction; and
  - ○ Clearly reduces safety margins.

(Source: Merritt & Klinect, University of Texas Human Factors Resource Project, 2006).

CONTROL STRATEGIES TO MANAGE THREATS AND ERRORS

Effective control strategies may be engineering-based tools associated with the aircraft such as cockpit automation, instrument displays, or warning devices. For example, the Ground Proximity Warning System (GPWS) is a so called "hard" defense mechanism to prevent controlled flight into terrain with visual and audio warnings to "pull up." A yoke "stick shaker" or stall warning horn would be another example of an engineering control strategy to warn pilots of low air speed and possible stall situation.

There are other control strategies which are more administrative in nature. The so-called "soft" defenses such as checklists, rules and regulations, standard operating procedures, etc., are used to direct crew members to take appropriate action under a given set of circumstances. Using CRM, LOSA, and TEM training concepts, modern airline crews are now taught the newest techniques and communication skills to either avoid the human error entirely, trap it from spreading into an undesired aircraft state, or finally, mitigate the consequences of the error through anticipation, error recognition, and ultimate recovery from the undesirable human error problem.

AIRBUS AND BOEING DESIGN STRATEGIES

In the past three decades, with the rapid growth in microprocessor technology, there has been a temptation on the part of some designers to build very complex systems based on the rationale that the systems could operate automatically. There are two fallacies in this argument. First, almost no major system on an aircraft truly operates fully automatically. Systems must be initialized or set up by the human, decisions about operating modes must be made, and then the systems must be monitored by humans for obvious reasons. Second, in the event of the failure of automation, it falls to the human to operate the system. This responsibility cannot be avoided or designed away. If the complexity of the system is unbridled, then the crew may not be able to perform their duties effectively or take over in the event of equipment failure.

In response, many design engineers with human factors sophistication have recognized that simplification offers an alternative to automation. If the system can be simplified, there may be no need for complex automation, and the same goal can be achieved without placing the human in a potentially hazardous position. An example is the fuel system on a multiengine aircraft. Those favoring automation would find no problem with creating a complex tank-to-tank and tank-to-engine relationship, as long as its management could be automated. If, for example, a fuel imbalance were created, automatic devices would detect the imbalance, determine a remedy, open the required transfer valves, and turn on the appropriate pumps to restore the proper balance. No human intervention would be required.

This example represents a philosophical difference between two major aircraft manufacturers—Airbus and Boeing. The Airbus general approach has been to remove the pilot from the loop and turn certain functions over to sophisticated automation. Compensation is automatic—the systems do not ask the crew's approval. Boeing's approach is to never bypass the crew: Sophisticated devices inform the crew of a need and, in some cases, a step-by-step procedure; but in the end, it is the crew who must authorize and conduct the procedure. Boeing is a strong advocate of simplification before automation. Their designers would look to a less complex relationship. An example would be fewer tanks to feed the engines, creating fewer tank-to-tank and tank-to-engine requirements, requiring less management by the crew and fewer opportunities for human error.

One of the potential difficulties with highly automated systems is that onboard computers may, unknown to the crew, automatically compensate for abnormal events. Efficient automatic compensation for abnormal events and conditions sounds attractive, but there is always a limit to the system's capacity to compensate. When automation is compensating for some worsening condition without the crew's knowledge, this can lead to a situation where it may be too late for the crew to override the system and prevent a catastrophe.

COCKPIT STANDARDIZATION

Between-fleets standardization of hardware is considered desirable to reduce training and maintenance costs as well as to prevent human error that may occur as a result of the pilots moving from one aircraft to another. During periods of rapid expansion of aircraft inventories and pilot personnel, there is frequent movement between aircraft as pilots bid for more lucrative assignments, more modern aircraft, or desirable bases. Some airline labor contracts limit the rapidity with which pilots may bid a new seat; others do not.

Most cockpit hardware is peculiar to the type of aircraft. However, certain cockpit hardware could be common to most or all models operated by a carrier; examples are radios, flight directors, certain displays, area navigation equipment, and weather radar. Other examples would be devices added after the original manufacture (e.g., TCAS, ACARS). When the carrier has the opportunity to purchase these add-on units, a common model will most likely be chosen for all the reasons stated above.

Where differences already exist between fleets, the airline may intervene by standardizing throughout the airline. For example, some airlines have invested in a common airlinewide model of the flight director.

Between-fleet standardization, if it involves retrofit rather than new equipment purchase, will be extremely costly, and its safety benefits may be modest compared to within-fleet standardization. Nonetheless, when pilots move rapidly through the seats of various aircraft or complete training for one aircraft and then return to another while awaiting assignment to the new aircraft, between-fleet standardization of cockpit hardware deserves inclusion in the list of intervention strategies.

Within-fleets standardization is far more critical. Long before the Airline Deregulation Act of 1978, carriers purchased aircraft from one another, thus generating mixed configurations within fleets. With the coming of deregulation, the pace of mergers and acquisitions, as well as used equipment purchases and leases, accelerated rapidly, producing fleets of traditional aircraft, such as B-727s, 737s, and DC-9s, that varied greatly with respect to cockpit configuration. These differences included different displays (e.g., various models of flight directors), warning and alerting systems (e.g., a host of altitude warning systems with various trigger points), every imaginable engine configuration, controls in different locations, various directions of movement of switches, and various operating limitations. One carrier, which had been through a number of mergers and acquisitions of other DC-9 operators, had eight different models or locations of altitude alerters. It later invested a very considerable sum to standardize the cockpits of its DC-9 fleet. Within-fleet standardization is considered a high-priority item by the line pilots and their safety committee. Southwest Airlines, which currently has a modern Boeing 737 (only) fleet, is a good example of a successful cockpit standardization program.

COCKPIT AUTOMATION AND PRECISION NAVIGATION

In today's modern commercial aircraft, computer technology, and the "glass cockpit" have become the norm. Navigation is accomplished electronically on moving maps using the latest techniques to enhance the situational awareness of the pilot. Nevertheless, as we have learned, the human error element is always present and must always be actively monitored by the crew. A good example of a human error situation occurred onboard a B-767 preparing to depart Atlanta for Miami. The clearance included as a waypoint the TEPEE (note spelling) intersection near Tampa. The captain entered TEEPE (note spelling) into the route page of the control display unit (CDU). Because there is a TEEPE intersection (near Waco, Texas), the CDU dutifully accepted the erroneous spelling and established it as a waypoint on the route from Atlanta to Miami. The sudden shift in course to the west-southwest toward TEEPE from the southward course toward TEPEE was immediately evident to the crew. A non-EFIS aircraft with the same CDU-FMS (such as some models of the B-737-300) would not have provided this form of error detection capability. The crew would have had to detect the error by some other check.

Once the aircraft system properly displays an abnormal condition, there must be an effective means of removing it and allowing the system to recover. The system must not permit irreversible errors. With traditional aircraft, this was usually not a problem. Working with less sophisticated systems, the pilots were closely coupled to the machine; an error, once detected, could usually be reversed quite easily. The advent of highly sophisticated automation raises the question of escape from error and system recovery. Generally, the problem is not that the error is irreversible but that a complicated recovery process can be difficult, time-consuming, and possibly error-inducing itself.

The modern Flight Management System (FMS) offers some novel features for protection against human error. It can store and process a vast amount of information typically contained in manuals, checklists, performance charts, flight plans, weather reports, and documents and paperwork of all sorts. This information can be displayed to the crew in text, numeric, and graphic forms on selected pages of the control display unit, the glass instrument panels, and elsewhere. Some of the information is automatically displayed, requiring no request from the crew (e.g., the wind vector on the navigation display); other information is available in the FMS on demand through pilot selection of the correct CDU page. The display of certain valuable information, such as suitable emergency airfields, is switch-selectable. Finally, if the FMS detects an abnormal computer condition, a brief message can be displayed in the "scratch pad" line of the CDU, and the pilot is alerted on two other displays that an FMS message is waiting. An example would be a request for a waypoint "not in the database."

The B-767/757 and the glass cockpit aircraft that followed possessed rudimentary forms of computer-based error elimination and protection. The Airbus A-320, introduced in 1988, took error protection a step further. The fly-by-wire feature offered the opportunity to fly maneuvers, such as maximum safe angle of attack (AOA) for wind shear escape, with no danger of entering a stall. The computer would simply stop the aircraft's increase in pitch short of its computed safe AOA. If the pilot continued to pull back on the stick, no more nose-up pitch would be commanded. An intelligent computer interposes an electronic line of defense between the pilot's control and the aircraft's control surfaces. Incidentally, such a system that has the capability of controlling and correcting an error is referred to as an *error-resistant* or *error-tolerant system*.

Other EFIS aircraft, such as the Boeing 757/767, offer escape guidance on the *altitude indicator* in the form of a target line for optimal nose-up pitch. In contrast with the approach taken in the A-320 design, the pilot remains in the loop in the Boeing aircraft. The pilot controls the pitch angle; the computer merely computes and displays the commanded nose-up pitch.

These two approaches emphasize not only disparate views of cockpit design but basic philosophical design differences: The A-320 essentially allows the pilot to pull the control stick all the way back and let the computer find the maximum angle of attack that will avoid a stall. Other EFIS aircraft depend on the pilot to follow the wind shear escape guidance cues. It is impossible to say which approach is more effective. Only time, experience, and years of accident investigation statistics will settle that question.

COMMUNICATION ISSUES

Verbal communication remains the weakest link in the modern aviation system; more than 70 percent of the reports to the Aviation Safety Reporting System involve some type of oral communication problem related to the operation of an aircraft. The ground collision between two B-747 aircraft in Tenerife in 1977, resulting in the greatest loss of life in an aviation accident, occurred because of a communication error. Technologies, such as airport traffic lights or data link, have been available for years to circumvent some of the problems inherent in ATC stemming from verbal information transfer.

One potential problem with ATC by data link is that the loss of the "party line" effect (hearing the instructions to other pilots) would remove an important source of information for pilots about the ATC environment. However, the party line is also a source of errors by pilots who act on instructions directed to other aircraft or who misunderstand instructions that differ from what they anticipated by listening to the party line. Switching ATC communication from hearing to visual also can increase pilot workload under some conditions. Further human factors study is necessary to define the optimum uses of visual and voice communications.

The term *communication* usually includes all facets of information transfer. It is an essential part of teamwork, and language clarity is central to the communication process. Adequate communication requires that the recipient receive, understand, and can act on the information gained. For example, radio communication is one of the few areas of aviation in which complete redundancy is not incorporated. Consequently, particular care is required to ensure that the recipient receives and fully understands a radio communication.

There is more to communication than the use of clear, simple, and concise language. For instance, intelligent compliance with directions and instructions requires knowledge of why these are necessary in the first place.

Trust and confidence are essential ingredients of good communication. For instance, experience has shown that the discovery of hazards through incident or hazard reporting is only effective if the person communicating the information is confident that no retributory action will follow her or his reporting of a mistake.

Communications within the cockpit can be affected by what some psychologists call the *transcockpit authority gradient (TAG)*, which is an expression of the relative strength and forcefulness of the personalities involved. For safe operations, the gradient between the captain and copilot should be neither too steep nor too shallow, thus encouraging free communication between the pilots, leading to improved monitoring of the aircraft operation. For example, when the gradient is too steep, the copilot may be afraid to speak up, thereby failing in his or her role of monitoring the captain's actions. When it is too shallow, the captain may not adequately exercise his or her authority.

Miscommunication between aircrews and ATC controllers has been long recognized as a leading source of human error. It has also been an area rich in potential for interventions. Examples are the restricted or contrived lexicon (e.g., the phrase *say again* hails from military communications, where it was mandated to avoid confusing the words *repeat* and *retreat*); a phonetic alphabet ("alpha," "bravo," etc.); and stylized pronunciations (e.g., "niner" due to the confusion of the spoken words *nine* and *five*).

As a result of the tragic ground collision between two B-747s at Tenerife in 1977, blamed largely on miscommunications between the tower and the two aircraft, the FAA encouraged controllers to restrict the word *cleared* to two circumstances—*cleared to take off* and *cleared to land*—although other uses of the word are not prohibited. In the past, a pilot might have been cleared to start engines, cleared to push back, or cleared to cross a runway. Now the controller typically says, "Cross runway 27," and "Pushback approved," reserving the word *cleared* for its most flight-critical use.

The need for linguistic intervention never ends, as trouble can appear in unlikely places. For example, pilots reading back altimeter settings often abbreviate by omitting the first digit from the number of inches of barometric pressure. For example, 29.97 (inches of mercury) is read back "niner niner seven." Since barometric settings are given in millibars in many parts of the world, varying above and below the standard value of 1013, the readback "niner niner seven" might be interpreted reasonably but inaccurately as 997 millibars. The obvious corrective-action strategy would be to require full readback of all four digits when working in inches.

A long-range intervention and contribution to safety would be to accept the more common (in aviation) English system of measurement, eliminating meters, kilometers, and millibars once and for all. Whether English or metric forms should both be used in aviation, of course, is arguable and raises sensitive cultural issues. At this time, the English system clearly prevails, since English is the ICAO mandated international language of aviation effective January 1, 2008.

PROFESSIONALISM IN AVIATION

After a series of incidents in the last decade, the NTSB conducted a 3-day public forum in May 2010 to focus on pilot and air traffic controller professionalism as a safety issue. Among the airline incidents discussed were the following:

- Comair Flight 5191, Lexington, Kentucky, where the aircraft took off on the wrong (short) runway. (August 2006)
- Colgan Airlines Flight 3407, Buffalo, New York, where the aircraft stalled and crashed short of the airport. (February 2009)
- Fatal midair collision over the Hudson River between an air tour helicopter and a fixed wing general aviation aircraft. The air traffic controller was engaged in personal business at the time. (August 2009)

- Northwest Airlines Flight 188 in which the aircraft overflew its destination (Minneapolis) while the pilots were distracted by a conversation while using their personal computers in the cockpit. (October 2009)

STERILE COCKPIT RULE (FAR 121.542)

In several incidents discussed at the NTSB forum, the pilots were engaged in a violation of the *sterile cockpit rule* which provides that pilots shall not engage in casual, non-pertinent conversation during critical phases of flight (taxiing, takeoff, and landing, and all other flight operations below 10,000 feet except cruise flight). These aircrew conversations were captured on the cockpit voice recorder, and a lack of professionalism or distractions were noted as a causal factor on several NTSB accident investigations in recent years.

Other key points discussed during the NTSB professionalism forum include:

- Structured development and training of professional pilots
- Developing excellence and professionalism in air traffic controllers through improved screening, selection, and training
- Ensuring effective pilot—controller communications
- The captain's role in ensuring professionalism

Complete proceedings on all the NTSB public forums and symposia for the past decade can be found on its Web site at www.ntsb.gov/.

CONCLUSION

People are pivotal to aviation safety. Although humans are largely responsible for commercial aviation's excellent safety record, human errors nonetheless cause or contribute to most accidents. Moreover, the rate of pilot-error accidents shows no sign of abating, while weather-related crashes are declining and aircraft component failures are rarely the sole factor in serious mishaps. Furthermore, accident and incident data analyses indicate that if only a portion of human-error problems can be solved, substantial reductions in accident risk can be attained.

Another concern among pilots and human-performance experts is that the increased level of cockpit automation may create a generation of pilots whose basic flying skills ("stick and rudder") deteriorate from lack of practice. If manual skills ever become needed because of automation failure/degradation or unusual plane attitudes and conditions that automation cannot handle, the pilot may not be up to the challenge. Manual piloting skills may have degraded because of the (over)use of automatic flight systems in lieu of hand flying and/or because of the lack of training and practice on certain maneuvers and skills.

Human error on the flight deck can never be totally eliminated. However, through judicious design; constant monitoring of accidents, incidents, and internal reports; and the aggressive use of reporting systems such as NASA's ASRS, the means of corrective action can be found. Air transportation enjoys an excellent safety record today largely because no part of the system is ever allowed to rest.

Lack of situational awareness has been identified as a contributing factor in many accidents and incidents. Modern airline training tools can greatly improve pilot professionalism through enhanced aeronautical decision making and a clear understanding of human factors principles using models such as HFACS to identify and control inevitable human error. Advanced Crew Resource Management training combined with LOFT, LOSA and Threat Error Management can significantly reduce commercial aviation accidents and incidents. Furthermore, the full implementation of ICAO mandated Safety Management System (SMS) throughout the world will enable aviation organizations to greatly improve the safety culture in coming years. More details on the SMS process are included in Chapter 13.

KEY TERMS

Human factors

Human performance

Emotion

Awareness

Motivation

Memory

Complacency

Attention

Attitude

Perceptions

Judgment

The Decide Model

Aeronautical Decision Making (ADM)

Human Factors Analysis and Classification System (HFACS)

Human Factors Intervention Matrix (HFIX)

Psychosocial factors

Transcockpit authority gradient (TAG)

Peer pressure

Ego and pride

Hardware factors

Task factors

Environmental factors

Fly-by-wire technology

Advanced Qualification Program (AQP)

Cockpit Resource Management (CRM)

Line-Oriented Flight Training (LOFT)

Line Operations Safety Audit (LOSA)

Threat and Error Management (TEM)

Automation

Electronic flight instrument system (EFIS)

Inertial reference system (IRS)

Flight Management System (FMS)

Control display unit (CDU)

Ground-proximity warning system (GPWS)

Controlled flight into terrain (CFIT)

Sterile cockpit rule

REVIEW QUESTIONS

1. Discuss the significance of human error in major aircraft accidents.
2. What are the seven major factors that affect human performance? Give examples of each.
3. Define human error, and explain the HFACS method used to classify human error.
4. Explain what is meant by management of human error.
5. Discuss the concepts of CRM, LOSA, and TEM in modern commercial aviation.
6. Discuss the role of the following in human error management:
   - Cockpit standardization
   - Cockpit automation
   - Warning and alerting systems
   - Flight Management System
   - Air-to-ground communication
7. Explain the differences in the Airbus and Boeing aircraft design strategies.

8. Discuss the importance of resolving communications issues in air-ground voice transmissions.

9. What was the purpose of the NTSB conference "Professionalism in Aviation" in May 2010?

REFERENCES

Alkov, Robert A. 1997. *Aviation Safety—The Human Factor.* Casper, Wyo.: Endeavor Books.

Duke, Thomas A. 1991. "Just What Are Flight Crew Errors?" *Flight Safety Digest.* Arlington, Va.: Flight Safety Foundation, July 1–15.

Garland, Daniel J., John A. Wise, and V. David Hopkin. 1999. *Handbook of Aviation Human Factors.* Mahwah, N.J.: Laurence Erlbaum Associates, Publishers.

Green, G. G., H. Muir, and James M. Gradwell. 1991. *Human Factors for Pilots.* Aldershot, U.K.: Gower Publishing Co., Ltd.

Helmreich, R.L., Merritt, A.C., and Wilhelm, J.A. 1999. The Evolution of Crew Resource Management in Commercial Aviation. International Journal of Aviation Psychology, 9(1), 19–32.

Hunt, Graham J. F., ed. 1997. *Designing Instruction for Human Factors Training in Aviation.* Brookfield, Vt.: Ashgate Publishing Co.

ICAO. 2009, *Safety Management Manual* (SMM) Doc 9859 AN/474. www.icao.int

Jensen, A. D., and J. C. Chilberg. 1991. *Small Group Communication: Theory and Application.* Belmont, Calif.: Wadsworth Publishing Co.

Lintern, G., S. N. Roscoe, and J. Sivier. 1990. "Display Principles, Control Dynamics, and Environmental Factors in Pilot Performance and Transfer of Training." *Human Factors.* 32: 299–317.

Maurino, Daniel E. 1990. "Education Is Key to ICAO's Human Factor Program." *ICAO Journal.* Montreal. October, pp. 16–17.

Merritt, A., and Klinect, J. 2006. *Defensive Flying for Pilots: An Introduction to Threat and Error Management.* The University of Texas Human Factors Research Project.

National Transportation Safety Board. 2010. "Professionalism in Aviation" www.NTSB.gov

O'Hare, David, and Stanley Roscoe. 1990. *Flight Deck Performance: The Human Factor.* Ames, Iowa: State University Press.

Orlady, Harry W., and Linda M. Orlady. 1999. *Human Factors in Multi-Crew Flight Operations.* Brookfield, Vt.: Ashgate Publishing Co.

Reason, J. 1990. *Human Error.* New York: Cambridge University Press.

Salas, E., and Maurino, D. 2010. *Human Factors in Aviation, Second Edition.* Burlington, MA: Academic Press.

Shappell, S. and Wiegmann, D. 2009. A Methodology for Accessing Safety Programs Targeting Human Error in Aviation. The International Journal of Aviation Psychology, 19(3), 252–269.

Shappell, S., Detwiler, C., Holcomb, K., Hackworth, C., Boquet, A., and Weigmann, D. 2007. Human Error and Commercial Aviation Accidents: An Analysis Using the Human Factors Analysis and Classification System. Human Factors, Vol 49, No 2 April 2007, pp. 227–242.

Villaire, N. E. 2009. *Aviation Safety: More Than Common Sense.* 2nd edition Casper, Wyo.: Endeavor Books.

Weigmann, D. A., and Shappell, S. A. 2003. *A Human Error Approach to Aviation Accident Analysis.* Burlington, VT.; Ashgate Publishing Company.

AIR TRAFFIC SAFETY SYSTEMS

LEARNING OBJECTIVES

After completing this chapter, you should be able to:

- Describe the mission of the FAA and discuss major milestones of Air Traffic Control history.
- Explain the basic components of the ATC system and ATC services available.
- Discuss the advantages of GPS and satellite-based navigation.
- Explain the purpose of the WAAS and GBAS GPS augmentation systems.
- Give examples of recent operational planning improvement programs.
- Describe the features of the NextGen Air Transportation System.
- Identify critical NextGen Transformational Programs.
- Discuss NextGen implementation challenges.

INTRODUCTION

The mission of the Air Traffic Control (ATC) system is to promote the safe, orderly, and expeditious flow of aircraft through the nation's airspace. Today's National Airspace System (NAS) has evolved over several decades into a complete infrastructure involving thousands of employees and billions of taxpayer dollars to administer. A short history of the U.S. ATC system is helpful to understand its current state.

MAJOR MILESTONES OF ATC HISTORY

- June 1956—Midair collision over the Grand Canyon in Arizona, killing all 128 occupants of the two airplanes, a TWA Super Constellation and a United Air Lines DC-7. The accident happened in clear weather under visual flight rules (VFR) in uncontrolled airspace. Such sightseeing detours over the Grand Canyon were a common practice at the time.

- December 1960—The FAA was only 2-years old when another disastrous midair collision occurred over New York City. A TWA Super Constellation and a United DC-8, both in a holding pattern under instrument flight rules (IFR) collided, killing all aboard. The United aircraft had navigation problems and excessive speed, which could not be detected by New York ATC towers lacking proper surveillance radar equipment.

- August 1981—Over 12,000 U.S. air traffic controllers (ATC) went on strike over labor conditions and were fired by the FAA. It took almost 10 years before overall ATC staffing levels returned to normal.

- January 1982—FAA released the first annual NAS Plan, a comprehensive 20-year blueprint to modernize the ATC and air navigation systems in the United States.

- April 2000—Creation of the FAA Air Traffic Organization (ATO), a performance-based department focusing on efficient operation of the ATC system. This was a major reorganization headed up by an FAA ATO Chief Operating Officer.

- September 11, 2001—Two terrorist airline attacks at the World Trade Center in New York City, plus hijackings which damaged the Pentagon in Washington, D.C., and ended in the crash of a fourth aircraft in a Pennsylvania field, caused the FAA to ground all aviation traffic in the country. The Transportation Security Administration (TSA) and Department of Homeland Security were created soon thereafter. Airline traffic decreased for 2 years but eventually recovered.

- December 2003—Next Generation Air Transportation System (NextGen) concept was authorized; a new, ambitious, multiagency effort to develop a modern air transportation system for the 21st century. This system faces significant challenges changing ATC from a radar-based to a satellite-based navigation and communication system.

The following section will discuss the fundamental components of the U.S. ATC system. A detailed discussion of this subject would require many chapters and therefore is outside the scope of this textbook. An excellent resource in this area for more information is *Fundamentals of Air Traffic Control*, 5th Edition (2011) by Michael S. Nolan referenced at the end of this chapter.

BASIC COMPONENTS OF THE ATC SYSTEM

AIRSPACE CLASSIFICATION

In the United States, the FAA has designated four categories of airspace as follows:

- Positive controlled airspace—These are areas in which FAA is responsible for separation of all aircraft, whether under Visual Flight Rules (see and avoid) or Instrument Flight Rules. Examples of positive control areas are high-density airports and very high altitude flights above 18,000 feet mean sea level.
- Controlled airspace—Areas in which ATC separates IFR traffic, but VFR pilots provide their own separation, weather permitting.
- Uncontrolled airspace—The pilots themselves provide all aircraft separation; ATC services are not provided by FAA.
- Special use airspace—Areas where air traffic may be prohibited or restricted with special rules such as the airspace around the White House in Washington, D.C.

Airspace classifications and communication requirements are taught in basic flight training, thus, are beyond the scope of this textbook. A chart illustrating U.S. airspace classifications, communication requirements, and weather minimums is provided in Fig. 8-1.

AIR TRAFFIC CONTROL SERVICES

The basic communication and Navigational Aids (NAVAIDS) to aviation have changed little since the 1970s. Air traffic controllers operate primarily from three types of facilities as indicated in Fig. 8-2, which is a general picture of how U.S. airspace is managed. These basic ATC facilities are as follows:

- Airport traffic control towers (ATCTs)
- Terminal radar approach control (TRACON) facilities
- Air route control centers (ARTCCs)

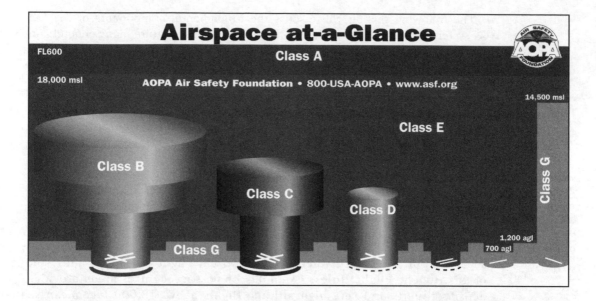

FIGURE 8-1   Airspace at-a-glance. (*Source: http://www.aopa.org/*)

Some of the basic NAVAID and radar surveillance equipment used in this system are the following:

- VHF Omni-directional Range (VOR) and Distance Measuring Equipment (DME) stations which transmit signals along "airways," highways in the sky, to route air traffic.

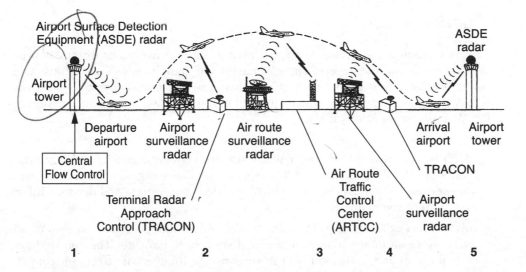

KEY:

1. The airport tower controls the aircraft on the ground before takeoff and then to about 5 miles from the tower, when the tower transfers aircraft control to a Terminal Radar Approach Control facility (TRACON). Controllers in the airport tower either watch the aircraft without technical aids or use radar–Airport Surface Detection Equipment for aircraft on the surface and airport surveillance radar for those in the air. Central Flow Control (in Washington, DC) can order the tower to hold flights on the ground if demand exceeds capacity at the arrival airport.

2. The TRACON, which may be located in the same building as the airport tower, controls aircraft from about 15 miles to about 30 miles from the airport, using aircraft position information from the aircraft surveillance radar. The TRACON then transfers control to an Air Route Traffic Control Center (ARTCC).

3. ARTCCs control aircraft that are en route between departure and arrival airports. Each ARTCC controls a specific region of airspace and control is handed off from one ARTCC to another when a boundary is crossed. Aircraft positions are detected by the air route to an Air Route Traffic Control Center (ARTCC).

4. The TRACON controls the arriving aircraft until it is within about 5 miles of the arrival airport tower, when control is transferred to the tower.

5. The airport tower controls the aircraft on the final portion of its approach to the airport and while it is on the ground.

FIGURE 8-2   General picture of how airspace is constructed and controlled.

- Instrument Landing System (ILS), which is a precision approach and landing aid that normally consists of a localizer, a glide slope, marker beacon, and an approach light system.

- Airport Surveillance Radar (ASR), used in conjunction with a transponder radar beacon device, is an approach control radar system used to separate aircraft within the immediate vicinity of an airport; it normally has a maximum range of 60 nautical miles.

AREA NAVIGATION (RNAV) EQUIPMENT

Area navigation is a collective term that permits aircraft navigation on any desired course within the coverage of specific navigation signals. RNAV may consist of station-to-station navigation, or it may be between random waypoints offset from published routes depending upon the type of equipment used. The primary types of RNAV equipment are as follows:

- *VORTAC* (co-located VOR and TACAN station) systems account for the greatest number of RNAV units in use. A TACAN (Tactical air navigation) system is a military version of VOR equipment, which provides both bearing and distance information to military aircraft.

- *Inertial navigation systems (INS)* are totally self contained aboard large aircraft and require no information from external sources to navigate. They provide aircraft position and navigation data in response to the inertial effects on components within the INS system.

- *LORAN-C* is a long-range radio navigation system that uses ground waves transmitted at low frequency to provide aircraft position at ranges up to 1,200 nautical miles.

- *Global Positioning System (GPS)* is a space-based navigation system providing very accurate position and velocity information and precise time on a continuous global basis. The GPS system is unaffected by weather and provides worldwide navigation services.

GPS ENHANCEMENTS

The current U.S. aviation navigation system is comprised of more the 4,300 ground-based systems whose signals are used by aircraft avionics for en route navigation and landing guidance. Despite the large number of ground systems, navigation signals do not cover all airports and airspace. Over the next several years, the navigation system is expected to increase its use of GPS satellites, augmented by ground monitoring stations, to provide navigation signal coverage throughout the NAS. Reliance on ground-based navigation aids is expected to decline as satellite navigation provides equivalent or better levels of service.

A transition to satellite navigation significantly expands navigation and landing capabilities, improving safety and efficient use of airspace. In addition, it will reduce the FAA's need to replace many aging ground systems, decrease the amount of avionics required to be carried in aircraft, and simplify navigation and landing procedures. The transition to satellite-based navigation consists of the following:

- Use of the Global Positioning System for en route-terminal navigation and nonprecision approaches (provided that another navigation system is onboard the aircraft). GPS is being described as the greatest aviation achievement since radios were introduced 50-years ago and is scheduled to revolutionize today's air traffic system. GPS is a radio navigation system composed of 24 orbiting satellites that provides extremely accurate three-dimensional position, velocity, track, and time, at low cost so that even general-aviation planes can use it. By picking up signals from four or more satellites, GPS receivers on the ground or in the aircraft can determine the location within approximately 15 meters of the exact position. By adding a datalink to the plane, its position, velocity, and track can be conveyed via satellite to communicate data to ground control units worldwide. Over-ocean surveillance can now be commonplace, providing additional flight paths and reduced separation for improved utility of the sky. Since all the traffic can now be determined accurately, each airplane's position can be uplinked or directly linked to planes, to display those within close proximity and with potential conflict. We will have not only communications, navigation, and surveillance (CNS) technologies, but also the potential for collision avoidance at a price that all aviators can afford. GPS alone does not meet the accuracy, availability, and integrity requirements critical to safety of flight. Signal augmentation is required for most landing operations.

- Deployment of the *wide-area augmentation system (WAAS)* to provide en route/terminal navigation and category (CAT) I precision approaches. Using WAAS, aircraft can now access over 2,300 U.S. runways in poor weather conditions with precision approach minimums as low as Category I (200 feet). WAAS enhances GPS signals to provide more precise location information to an accuracy of approximately 3 meters. WAAS is designed to use reference stations covering wide areas throughout the United States to cross-check GPS signals and then relay integrity and correction information to aircraft via geostationary communication satellites. WAAS enhances availability by using these satellite to provide a GPS-like navigation signal (Fig. 8-3)

- Deployment of the *ground-based augmentation system (GBAS)* to augment GPS for CAT I, II, and III precision approaches. GBAS provides precise correction data to airborne and surface receivers that results in navigation accuracy of less than 1 meter to distances of a 20 to 30 mile radius around the airport (Fig. 8-4).

Precision approaches are categorized in terms of decision heights (ceilings) and minimum visibility. Listed below are FAA definitions and requirements for instrument approaches.

- *Nonprecision approach (NPA).* This is an instrument approach procedure based on a lateral path and no vertical guide path.

- *Precision approach (PA).* This is an instrument approach procedure based on lateral path and vertical guidance.

- *Category I.* Category I operation is a precision instrument approach and landing with a decision height that is not lower than 200 feet (60 m) above the threshold and with either a visibility of not less than ½ statute mile (800 m) or a runway visual range (RVR) of not less than 1800 feet (550 m).

Wide Area Augmentation System

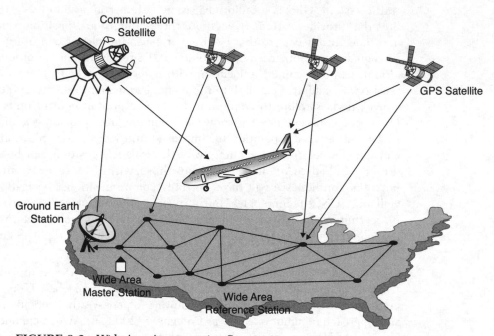

FIGURE 8-3   Wide Area Augmentation System. (*Source: NASA www.aeronautics.nasa.gov/*)

- *Category II*. Category II operation is a precision instrument approach and landing with a DH lower than 200 feet (60 m), but not lower than 100 feet (30 m), and with a RVR of not less than 1200 feet (350 m).

- *Category III*. Category III operation is a precision instrument approach and landing with a DH lower than 100 feet (30 m) or no DH, and with an RVR less than 1200 feet (350 m).

ADVANTAGES OF SATELLITE-BASED NAVIGATION

Satellite-based navigation enables significant operational and safety benefits. It meets the needs of growing operations because pilots will be able to navigate virtually anywhere in the NAS, including at airports that currently lack ground navigation and landing signal coverage. Satellite-based navigation will support direct routes.

With satellite navigation, the number of published precision approaches has greatly increased. FAA is adding 500 new WAAS approaches each year and the current number of nearly 2,400 approaches exceeds the number of traditional ILS approaches in the United States. In addition, combining GPS with cockpit

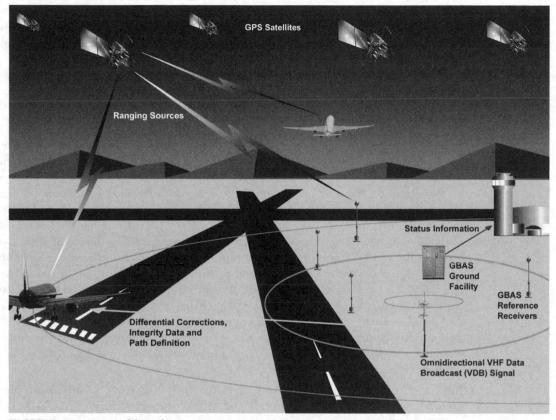

FIGURE 8-4   Ground-based augmentation system. (*Source: FAA http://www.faa.gov/*)

electronic terrain maps and ground-proximity warning systems can help pilots avoid controlled flight into terrain (CFIT) problems.

Satellite-based navigation also decreases the number of ground-based navigation systems, thereby reducing infrastructure costs. For a precision approach today, each runway end needs a dedicated instrument landing system. WAAS can provide the precision approach guidance for most of the runways in the NAS. A single GBAS will provide CAT II/III precision approach guidance at all runways at an airport in the future, thus reducing maintenance and logistics costs.

WEATHER AUTOMATION SYSTEMS

Weather conditions interfere with flight operations and contribute to aviation accidents more than any other factor. Given these major impacts, new FAA weather programs collect, process, transmit, and display weather information to users and service providers, both during flight planning and in flight.

The key to reducing weather-related accidents is to improve pilot decision making through increased exchange of timely information. Service providers and users receive depictions of hazardous weather, simultaneously enhancing common situational awareness.

As the aviation weather plan evolves from present-day separate, stand-alone systems to weather systems that are fully integrated into the NAS, the focus is on the following two key capabilities:

- Improved processing and display, with the key systems being the *integrated terminal weather system (ITWS)* and *weather and radar processor (WARP)*. ITWS is an automated weather system that provides near-term (0 to 30 minutes) prediction of significant terminal area weather for major terminal locations. ITWS integrates data from radar, sensors, National Weather Service models, and automated aircraft reports. It generates products, including wind shear and microburst predictions, storm cell hazards, lightning information, and terminal area winds. WARP is an integrated system that receives and processes real-time weather data from multiple sources and provides weather information for use by the ARTCCs and air traffic control system command center (ATCSCC) to support the en route environment. It also receives gridded forecast data from the National Weather Service and provides this information to other NAS automation systems. WARP has direct and indirect connections to the *next-generation weather radar (NEXRAD)* and prepares national and regional weather images for the controllers' displays.

- Improved sensors and data sources, featuring the NEXRAD, *terminal Doppler weather radar (TDWR)*, and ground- and aircraft-based sensors. NEXRAD is a national network of Doppler weather radar to detect, process, distribute, and display hazardous weather, providing more accurate weather data for aviation safety and fuel efficiency. This radar has a 250-mile range, and the network covers the majority of the domestic en route airspace. This weather detection system provides information about wind speed and direction in the areas of precipitation, convective activity, tornadoes, hail, and turbulence. TDWR detects localized microbursts, gust fronts, wind shifts, and precipitation in the immediate terminal area at key locations. The radar provides alerts of hazardous weather conditions in the terminal area and advanced notice of changing wind conditions to permit timely change of active runways.

The FAA is conducting research to improve the capability to predict weather hazards and to communicate these to NAS users. New wind, temperature, icing, and weather hazard modeling will be used to help diminish weather-related delays and improve safety.

Aviation weather research is being focused on the following areas: in-flight icing, aviation gridded forecast system, ground deicing operations, convective weather, short-term ceiling and visibility predictions, turbulence, and wake vortices.

OPERATIONAL PLANNING IMPROVEMENTS

To improve flight planning, the FAA has introduced new and improved information services in the areas of *traffic flow management (TFM)* and flight services that enable collaboration-service providers and users sharing the same data and negotiating to find the best solutions to meet operational needs.

Traffic flow management capabilities are centralized at the ATCSCC. Some functionality is distributed to traffic management units at ARTCCs, TRACON facilities, and the highest-activity ATCTs.

NAS-wide information service enables data exchange between users and the FAA to facilitate a collaborative response to changing NAS situations, rather than a local solution based on incomplete data. This gives users and service providers a common view of the NAS for improved decision making during all phases of flight, including flight planning.

In addition, the flight plan will be replaced by the flight object, which will be designed for dynamic updates and made available to authorized NAS service providers and users to manage flight operations collaboratively. The flight object will contain additional data such as the user's route and altitude preferences, the aircraft's weight, gate assignments, departure/arrival runway preferences, and location while in flight.

The goal of operational planning improvements is to integrate operational and business decisions to gain efficiency, predictability, and flexibility in flight operations. Each day, approximately 100,000 flights use the NAS, requiring many decisions to manage all the traffic. The TFM program provides tools to help users and service providers make collaborative decisions to prioritize and schedule flights and to better organize air traffic locally and nationally.

During flight planning, improved tools are used to predict locations and the impact of traffic demand and weather along planned routes and at the destination. As the flight progresses, additional updates on weather, NAS status, and other user-specific data are provided to the airline and other operational centers as appropriate. These new tools will eventually help plan direct flight paths, sequence departures and arrivals, change routes, and balance capacity and demand throughout the NAS.

The *enhanced traffic management system (ETMS)* is an upgraded program to replace hardware and software. ETMS is an existing traffic flow management computer system used by specialists to track, predict, and manage air traffic flows. The ETMS upgrade replaces controller workstations, computers, peripherals, and proprietary software to sustain current traffic flow management capability and meet the need for continued improvements in collaboration.

AUTOMATED FLIGHT SERVICE STATIONS (AFSSs)

Automated FAA Flight Service Stations (AFSSs) provide planning assistance, aviation weather, and aeronautical information to commercial, general-aviation, and military pilots. A new AFSS contract was awarded in September 2010 for a 3-year period.

The AFSS modernization plan replaced outdated automation systems that have limited capabilities with a new *operational and supportability implementation system (OASIS)*. OASIS incorporated the functions provided by the direct user access terminal (DUAT) service and the graphics weather display system (GWDS). Like the DUAT service, OASIS allows pilots to self-brief and file flight plans on the worldwide web or over the telephone.

AIRPORT SURFACE DETECTION EQUIPMENT, MODEL X (ASDE-X)

As frequently noted by the FAA and NTSB, the potential for runway incursions and collisions on taxiways increases each year. To combat this problem, FAA has deployed ASDE-X at 35 major U.S. airports. This system allows controllers to detect potential runway conflicts by providing detailed coverage of movement from surface radar, ADS-B sensors aboard aircraft, and aircraft transponders, among other sources. This technology is especially important to controllers at night or in weather conditions of poor visibility.

DEPARTURES AND ARRIVALS

Resolving congestion at the busiest U.S. airports requires a combination of modern technology and additional runways. The FAA is working with airport operators to help plan and develop new runways to accommodate increased aircraft operations and use new technologies, while meeting environmental requirements.

Arriving and departing aircraft are sequenced in and out of the airport by traffic controllers at the TRACON facilities. Maintaining a steady flow of aircraft, particularly during peak periods, can be improved by providing controllers with tools for sequencing and spacing aircraft more precisely. The objective is to reduce variability in services and optimize use of airspace and available runways.

The terminal modernization plan specifically includes installation of the new *standard terminal automation replacement system (STARS)* and aircraft sequencing tools. STARS is an all-digital, integrated computer system with modern color displays and distributed processing networks. STARS can be easily upgraded and supports current and future surveillance technology, traffic and weather information, and sequencing and spacing tools. The new STARS workstation will display air traffic, weather overlays, and traffic flow management information for controllers. STARS will interface with advanced communications, navigation, surveillance, and weather systems planned for the NAS modernization. It will replace the en route automated radar tracking system used in Alaska and at off-shore locations.

EN ROUTE AND OCEANIC OPERATIONS

The evolution toward a NextGen environment requires significant improvements in en route and oceanic computer systems and controller decision support tools.

The aging automation infrastructure must be replaced before new applications and improved services can be provided.

Currently, en route and oceanic facilities are colocated but do not share common systems, primarily because of the lack of surveillance and direct communications services over the ocean. The addition of oceanic surveillance and real-time direct communications will enable oceanic services to gradually become comparable with en route services, and oceanic and en route systems will evolve to a common hardware and software environment. Within both the Atlantic and Pacific airspaces, FAA controllers monitor aircraft position using Advanced Technologies and Ocean Procedures (ATOPs) equipment.

In the domestic airspace, aircraft are monitored by radar and typically follow the fixed route structure of airways, preventing pilots from flying the most direct route or taking advantage of favorable winds.

In oceanic airspace, aircraft follow "tracks" that are aligned each day with prevailing winds. Lack of radar surveillance and direct controller-pilot communications requires oceanic separation standards to be 20 times greater than those in domestic airspace. The large separations limit the number of available tracks. Therefore, some flights are assigned a less than optimum altitude, and there is insufficient opportunity to adjust altitudes to conserve fuel. Additional tracks and access to optimum altitudes would reduce fuel consumption and costs substantially. When fully integrated with ADS-B, ATOPs will reduce the required aircraft separation from 50 nautical miles to 30, and possibly to as little as 10 nautical miles of separation between aircraft. This will improve the efficiency of oceanic routes and reduce controller workload.

THE NEXT GENERATION AIR TRANSPORTATION SYSTEM (NextGen) (www.faa.gov/)

The FAA defines NextGen as an umbrella term for the ongoing and wide-ranging transformation of the National Air System (NAS). It represents an evolution from a ground-based system of air traffic control to a satellite-based system of air traffic management (ATM). NextGen is designed to relieve airspace congestion problems at high-density airports, especially in areas such as New York City with three large commercial airports which is considered a metroplex by the FAA. The National Airspace System currently handles almost 50,000 flights per day and more than 700 million passengers per year. New technologies such as the Global Positioning System (GPS) and the introduction of new transformational networking programs should allow more direct routing of aircraft, thus saving substantial time, fuel, and money for the airline industry and its customers. As originally planned, NextGen would be implemented over a multiyear period ending in the year 2025.

The NextGen 2010 implementation plan focuses on operational improvements in the "midterm" period through the year 2018. In Fig. 8-5, FAA illustrates the top level advantages of NextGen using the routine phases of flight in order to make the concept more easily understood.

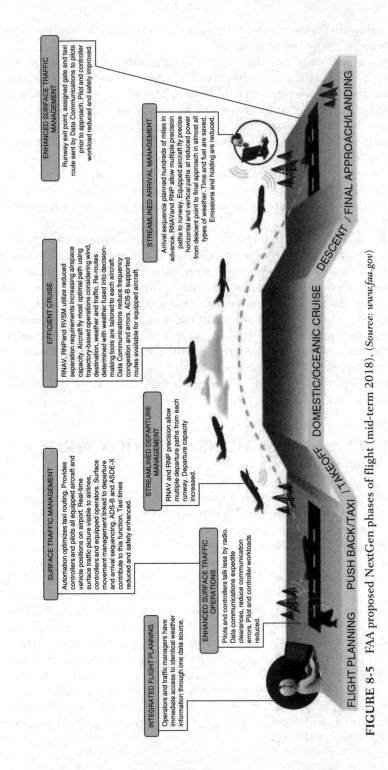

ENHANCED SURFACE TRAFFIC MANAGEMENT

Runway exit point, assigned gate and taxi route sent by Data Communications to pilots prior to approach. Pilot and controller workload reduced and safety improved.

STREAMLINED ARRIVAL MANAGEMENT

Arrival sequence planned hundreds of miles in advance. RNAV and RNP allow multiple precision paths to runway. Equipped aircraft fly precise horizontal and vertical paths at reduced power from descent point to final approach in almost all types of weather. Time and fuel are saved. Emissions and holding are reduced.

EFFICIENT CRUISE

RNAV, RNP and RVSM utilize reduced separation requirements increasing airspace capacity. Aircraft fly most optimal path using trajectory-based operations considering wind, destination, weather and traffic. Re-routes determined with weather fused into decision-making tools are tailored to each aircraft. Data Communications reduce frequency congestion and errors. ADS-B supported routes available for equipped aircraft.

SURFACE TRAFFIC MANAGEMENT

Automation optimizes taxi routing. Provides controllers and pilots all equipped aircraft and vehicle positions on airport. Real-time surface traffic picture visible to airlines, controllers and equipped operators. Surface movement management linked to departure and arrival sequencing. ADS-B and ASDE-X contribute to this function. Taxi times reduced and safety enhanced.

STREAMLINED DEPARTURE MANAGEMENT

RNAV and RNP precision allow multiple departure paths from each runway. Departure capacity increased.

INTEGRATED FLIGHT PLANNING

Operators and traffic managers have immediate access to identical weather information through one data source.

ENHANCED SURFACE TRAFFIC OPERATIONS

Pilots and controllers talk less by radio. Data communications expedite clearances, reduce communication errors. Pilot and controller workloads reduced.

FLIGHT PLANNING   PUSH BACK/TAXI   TAKEOFF   DOMESTIC/OCEANIC CRUISE   DESCENT / FINAL APPROACH/LANDING

FIGURE 8-5   FAA proposed NextGen phases of flight (mid-term 2018). (*Source: www.faa.gov*)

NextGen Transformational Programs

On its Web site, FAA details six (6) NextGen portfolio programs which are needed to transform the current NAS structure into a satellite-based system of the future. Each of these programs will be discussed in the following pages:

1. *Automatic Dependent Surveillance Broadcast (ADS-B).* The components of ADS-B are
   - Automatic—Periodically transmits information with no pilot or operator input required.
   - Dependent—Aircraft position and velocity vectors are derived and dependent on the Global Positioning System (GPS) satellite network.
   - Surveillance—A new method of determining position instead of using surveillance radar.
   - Broadcast—Information is broadcast (transmitted) to anyone with the appropriate ADS-B equipment.

ADS-B uses GPS signals along with aircraft avionics to transmit the aircraft's location to ground receivers. The ground receivers then transmit that information to controller ATC screens and to aircraft cockpit displays so the aircraft can "see" each other's relative position. The improved accuracy and reliability of satellite signals over existing radar surveillance stations means that FAA controllers will be able to safely reduce the mandatory IFR separation between aircraft. Relying on satellite signals instead of ground-based navigation aids also means that aircraft can fly more directly from point A to point B, saving time and money, and reducing fuel consumption. United Parcel Service (UPS) voluntarily equipped 107 of its cargo aircraft with ADS-B avionics knowing that it will recoup its investment quickly on daily flights to and from its hub airport, Louisville, Kentucky. ADS-B now also covers the Gulf of Mexico, which brings new air traffic surveillance services to low altitude helicopters flying to oil platforms beyond the range of conventional radar. The FAA plans to have ADS-B coverage nationwide in 2013.

2. *Collaborative Air Traffic Management (CATM).* This program is designed to accommodate aircraft flight operational preferences to the greatest extent possible in a collaborative fashion working closely with ATC. CATM supports improved flight planning and a more flexible ATC capable of in-flight adjustment to obtain more favorable routings and altitudes to save time and money. An important core program required for CATM implementation is known as the En Route Automation Modernization (ERAM) program, which is scheduled to replace aging, 30-year-old ATC equipment at all 20 FAA Air Route Traffic Control Centers (ARTCCs) located across the continental United States. Unfortunately, the ERAM program has experienced significant cost overruns, software problems, and schedule delays at key locations, which jeopardizes CATM and other subsequent NextGen programs.

3. *System Wide Information Management (SWIM).* SWIM is a network centric program that will essentially provide a "NAS-wide" web for aviation data. SWIM will manage surveillance, weather, flight data, aeronautical, and NAS status

information to seamlessly provide it to ATC customers in a simple, useable format. SWIM will use commercial off-the-shelf (COTS) hardware and software to support common situational awareness, connecting a myriad of NAS information systems. In 2005, ICAO adopted the SWIM concept and it is now an integral part of development projects in the United States and Europe.

4. *Data Communications Program (DATACOMM).* At present, most communications between ground controllers and aircraft are made by analog voice communications. The use of such two-way voice communication is very inefficient and labor intensive, and voice messages are often easily misunderstood. Initially, the new DATACOMM program using text messages will be a supplemental means for two-way exchange for ATC clearances, instructions, flight advisories, and flight crew requests and reports. As DATACOMM becomes the new method of operation, the majority of air/ground exchanges will be handled by digital data exchange to the crew or directly to the aircraft's flight management system (FMS), providing more efficiency and safety in the process.

5. *NAS Voice Switch (NVS) program.* This is a program to replace NAS voice switches more than 20-years old with a new technology switching system capable of supporting future NextGen requirements. Currently in the system there are 13 different NAS voice switches, each with different training and logistics requirements. The goal of NVS is to enable voice switch flexibility, and allow the airspace sectors to be dynamically reconfigured according to controller workload without the physical movement of ATC personnel. This common voice switch network should simplify the system and save significant costs of training and logistics/support.

6. *NextGen Network Enabled Weather (NNEW) program.* This is the last transformational program in concept development by FAA. Adverse weather has a considerable impact on flight operations. The NNEW program provides universal access to current weather information to enable collaborative weather decision making. The key to this program is the interagency effort to develop a "four dimensional weather data cube" (4D WX Data Cube) which will provide common, universal access to all aviation weather data. The stated purpose of this program is to provide better weather information integrated into ATC decision support tools to improve the quality of controller decisions and greatly reduce controller workload during adverse weather conditions.

NEXTGEN IMPLEMENTATION CHALLENGES

According to the FAA, NextGen is one of the most complex systems ever developed by the U.S. Government. Many significant challenges to success lie ahead to fill the gaps that remain in this ambitious and expensive system. According to the GAO and DOT Inspector General's offices, among these challenges are the following:

- FAA establishment of firm NextGen technical requirements, especially for transitional research and development programs.

- Identification of critical path decisions needed in terms of airspace development, and changes in the roles and responsibilities of pilots and controllers in the new system.

- Assessment of the safety and human factor risks of mixing old NAS equipment with new NextGen systems and concepts, and development of corresponding mitigation strategies to meet these risks.

- Solving the technical problems of introducing unmanned aircraft systems into the NAS and NextGen environment.

- Management of cost and schedule growth during the extended development of NextGen. The initial cost estimates of $40 billion by 2025 could go as high at $160 billion at the highest performance levels of NextGen.

The FAA is dependent upon Congress to appropriate funds on a year-to-year basis. The agency's budget request, part of the Department of Transportation package, is included in the President's January budget submittal for the following fiscal year. Before the agency's request is included in the Presidential submittal, it has already gone through review by the Department of Transportation and the Office of Management and Budget.

After Presidential submittal, the agency's request is submitted to congressional committee scrutiny through study, hearings, questions and responses, and finally to the full Senate and House for approval on a department-by-department basis. Unfortunately, the FAA has had some spectacular program failures in the past, and Congress has a long memory for such embarrassments. Another factor is that the FAA budget is considered in the context of the entire budget and national priorities, placing the agency in the position of competing with social, space, and defense programs. Only after the budget process is complete, taking nearly 2 years from initial work to approval, can the agency start procurement or construction of new NextGen systems.

The question at this point is, "Will the FAA be successful under the present technical and funding challenges facing this critical myriad of programs?"

KEY TERMS

National Airspace System (NAS)

Next Generation Air Transportation System (NextGen)

Navigational Aids (NAVAIDS)

Very high frequency omni-directional range (VOR)

Distance-measuring equipment (DME)

VORTAC

Instrument Landing System (ILS)

Airport Surveillance Radar (ASR)

Area Navigation (RNAV)

Instrument flight rules (IFR)

Visual flight rules (VFR)

Airport Traffic control towers (ATCTs)

Terminal radar approach control (TRACON)

Air route traffic control centers (ARTCCs)

Global Positioning System (GPS)

Wide-area augmentation system (WAAS)

Ground-based augmentation system (GBAS)

Collaborative Air Traffic Management (CATM)

System Wide Information Management (SWIM)

Data Communications (DATACOMM)

NAS Voice Switch (NVS)

Nonprecision approach (NPA)

Precision approach (PA)

Decision height (DH)

Runway visual range (RVR)

Category (CAT) I, II, and III approaches

Automatic Dependent Surveillance—Broadcast (ADS-B)

Integrated terminal weather system (ITWS)

Weather and radar processor (WARP)

Next-generation weather radar (NEXRAD)

Terminal Doppler weather radar (TDWR)

Airport Surface Detection Equipment, Model X (ASDE-X)

Traffic flow management (TFM)

Next Gen Network Enabled Weather (NNEW)

Automated Flight Service Station (AFSS)

Enhanced traffic management system (ETMS)

Operational and supportability implementation system (OASIS)

Standard terminal automation replacement system (STARS)

REVIEW QUESTIONS

1. Describe how air traffic is controlled.
2. What is the mission of the FAA's air traffic control system?

3. Identify some of the major milestones in ATC history.

4. What are some of the key components of the ATC system?

5. Discuss the basic differences between visual and instrument flight rules (VFR vs. IFR).

6. What is the purpose of a terminal radar approach control (TRACON)?

7. How does GPS work?

8. Compare and contrast WAAS and GBAS.

9. Distinguish between category I, II, and III approaches.

10. What are some of the advantages of satellite-based navigation?

11. What is automatic dependent surveillance broadcast (ADS-B)? Distinguish between ADS-B and ASDE-X.

12. What are the key components of the en route/oceanic ATC system?

13. What is NextGen and what are its goals?

14. Describe the six NextGen transformational programs.

15. Discuss the implementation challenges facing the ambitious NextGen Air Transportation System

REFERENCES

FAA. 2010. *Action: Timely Actions Needed to Advance the Next Generation Air Transportation System,* FAA Report Number AV-2010-068. U.S. Department of Transportation, Office of the Inspector General, Washington, D.C. June 16, 2010

FAA. 2010 *Delivering NextGen*, Federal Aviation Administration, Washington, D.C. www.faa.gov April 8, 2010.

FAA. 2011 *FAA NextGen Implementation Plan* 2011. Federal Aviation Administration, Washington, D.C. www.faa.gov/

FAA. 2011 Fact Sheet—*Automatic Dependent Surveillance—Broadcast (ADS-B)*, Federal Aviation Administration, Washington, D.C. www.faa.gov/ 2011.

ICAO. 2011 International Civil Aviation Organization, Montreal, Canada www.icao.int

Nolan, M. S. 2011. Fundamentals of Air Traffic Control, 5th ed. Clifton Park, NY. Delmar Cengage Learning.

U.S. Congress, General Accounting Office, 1995. *National Airspace System: Comprehensive FAA Plan for Global Positioning System is Needed*. Washington, D.C.: U.S. Government Printing Office, May.

USDOT. 2010 U.S. Department of Transportation, Office of Inspector General Testimony Number CC-2011-001 dated December 21, 2010; retrieved March 2011, from www.oig.dot.gov

U.S. Government Accountability Office, 2010 *Integration of Current Implementation Efforts with Long-term Planning for the Next Generation Air Transportation System. Washington,* D.C. November 22, 2010.

AIRCRAFT SAFETY SYSTEMS

LEARNING OBJECTIVES

After completing this chapter, you should be able to

- Recognize the importance of jet engine development and advances in solving problems of fuel consumption, noise, reliability, durability, stability, and thrust.
- Describe several breakthroughs in technology evolving from development of the B-47.
- Explain how improvements in high-lift systems have resulted in safer aircraft.
- List and briefly describe advances that have taken place in five stopping systems.
- Discuss the importance of stability and control characteristics in relation to safety; give several examples of developments in powered controls and low-speed stall characteristics.
- Understand how criteria and procedures used in aircraft design over the years have produced long-life, damage-tolerant structures with excellent safety records.
- Explain the concept of fail-safe design.
- List and discuss the parameters that define aircraft aging.
- Describe several approaches and new technologies designed to address the problems of wind shear, volcanic ash, ice, and precipitation.
- Summarize some of the flight-deck technology changes that have made significant contributions to improving safety.
- Explain how computational fluid dynamics (CFD) has enhanced the study of wing design and engine/airframe integration.
- Discuss the importance of the following technologies to aircraft development and safety: wind tunnel, flight simulator, structural tests, flight tests, flight data recorder, and crew voice recorder.

INTRODUCTION

Rapid advances in technology have led to the development of extremely complex and highly sophisticated commercial aircraft. Major portions of tasks that were demanding and performed manually are now automated. New fly-by-wire concepts, for example, sever the mechanical connection between the pilot and the aircraft wing and tail. *Fly-by-wire* refers to the electronic linkage from the sidestick (or control yoke) to flight control computers to activate flight controls. This type of system can come in a number of forms. In many aircraft, such as the Airbus A-320, it incorporates *protections*, which prevent the pilot from exceeding certain limits.

Pilots often perceive that they can fly closer to the margins with a conventional airplane and get the last ounce of performance. In reality, the bigger danger may lie in overreacting and getting into an unrecoverable attitude or being too timid and not getting the required performance. For example, close traffic may require rapid pitch and roll inputs. This is no time to consider limits. With a conventional aircraft, there is a real risk of excessive bank or overstressing the aircraft, while the fly-by-wire system will automatically limit bank to 67 degrees and prevent more than 2.5 G's as in the case of a wind shear escape maneuver during approach. Older, conventional aircraft may have no recovery tools, other than a target pitch attitude, with additional pitch increases up to stick-shaker actuation. Others have more accurate flight director guidance for pitch target but still have no stall protection. The fly-by-wire aircraft combination of an excellent speed reference mode for the flight director and stall protection, even with full aft stick, is a distinct advantage. This allows the pilot to pitch up to the command bars, yet not be concerned with overdoing it. This is a valuable feature. In other words, it really does not matter that a conventional airplane may have a theoretical extra margin of performance by pressing the limits. What matters is the level of performance that is readily available to the pilot.

Pilots soon gain confidence that fly-by-wire will not allow limits to be exceeded, which promotes maneuvering to the edge of the safe flight envelope, probably closer than they would dare with a non-fly-by-wire airplane.

Much of the technology of the airplane can be attributed to the many new and improved design tools that the engineer has available. Wind tunnels, simulators, and analog and digital computers have each contributed to the remarkable safety-record improvements achieved by today's jet transports. The Wright brothers' achievement was aided (maybe even made possible) by the wind tunnel they built to verify the density of air.

The purpose of this chapter is to relate how this wealth of technology and the technology of design tools have contributed to the excellent safety record of today's jet transport fleet. In addition, a discussion of new technologies that should further improve the safety and efficiency of tomorrow's airplane is included.

World War II hastened the development of several different technologies used directly in the development and use of the jet transport. The jet engine, radar, and wing sweep are three of those major technologies. The only airplanes available to the airlines in the late 1940s were propeller-driven, powered by large reciprocating engines. Efforts to increase their size and performance came at the expense of engine reliability. In many accidents, the loss of an engine was a contributing factor.

After initial development, it was realized that the jet engine offered improved reliability, safety, and performance. The jet engine's greatest challenge was its propensity for high-speed, high-altitude operation. This led to a swept-wing design, which favored a much higher wing loading. Early operation at high speed at high altitude near the airplane's critical Mach number was an adventure in pilot technique, clear air turbulence, and jet streams. The impact of these factors and the technical solutions found to make the jet the safe and efficient mode of travel it is today make a very interesting story.

The jet transport era was ushered in with the advent of the Boeing B-47. It was followed in 7 years by the Boeing Dash 80 prototype. There were many safety-of-flight issues associated with these large swept-wing jet aircraft. This chapter focuses on those technologies that were pivotal to safety as the jet frontier was opened.

JET ENGINE DEVELOPMENT

Credit for the jet engine goes to two individuals who worked independently of each other just prior to and during World War II: Frank Whittle of England and Hans von Ohain of Germany. Through the efforts of General Hap Arnold, the British made Frank Whittle's work available to U.S. engine companies. General Electric (GE) started jet engine development for the Air Force, and Westinghouse started it for the Navy. The effort at GE first produced the axial flow J-47 engine, which was used in quantity in the F-86 and B-47 production airplanes. The work of Pratt & Whitney (P&W) led to the J-57 engine used on the B-52 and 707 series aircraft.

Very rapid progress has been made through the years by the engine manufacturers in solving problems of fuel consumption, noise, reliability, durability, stability, and thrust. The resulting improvements in jet engine performance and reliability rapidly overtook those of the piston engines which were experiencing significant problems with increased size. The reduced complexity and frequency of maintenance of the jet engine lessened the chances of human error. As is discussed later, simplification in controls improved layout and display conditions in the flight deck that subsequently served to reduce the chances for human error.

Jet airplane performance resulted directly from jet engine propulsion efficiency, increasing as airspeed increased. Propeller efficiency falls off as airspeed increases, making it impractical for high-speed flight. Efforts to improve jet engine capability have been nothing short of spectacular. This progress has been made while meeting diverse engine-thrust levels asked for by the aircraft manufacturers to satisfy ever-changing payload-range requirements.

These engines also had to satisfy more stringent noise requirements. Much of the improvement in the last 40 years has been due to the engine and its installation when measured in terms of increased passenger-miles per pound of fuel burned. This jet engine technology was even more impressive when one notes that this performance improvement was achieved with a corresponding improvement in the engine's reliability and safety. But the *in-flight shutdown* (IFSD) rate and the resultant loss of thrust and systems were only a part of the safety concern, as the airplane was designed for this event. The other concern was the failure mode (passive failure or a disk rupture), which had to be treated differently. This was the primary reason for going to the pod-mounted engine installation.

RECENT DEVELOPMENTS IN JET ENGINE DESIGN

The air transport industry has made excellent progress in jet engine developments in the first decade of the 21st century through computer-aided design and testing. Next generation aircraft are making maximum use of the latest composite materials and design processes to reduce weight, improve performance and lower maintenance costs. Lightweight composites such as graphite, kevlar and fiberglass are being used throughout the aircraft main structure and as fan blades in jet engines where the operating temperatures are low. The use of composites in the Boeing 757 aircraft is shown in Fig. 9-1. Additionally, the fuel performance of the latest Airbus and Boeing aircraft is improving, and reduced fuel consumption results in a smaller carbon footprint including less CO_2 emissions on an aggregate level. Likewise, engine noise improvements have been achieved through the slower fan speeds and lower jet velocity of today's modern jet engines.

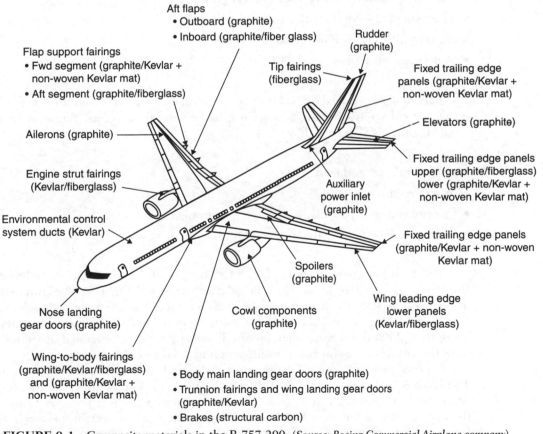

FIGURE 9-1 Composite materials in the B-757-200. (*Source: Boeing Commercial Airplane company*)

LONG-RANGE COMMERCIAL JET TRANSPORT ERA

A great leap forward for commercial transport airplanes came with the Boeing B-47, a military jet bomber that first flew on December 17, 1947, the 44th anniversary of the Wright brothers' first flight. It all started with the vision of a few designers to adapt this new jet propulsion system to the airplane in such a way as to take full advantage of the jet engine's performance characteristics. The Boeing Transonic Wind Tunnel (BTWT) was an invaluable tool to establish the successful configuration. A number of radical airframe differences characterized the XB-47 design:

- Highly swept wing (35°)
- High-aspect-ratio plan form (9.43)
- Very wide speed range
- Long-duration, high-altitude operation
- High wing loading (double that of previous designs)
- Thin wing (12% constant-thickness ratio)
- An extremely clean aerodynamic design
- Pod-mounted engines

This design produced a revolutionary performance advantage but also presented some real safety challenges requiring technological solutions. Some of these challenges involved are as follows:

- How to take off and land
- Stopping-distance considerations
- Control system capability over a large speed range, and flutter
- Structural integrity for this wing plan form and speed range

These challenges were, but a few, encountered by the test program following the first flight on December 17, 1947. An experimental flight test program was undertaken, during which many design decisions were validated and valuable lessons learned (Fig. 9-2). Each of these contributed to today's jet transport safety. More than 2000 B-47s were manufactured, and the airplane remained operational until the late 1960s, again contributing much to today's airplane safety.

Another great leap forward came with the Boeing B-367-80 (Dash 80) prototype. The Dash 80 took full advantage of the B-47 experience. One significant difference from the B-47 was the return to a tricycle gear, which made it possible to tailor its high-lift system for takeoff and landing and to increase the weight on its wheels for improved stopping capability. These changes are discussed along with

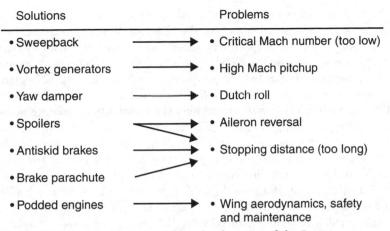

Solutions	Problems
• Sweepback →	• Critical Mach number (too low)
• Vortex generators →	• High Mach pitchup
• Yaw damper →	• Dutch roll
• Spoilers →	• Aileron reversal
• Antiskid brakes →	• Stopping distance (too long)
• Brake parachute →	
• Podded engines →	• Wing aerodynamics, safety and maintenance

FIGURE 9-2   Lessons learned from development of the B-47.

improvements in the control system, structure, and other technologies that evolved from the Dash 80 prototype.

HIGH-LIFT SYSTEMS

Development of *high-lift systems* had to keep pace with the transition from piston engine to jetliner operations. Early jet bombers used simple trailing-edge flaps and explored but did not employ leading-edge flaps. These trailing-edge flaps provided a low-drag solution to the takeoff climbout problem but created another problem for the approach to landing. Since flap drag was low in the approach phase, the power setting of the engines was low, and it became difficult to quickly accelerate the engines to go-around power for a missed approach. Glide path control was also difficult with these low-drag flaps. With the B-47 in the landing configuration, the engines took 13s to accelerate to 60 percent rpm. Part of the solution was to incorporate multielement slotted trailing-edge flaps, which provided the required higher drag and power settings. The remainder of the solution was to reduce the acceleration time for the engines.

Incorporation of leading-edge flaps resulted in greater safety due to better takeoff and landing field length performance and greater stall margin. Leading-edge flaps were incorporated on some of the early jet transports primarily because of the different takeoff characteristics of jetliners versus piston-powered airplanes. Prop wash over the wing of propeller aircraft provides a built-in factor of safety at low speeds, enabling the airplane to continue to climb at airspeeds less than the power-off stall speeds. Leading-edge flaps on jets extend the margin of safety to even lower speeds and permit takeoff and climbout margins similar to those in prop airplanes. Some overrun accidents with early jets not equipped with leading-edge devices were attributed to early rotation of the airplane.

The early jetliners operated into and out of a limited number of airports, and their high-lift systems were adequate for the range of altitudes and temperatures encountered at these airports. As jet travel became more economical, an increased diversity of airports was served, and the technology of flap systems changed to accommodate them. In the interest of shorter field-length performance, the Boeing 727 program developed a high-camber airfoil by having a highly deflected slat and a three-element trailing-edge flap system. This system provided the capability to operate at lower approach and landing speeds, giving the 727 the ability to land on shorter runways. Today's jetliners incorporate two-position leading-edge slats to provide low-deflection setting for low-drag takeoffs, a higher-deflection position for landing at lower speeds, and better visibility from a lower deck angle.

A further merit of a good high-lift system is to have higher leading-edge-device camber on the outboard wing to provide a more stable stall plus better stall characteristics, which provide enhanced safety by making for a more forgiving airplane. To enhance safety and improve stall characteristics while maintaining the low-camber leading edge for takeoff, the Boeing 757 added the auto slat to its configuration. This feature provides an extra margin from stall when the auto slat actuation angle of attack is reached. This feature also provides additional maneuvering margin in the case of a wind shear encounter.

STOPPING SYSTEMS

A number of improvements in aircraft stopping systems have taken place over the years, greatly enhancing safety. They include antiskid, fuse plugs, autobrakes, speed brakes, and thrust reversers.

ANTISKID.   The higher takeoff and landing speeds for jetliner operations compared with prop airliners created the need for more efficient stopping devices. An *antiskid system* was recognized as a necessity early in the jet age. Most of the time pilots do not need antiskid at all. The runway is dry and long, and the brakes are not applied hard enough to skid the tires. However, when the runway is slippery and short, the ability of the antiskid system to maximize braking effectiveness becomes very important.

Early antiskid systems were developed and tested successfully on the dual-wheel main landing gear assemblies of large propeller-driven aircraft. However, the design of antiskid systems soon became a new technological challenge because of the incorporation of multiple-wheel main landing gears on the first jetliners. The early antiskid systems for the four-wheel truck main gears controlled tandem pairs of wheels. Later the technology to control each wheel independently was incorporated to maximize the effectiveness of the antiskid system. Digital antiskid systems incorporating microprocessors enhance the reliability and effectiveness of today's antiskid systems.

ENGINEERED MATERIALS ARRESTOR SYSTEM (EMAS).   FAA requires that commercial airports under FAR Part 139 have a Runway Safety Area of 1000 feet beyond the end of the runway if possible. Since the standard 1000 feet is not practical at some airports, EMAS systems, a bed of lightweight crushable concrete, may be installed at the end of a runway to stop the aircraft. Currently, EMAS is installed at 51 runway ends at 35 U.S. airports, and has been credited with several successful arrestments preventing injury and aircraft damage.

FUSE PLUGS. Air pressure buildup in tires after braking during early high-speed refused-takeoff tests caused the tires to explode, sending chunks of rubber flying into the airframe. Incorporation of low-melting-point fuse plugs in the wheels allowed the tires to deflate before severely heating and exploding.

AUTOBRAKES. The incorporation of an *automatic braking system* is a recent enhancement to safety. This system enables automatic brake application on landing or during a *refused takeoff* (RTO). The landing autobrake system controls brake pressure to maintain aircraft deceleration at one of five pilot-selected values, provided that sufficient runway friction is available to maintain this level. The RTO autobrake system applies full braking upon closing throttles above a fixed speed (for example, 85 knots). Using autobrakes frees the pilot to concentrate on other activities, such as applying reverse thrust and guiding the airplane to a smooth, safe stop.

SPEED BRAKES. While main wheel brakes remain the primary method of stopping aircraft on the runway, technological development of ancillary stopping devices has kept pace with development of wheel brake systems. The need to apply a download on the wheels was recognized as a requirement to enhance the effectiveness of wheel brakes. The use of *wing spoilers* (*speed brakes*) to increase download on the main wheels was incorporated on the early jets. The percentage of wingspan covered by these devices has increased on later models, increasing their effectiveness. Implementation of automatically deployed spoilers on later models has enhanced stopping capability significantly. Sequencing of the spanwise spoilers after touchdown eliminates pitchup, which can occur when all the spoilers are deployed simultaneously.

THRUST REVERSERS. Commercial jet transport operation into regional airports would not have been possible without *thrust reversers*. Although FAA regulations did not require thrust reversers, they were a must for most airline customers. It was very desirable that jetliners be able to land on slippery runways within the FAR-required field length. Early experience revealed that some types of thrust reversers lost their effectiveness at high landing speeds, which required the airframe manufacturers to conduct tests to verify early reverser concepts.

Designs of the 737-100 and 737-200 presented The Boeing Company with a difficult challenge in integrating the thrust reversers, because the engines were mounted very close to the lower surface of the wing and very close to the ground. The initial design decision was an economic one: Use the entire power package and nacelle from the Boeing 727. The resulting reverser configuration partially trapped the reverser efflux between the trailing-edge flaps and the leading-edge devices, creating a "bubble" of air on which the airplane floated in ground effect, greatly reducing stopping effectiveness. A complete redesign of the reverser, including a lengthening of the nacelle to accommodate a target reverser aft of the flaps, resulted in a tremendous improvement in stopping capability. In fact, the 737-200 has such an effective reverser that the airplane can stop within its FAR scheduled wet distance using only thrust reversers.

FLYING QUALITIES

Before it flew, the flying qualities of the B-47 jet airplane were known to be significantly affected by its swept wing, high speed, and high wing loading. A number of technological advances enhanced the handling qualities of this configuration so that today the flying qualities of the Boeing 757, 767, and 777 and Airbus 320, 330, and 340 are the standards of the industry.

The stability and the control characteristics designed into an airplane have one of the most significant impacts on the airplane's flight safety. The airplane's response to an engine-out, system failure, or atmospheric disturbance has to be controllable within the trained commercial pilot's ability. To ensure this controllability, the technology for modeling airplane dynamic response has been continually advanced along with improved modeling of atmospheric disturbances. These models are then incorporated into flight simulators, and piloted evaluations of the designs are undertaken to give direction as to the best flying qualities and pilot techniques. Out of these simulations, technology was developed for determining airplane configuration, including the best control laws for automatic flight control and stability augmentation systems. The computer's and simulator's parts in enabling this progress, and thus improving aircraft safety, are discussed later.

POWERED CONTROLS.   The B-47 was ahead of its time with closed-loop hydraulic position servos on all control surfaces. Each axis had manual backup through servo tabs and aerodynamic balance cavities. Hydraulic-powered elevator force feel was provided with a Q-spring. It sensed an increase in dynamic pressure and adjusted the pilot's force feel system to protect against an inadvertent maneuver that could exceed the airplane's structural limit. The technology that followed on jetliners greatly improved the state of the art of hydraulic systems and actuators by providing more reliable, redundant hydraulic supply and smooth transition from boost to manual control. It also prevented a jammed control failure mode.

The B-52 design proceeded before the B-47 hydraulic power control system could be perfected. The B-52 was designed with manual controls on all axes. The elevator and rudder controls had 10 percent chord surfaces, and a variable-incidence horizontal stabilizer was used for longitudinal trim and control. Aileron control was supplemented by wing spoiler controls that were flight-tested on the B-47. On commercial designs, the rudder and elevator sizes were increased to provide more engine-out control and maneuvering capability, stabilizer mistrim, and dive recovery capability on the longitudinal axes.

Spoiler lateral controls were explored on the B-47 after it was discovered that at high airspeeds aileron control reversed due to the resultant elastic torsion of the wing from an aileron-developed lift component opposite to and of greater magnitude than that of the deflected aileron.

The Boeing Dash 80 prototype perfected the combination aileron and spoiler control further so that essentially no yaw or pitch coupling occurred following a control input. The adverse yaw associated with a lateral control input is balanced by yaw from the spoilers.

The B-47 also encountered adverse shock interactions at high speeds. The swept wing greatly delays this onset, but as Mach 1 is approached, they occur. This problem was investigated thoroughly in flight. The solution turned out to be rather simple. Small vanes called vortex generators were installed on the wing, altering the boundary-layer shock interaction to reduce pitchup and increasing the buffet margin to allow greater turn performance at altitude. These devices have proved valuable on subsequent transport designs.

The other flying quality discovery of the B-47 program was that the swept wing has significantly different Dutch-roll characteristics. In a straight wing, it is primarily a nose oscillation, but with wing sweep, it has significant roll that is slightly out of phase with the yaw. It was computed prior to flight that with no dihedral the Dutch-roll mode would be sufficiently damped, but flight test showed that not to be the case. Again, technology advanced quickly with the invention of the electromechanical yaw damper to solve this problem. It senses the yaw motion that occurs in this mode and applies opposite rudder, thus relieving the pilot of extra workload or the possibility of the aircraft becoming upset in turbulence. The electromechanical yaw damper proved to be a solution with a long life. Electronic yaw dampers are in all present-day commercial jet transports.

LOW-SPEED STALL CHARACTERISTICS. Design of the high-lift systems was a challenge in that these systems had to provide commercial jets with safe takeoff and landing margins, in all places and conditions. The system also has to exhibit satisfactory flying qualities in the extremely unlikely event of a stall.

Early swept-wing designs exhibited poor stall characteristics because the wing tip would stall first. The remaining inboard wing lift, being ahead of the center of gravity, would give the aircraft an undesirable pitchup tendency. Wing technology involving the selection of airfoils, wing twist, and tailoring of the leading-edge and trailing-edge flaps was initially advanced through the use of wind tunnels, then flight testing. Later, the understanding of this problem was greatly advanced through the use of *computational fluid dynamics* (CFD).

Two technologies, the stick-shaker and auto slat gapper, have been effective in providing good stall characteristics for both takeoff and landing. The best design philosophy is to first work on avoiding the stall. Some airfoil-wing combinations provide adequate stall warning to the pilot by virtue of buffet that occurs as the stall angle of attack is reached. As high-performance wings were developed, this buffet margin was found inadequate and a "stick-shaker" was provided. The technology of this system involved finding the best sensor and location to key on. The auto slat was implemented when it was found that the characteristics of a sealed slat, which had improved takeoff performance, had poor flying qualities at stall. The solution was to automatically open the slat gap as stall was approached.

Stall safety was greatly improved by use of the simulator. Early in demonstrating airplane stall and recovery techniques, it was found that the situation was too unforgiving to a student applying improper technique. Today, stall avoidance and recovery are carried out safely in the simulator. Stall characteristics are thoroughly investigated during flight tests involving, in many cases, more than 700 stalls. The designer is anxious to get as low a stall speed as possible for overall safety in addition to reduced structural weight. However, finding an ample stability and control configuration makes it difficult, but necessary, to achieve.

STRUCTURAL INTEGRITY

The importance of structural integrity to commercial aircraft safety is obvious. What was not so obvious was the fracture-mechanics problems encountered and the durability of the jet airplane that extended its life well beyond anything previous. The B-47 and the de Havilland Comet were the first large jets to become operational, and both encountered fatigue problems. The Comet encountered fuselage skin fatigue problems that led to a series of accidents. Subsequent investigation into these and the B-47 fatigue problems pushed the state of the art for aircraft structural design technology forward very rapidly. The B-47 and B-52 fatigue problems came to light after the U.S. Air Force started flying low-level radar avoidance missions.

The real challenge to commercial aircraft was that they would experience much higher flight-hours at lower stress levels. One hour of low-level flight was equivalent to 80 hours at cruise. These lessons learned were shared with all, and manufacturers have been benefiting from these improvements in aircraft safety ever since.

One of the material fracture-mechanics properties came to light. Laboratory data compared the tensile strength of three different aluminum alloys as affected by the length of a crack. In the past, it was general practice to design engineering structures such as aircraft, bridges, buildings, and pipelines to a required new, uncracked strength including a factor of safety. This factor of safety was intended to provide for degradation by corrosion, fatigue, damage, etc. It is clear today that engineering structures should be designed to have adequate strength after they have sustained fatigue, corrosion, and use damage to an inspectable level.

A lot of credit has to be given to the electron microscope. By electron microscope examination on failure surfaces of failed structures, it became clear that each time the structure was loaded, there would be crack growth. The initiation and growth of cracks could be identified with their prior loading. It was also clear that the fatigue and corrosion cracks of most aircraft materials start at the exposed surface of the material, permitting inspection for fatigue and corrosion cracks for most installations.

The attention of aircraft designers became focused on the rate at which cracks grow, the strength of cracked structures, and the variation of these factors for different materials. It was found that the rate of crack growth was a function of loading. It was then that considerable attention was given to understanding the loading cycles commercial aircraft would see in service.

STRUCTURAL SAFETY. Criteria and procedures used in commercial airplane design over the last three decades have produced long-lived, damage-tolerant structures with excellent safety records. This has been achieved through diligent attention to detail design, manufacturing, maintenance, and inspection procedures. Structural safety has been an evolutionary accomplishment, with attention to detail being the key to this achievement. These design concepts, supported by testing, have worked well due to the system that is used to ensure that the fleets of commercial jet transports are kept flying safely throughout their service lives. This system has three major participants:

- The manufacturers that design, build, and support airplanes in service
- The airlines that operate, inspect, and maintain the airplanes
- The airworthiness authorities who establish rules and regulations, approve the design, and promote airline maintenance performance (Fig. 9-3)

Airplane structural safety depends on the diligent performance of all participants in this system. The responsibility for safety cannot be delegated to any single participant.

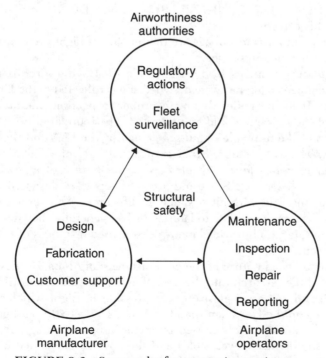

FIGURE 9-3 Structural safety system interaction.

All jet transports are designed to be damage-tolerant, a concept that has evolved from the earlier fail-safe principle. On the whole, service experience with fail-safe designs has worked very well with thousands of cases where fatigue and other types of damage have been detected and repaired. The question being debated among experts in the industry is whether the fail-safe design practices used in the 1950s and 1960s are adequate as these airplanes approach or exceed their original economic life objectives. Note that there is no limit to the service life of damage-tolerant-designed airplane structure (see the section on aging aircraft next), provided the necessary inspections are carried out along with timely repair and replacement of damaged structure or preventive modifications for airplanes exceeding economic design life objectives. Operational efficiency is affected by the cost and frequency of repair. Durability may, therefore, limit the productive life of the structure.

AGING AIRCRAFT.   One of the major problems facing the FAA and air carriers today is aging aircraft. By definition, aging aircraft are aircraft that are being operated near or beyond their originally projected design goals of calendar years, flight cycles, or flight hours. Nothing in the FARs pertains directly to the life of an airplane as measured in calendar years, but it was customary for designers to assume approximately 20 years as the calendar life of an airplane. For the first generation of jet transports, some designers believed that the aircraft would be technically obsolete within 20 years. Using this measure, a number of the earlier models of the DC-9s, B-737s, B-727s, B-747s, and DC-10s still flying today would fall into this category.

The two important measures of age are the number of flight cycles and the number of flight hours accrued in service.

Recent regulatory changes by FAA have clarified aging aircraft monitoring requirements. Commercial aircraft were designed and certificated in the United States after World War II in the belief that with proper inspection, maintenance, and repair, the life of the airframe could be unlimited. The foundation for this premise was the adoption of the principle of *fail-safe design* by the FAA and the industry in the early 1950s. This rule required that a specified level of residual strength be maintained after *complete failure* or *obvious partial failure* of a *single principal structural element*. The early U.S. jet and propjet fleets were designed, tested, and FAA-certified to this rule without a specified life limit. The service experience acquired by this fleet by the middle to late 1970s had generally shown a satisfactory level of structural safety and provided many documented instances of the validity of the fail-safe concept.

However, the unusual nature of the circumstances of Aloha Airlines Flight 243 greatly accelerated the research into the area of aging aircraft. On April 28, 1988, a Boeing 737-200 suffered an explosive decompression in flight which required an emergency landing on the Hawaiian island of Maui. The NTSB investigation report concluded that the accident was caused by metal fatigue. The age of the aircraft became a key issue (it was 19-years old) and the 737 had sustained a remarkable number of takeoff-landing cycles, 89,090, the second most such cycles for a plane

FIGURE 9-4   Aloha Airlines Flight 243 evacuation. (*Source: Associated Press photo*)

in the world at that time. Fig. 9-4 is a photo of the aircraft evacuation taken shortly after landing showing extensive damage to the plane's fuselage.

The Aloha Airlines accident of 1988 focused public and congressional attention on aging aircraft. This accident and other structural failures stimulated a reexamination of the current approaches to the structural integrity of aging aircraft. Fatigue-initiated damage is the primary cause of concern about *aging aircraft*. When aircraft are properly inspected and maintained, corrosion and accidental damage should be understood and controlled long before the design life is reached. On the other hand, the problem of fatigue-initiated damage increases with time or, more properly, with use as measured by flight hours or flight cycles or both.

Fatigue damage to the fuselage is caused primarily by the repeated application of the pressure cycle that occurs during every flight. Fatigue damage to the wings is caused by the ground-air-ground cycle that occurs during every flight and by pilot-induced maneuvers and turbulence in the air. Thus, the design-life goal of flight hours is more important for the wings than it is for the fuselage. Fatigue-initiated damage is a random phenomenon in which the probability of existence at any specific point in the structure increases with time. Because detecting the damage is also probabilistic, the success of the damage tolerance process in preserving airworthiness lies in an acceptably low probability of the presence of damage and high probability of timely detection. A large transport airplane is a complex structure in which a large number of points are susceptible to fatigue cracking that could propagate to the point at which the residual strength would be less than the damage tolerance requirement. The number of points in the structure at which cracks will initiate increases with the age of the structure. As the number of initiation sites increases, the probability increases of not detecting at least one before the strength degrades to the limit. In other words, at some time the risk of not having limit-load capability at some point in the structure may be too great for the airframe to be considered airworthy. Thus, even an airframe designed to be damage-tolerant may reach a time in its life when additional inspections, maintenance, and repair will be required to maintain airworthiness. A common scenario for the fuselage includes the following:

- Cracks initiate at the edges of the fastener holes in the center of the panels along a splice.
- With repeated flight cycles, these small cracks link to form a patch spanning several holes
- The patch becomes sufficiently long that it is detected during scheduled checks before rapid growth occurs.

Rapid growth, if it occurs, will be arrested by *crack turning* at the tear straps or the frame or both, resulting in a safe depressurization that permits the pilot to safely land the airplane. Although this scenario has actually occurred, it is not the only possible scenario. At least two other scenarios can be envisioned in which cracking may not be arrested by the tear straps (e.g., if the fastener holes in the tear straps already contain small cracks or if there is widespread cracking in several adjacent bays). In these circumstances, the crack may continue to propagate rapidly along the splice and lead to uncontrolled depressurization, in which case neither the tear straps nor the frame is fulfilling the original fail-safe function. This, in fact, is what occurred during the Aloha Airlines incident—the uncontrolled crack propagation resulted in the loss of a large portion of the fuselage. The structural condition previously described, wherein widespread cracking in tear straps and adjacent bays occurs, is called *multiple-site damage* (MSD). Clearly, MSD is more likely to exist in heavily used aircraft. A key factor in maintaining the safety of aging aircraft is the determination of age in flight cycles, flight hours, or both, of the onset of MSD.

It is assumed in analyses of this phenomenon that the airframe was designed and manufactured as intended, such that aging is related to the *wear-out* phase of the well-known *bathtub curve*. Because neither design nor manufacturing control is perfect, cracks occur in airframes long before the design goal life is reached. This is not an aging phenomenon but should be considered as the population of locations in which the design or construction was deficient (i.e., local "hot spots"). Some are revealed by the airplane fatigue tests and others during the early service experience with the aircraft. As cracks are detected and corrective action is taken, the rate of new hot-spot cracking decreases with time; the net result is that the total population of crack locations at any given time would be small. Experience has shown that the risk of undetected cracks during this phase of the operational life can be controlled to a safe level by appropriate inspection and maintenance programs.

In summary, the following primary technical issues are posed by the aging aircraft fleet:

- *MSD*, the undesirable condition caused by widespread cracking of the structure, negates fail-safety and damage tolerance to discrete sources. Neither multiple-load-path nor crack-arrest fail-safe features of commercial transport aircraft can be depended on to protect the structural safety of the aircraft after the onset of MSD.

- *Corrosion*, a time-dependent process, decreases the size of structural members, leading to higher stresses and lower structural margins. Corrosion also has undesirable synergism with the factors that lead to cracking of the structure, factors that are not well quantified and are not considered in the damage tolerance and fail-safe design of the structure.

- *Nondestructive inspection* is the key to assessing the health of the aging aircraft fleet. It should not be relied on for ensuring the continuing airworthiness of an aircraft that may be approaching the onset of widespread cracking, that is, the threshold of MSD as measured in calendar years, flight cycles, or flight hours.

- *Structural repairs*, which are more prevalent in the aging aircraft fleet, are generally made to regain static strength and may not adequately fulfill damage tolerance and fail-safe requirements.

- *Terminating actions*—the FAA language that denotes the structural actions necessary to eliminate MSD—do not have the database in terms of component and full-scale testing comparable with that on which the original structures were certified. Thus, neither the design life of the terminating actions nor the inspection intervals for continuing airworthiness can be established without further testing and analysis.

Fortunately, the action taken by the FAA in the wake of the Aloha Airlines accident has been noteworthy. In 1991, the U.S. Congress passed the Aging Aircraft Safety Act which required air carriers to demonstrate that maintenance of an airplane's age sensitive parts has been "adequate and timely enough to ensure the highest degree of safety". This legislation was codified by the FAA as the Aging Airplane Safety Rule (AASR) which requires airlines to ensure that repairs or modifications made to their airplanes are damage tolerant. On January 25, 2005, the FAA issued its final AASR regulation which requires repetitive inspections and record reviews every seven years for all transport airplanes greater than 14 years old. In November 2007, FAA issued Advisory Circular 120-93 to provide detailed guidance for "Damage Tolerance Inspections for Repairs and Alterations". Three years later, in November 2010, FAA issued its final rule which is considered a comprehensive solution to the problem of widespread fatigue on aging aircraft. This new rule seeks to prevent "widespread fatigue damage" (WFD) by requiring aircraft manufacturers to establish a specific number of flight cycles or hours an aircraft can operate and be free from WFD without additional inspections for fatigue. At least 4,000 U.S. registered aircraft are affected by this new rule, and if adopted worldwide, the number of affected airplanes could more than double.

CABIN SAFETY

Where is the safest place to sit in an airplane? The many people who posed that question in 1985 realized that surviving an airplane crash is often possible. But few of them also appreciated that survival and cabin safety depend on a great deal more than seating position, or that better emergency procedures and equipment, different cabin materials, and stronger seats can contribute to higher survival rates. In 1985, the NTSB addressed the issue of cabin safety on several fronts. It issued a study on emergency equipment and procedures relating to in-water air carrier crashes and found that equipment and procedures were either inadequate or designed for *ditchings* (emergency landings on water where there is time to prepare and that involve relatively little aircraft damage) rather than more common short (or no warning) in-water crashes. The NTSB recommended improvements in life preservers, passenger briefings, emergency-evacuation slides, flotation devices for infants, and crew postcrash survival training.

Later in the year, the NTSB completed a study on airline passenger safety briefings. The NTSB concluded that in the past, "The survival of passengers has been jeopardized" because they did not know enough about cabin safety and evacuation. The study also found wide variances and sometimes inaccuracies in oral briefings and in information on seatback-stored safety cards.

Effective evacuation of an aircraft also depends on flight attendants. In October 1985, the FAA issued a notice of proposed rulemaking to require protective breathing equipment for flightcrews and cabin attendants. The NTSB had recommended making this equipment available following its investigation of the fatal fire on the Air Canada DC-9 at Cincinnati International Airport in June 1983. The recommendation repeated one made a decade earlier after the NTSB participated in the investigation of a foreign accident involving a cabin fire. The NTSB believed that without protective breathing equipment, flight attendants could easily be incapacitated by fire and smoke, could not make effective use of fire extinguishers, and could be of no use during an evacuation. In 1985, the NTSB also supported a petition for rulemaking to set flight attendant flight-duty time limits.

In view of public attention to the issue of cabin safety, the FAA has taken a number of regulatory actions to increase the likelihood of passenger survivability in aviation accidents. Among these improvements are

- Fire blocking seat cushions to prevent fire or mitigate its effects. FAA Advisory Circular 120-80 dated 1/8/2004 provides detailed guidance on how to deal with in-flight fires, emphasizing the importance of crewmembers taking immediate and aggressive action.

- Emergency floor lighting to improve the chances and speed of evacuation by 20 percent under conditions where there is significant smoke in the cabin. Improved passenger safety briefings and cabin safety data cards have also made a positive impact.

- Stronger seats that must be able to withstand 16 times the force of gravity (16g) are required for Part 121 aircraft that were build after October 27, 2009. Using a test dummy, these seats must undergo dynamic testing and evaluation to ensure they are effective in the demanding aviation environment.

SAFETY DESIGN FOR ATMOSPHERIC CONDITIONS

Long-duration, high-altitude, and high-speed flight can be significantly impacted by many diverse atmospheric phenomena. Four of the more important areas of concern include

- Turbulence
- Wind shear
- Volcanic ash
- Ice and precipitation

The first line of defense should be to detect and avoid these hazards. When this strategy is impractical, solution must address safe operations and maneuvers in the event of an encounter. Great progress has been made toward both.

TURBULENCE

It became apparent as jet airplanes started operating at high altitudes for extended periods that the structure of atmospheric turbulence was different from that at the lower altitudes. One advance that helped was the velocity-load factor, in *G*'s, altitude (VGH) recorder. Tests showed the need for further research, one avenue being the *High Altitude Clear Air Turbulence* (*HICAT*) program. Data collected in this program markedly changed the design criteria of commercial and military jet airplanes.

Early in the jet era, there were a number of what were referred to as "jet upsets" from severe air turbulence encounters. Study of these encounters led to a better understanding of how to safely fly jet aircraft in such an encounter. Results indicated that it was important to not change trim, to disengage the early autopilot's autotrim feature and airspeed/Mach hold, if necessary, and to fly attitude and let the airspeed and altitude vary somewhat.

Between 1983 and 1997, the NTSB investigated 99 turbulence accidents and incidents that resulted in two fatalities and 117 serious injuries. Most of these injuries—many of which involved fractures of the spine, skull, and extremities—were completely preventable if the occupants had been restrained by seatbelts.

Since 1972, the Board has made several safety recommendations to the FAA and the National Weather Service (NWS) and has been involved in numerous investigations dealing with the hazards of turbulence. For example, in March 1993, the engine of a B-747 cargo flight separated from the airplane in severe turbulence conditions while departing from Anchorage, Alaska. As a result of this investigation, the Board recommended that the NWS develop in greater detail turbulence forecasts using data from the National Weather Service's Doppler weather radars (WSR-88D). The NWS has implemented this recommendation.

The Board investigated another severe turbulence encounter involving a United Airlines B-747 that occurred in the western Pacific on December 27, 1997. This encounter resulted in one fatality and many injuries to passengers and flight attendants. The Board's investigation focused on turbulence forecasting, flightcrew training, dissemination of information on turbulence, and crew procedures in areas where turbulence is forecast.

The NWS Aviation Weather Center and NOAA Earth System Research Laboratory are also making progress in improving its clear air turbulence and mountain wave forecast products. Further, the FAA Aviation Weather Research program has a multidisciplinary team addressing the turbulence problem, and researchers from the National Center for Atmospheric Research (NCAR) are working on new algorithms for using data from the NWS Doppler weather radars to detect turbulence. The NCAR researchers are also developing software that may be able to turn airborne commercial aircraft into a real-time turbulence-sensing platform. The software will use onboard sensors and computers to measure and analyze turbulence as the aircraft flies through it. The data will be transmitted to the NWS, where they will be used to create accurate, real-time turbulence maps.

WIND SHEAR

With the increased frequency of flights, wind shear–related takeoff and landing accidents increased. The first part of the investigation into wind shear involved isolating the wind profile that got the airplane in trouble. Improved onboard data recorders on test airplanes showed the profiles were more severe than previously known. Additionally, the downburst phenomenon associated with certain weather conditions was discovered (Fig. 9-5). A three-pronged approach was initiated. The first prong involved training crews on avoidance of the phenomena, the second related to better detecting the conditions that can produce wind shear and alerting the crew, and the third prong dealt with getting maximum performance from the airplane if the crew inadvertently encountered wind shear.

Good-fidelity simulators were used for this analysis. In evaluating means of helping the crew get the greatest performance from the airplane, it was possible to use the latest technologies that were being incorporated in advanced flight decks. A very successful task force composed of individuals from the FAA, airplane manufacturers, and airlines made a great contribution to safety by producing a wind shear training aid for flightcrew avoidance and procedures for getting maximum performance in the presence of wind shear. Additionally, algorithms were developed and displays were modified to provide guidance and situational awareness, which enhanced flightcrew performance.

Another winds-aloft hazard is mountain waves. The loss of a 707 near Mount Fuji was the result of such an occurrence. We still have a lot to learn about wind in the vicinity of mountains, and the supercomputer's contribution to modeling this phenomenon gives hope for further improving the safe operation of aircraft in this environment. Also needed is further research on sensors capable of detecting these adverse winds.

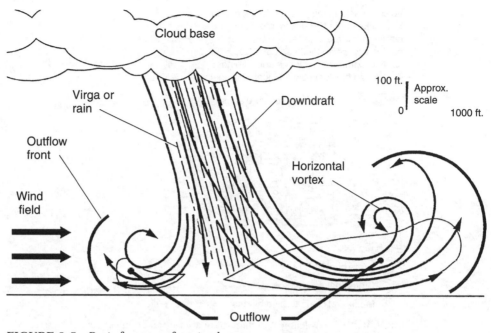

FIGURE 9-5 Basic features of a microburst.

VOLCANIC ASH

Volcanic activity has occurred throughout the world since the earliest of times. The first recorded impact on aviation was on March 22, 1944, when Mount Vesuvius did more damage to an airfield than any enemy activity. It was another 36 years before the next direct impact, and that was in 1980, when a civil Lockheed C-130 Hercules inadvertently penetrated a dense ash cloud from Mount St. Helens in western Washington following its second major eruption. The aircraft lost power on two of its four engines, but the crew was able to return to and land safely at McChord AFB. Since then, there have been seven major eruptions worldwide.

The most serious recent volcanic ash eruption was reported on April 14, 2010 when ICAO's authority in London issued an alert that a strong ash plume was moving from an eruption in Iceland towards northwestern Europe. This eruption resulted in the sudden closure of large areas of European airspace, around ten million passengers were affected, and total economic damage reached almost 5 billion dollars. The danger of engine flameout and severe damage to aircraft was significant for 2 weeks (See Fig. 9-6).

In the wake of the Icelandic volcanic crisis, ICAO is currently working with European aviation authorities to improve the forecasting of volcanic ash trajectory and dispersion. A mobile radar unit will soon be deployed in Iceland to improve the

Volcanic ash, small particles of glass and
tiny rocks could be sucked into jet engines,
jamming the mechanical systems. Jet
engines can also melt ash particles,
coating compressor blades and turbines.
This could block airflow and stop the engine.

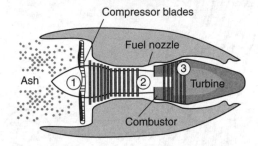

Effects of ash:

(1) Jams blades

(2) Clogs fuel and air flow

(3) Melts and coats turbine blades

FIGURE 9-6   Volcanic Ash damage to aircraft jet engines.
(*Source: USA Today www.usatoday.com*)

accuracy and speed of information on the height of any future ash cloud. An accurate assessment of the height of the initial ash plume is critical for predicting its subsequent movement and dispersion. A coordinated European approach to this problem is in process during 2011 under the "Single European Sky" initiative.

ICE AND PRECIPITATION

Ice and precipitation affect airplane operation in two areas. One is the effect they have on the airframe and engine, and the second is the effect on tire-ground contact. Unlike the previous atmospheric topics, these elements are daily occurrences, depending on the season and location. Safe operations depend on a coordinated team effort by airline maintenance crews, flightcrews, and airport authorities, as well as ground and air traffic control. The technologies that help are many and depend on specific icing or precipitation conditions. Again, procedures play an important role. Flight simulator training continues to be an effective tool in developing correct takeoff and landing procedures for ice-contaminated runways to keep safe operation possible.

Developing all these procedures involves extensive wind-tunnel, environmental laboratory, and flight tests to fully understand the problems and show performance of the airplane under these conditions. The contribution to safety in some of these cases is focused on understanding the limitations of the airplane under adverse weather conditions. Only then is safety improved when those responsible for each aspect of the operation implement the procedures that come from these tests.

In-flight icing is one of the FAA's top weather research priorities. Improved operationally available, high-resolution, accurate forecasts of atmospheric icing conditions are needed. Several safety recommendations dealing with in-flight icing were issued as the result of the NTSB's investigation of the ATR-72 accident that occurred at Roselawn, Ind., on October 31, 1994. These recommendations and the findings from the investigation provided the major impetus to the icing research efforts of the FAA.

Then again on January 9, 1997, Comair Flight 3272, an Embraer 120, crashed near Monroe, Mich., destroying the airplane and killing all 29 people on board. There were reports of moderate icing in the area at the time of the accident. In May of the same year, the Board issued four urgent safety recommendations to the FAA regarding icing. Almost concurrently, the FAA issued a notice of proposed rulemaking to modify operating procedures in icing conditions. Comair modified its operating procedures based on the FAA's proposed rule, and the FAA issued a final rule. The Board also worked with the National Center for Atmospheric Research and the NASA Lewis Research Center regarding weather issues, and in January 1998, Board personnel traveled to Brazil, where the airplane is manufactured, to review all pertinent test data on icing and to perform studies in the engineering simulator.

Issues examined regarding this accident included flightcrew training, operations in icing conditions, and aircraft performance. This accident resulted in several significant research activities in icing. These activities include the quantification of the performance loss due to small amounts of surface roughness on a wing's leading edge and the performance penalties associated with pneumatic deice boot intercycle ice.

In recent years, the NTSB and FAA have become more definitive in the area of ice detection and prevention. On August 3, 2009, the FAA published a final rule, applying to new transport aircraft designs, that became effective September 2, 2009. Although the new rule does not address existing airplane designs, the FAA is considering similar rulemaking that would apply to those designs. Under the revised certification standards, new transport aircraft designs must incorporate one of three methods to detect icing and to activate the airframe ice protection system:

- An ice detection system that automatically activates or alerts pilots to turn on the ice protection system;
- A definition of visual signs of ice buildup on a specified surface (e.g., wings) combined with an advisory system to alert the pilots to activate the ice protection system; or
- The identification of temperature and moisture conditions conducive to airframe icing that would alert pilots to activate the ice protection system.

The new FAA standards further require that, after initial activation, the ice protection system must operate continuously, automatically turn on and off, or alert the pilots when the system should be cycled.

FLIGHT DECK HUMAN–MACHINE INTERFACE

The technology topics discussed to this point have dealt primarily with the design aspects of the airplane as a machine and its capability to operate safely in the atmospheric environment. This section takes a look at the technology that involves the human–machine interface. It involves the flightcrew and one or more other parties: maintenance, ground operations, weather advisers, air and ground traffic control, and others. It all comes together on the flight deck. The human–machine interface often becomes the determining factor in the event of an emergency, where correct, timely decisions and execution make the difference between life and death. As discussed earlier in this text, human factors is the science that deals with the human–machine interface in an attempt to maximize the potential for safe, efficient operation while eliminating hazardous conditions resulting from human error. This technology was incorporated only into certain details of the early jet airplanes. As the state of knowledge of airplane design and human-factor research and understanding has advanced, so has the jet transport flight deck improved.

Technology in flight decks has improved continuously since the early days of aviation. Notable advancements are radio communication, radio and inertial navigation, and approach systems. The jet engine greatly simplified cockpit controls and displays. The following are some of the flight-deck technology changes that have made a significant contribution to improving safety:

- Crew alerting and monitoring systems
- Simple system designs
- Redundant systems
- Automated systems (when essential)
- Moving Map display
- Engine-indicating and crew-alerting system (EICAS)
- Glass cockpit Displays with color enhancement
- Ground-proximity warning system (GPWS)
- Traffic Collision Avoidance System (TCAS)
- Aircraft Communications Addressing and Reporting System (ACARS)
- Flight management system (FMS)

EARLY COCKPIT DEVELOPMENT

The 707, 727, DC-8, and early 747, DC-10, and L-1011 flight decks used a standard arrangement of pilot, copilot, and flight engineer. Much like the previous aircraft, the airplane systems were designed for the flight engineer to be the systems operator. Large instrument panels were mounted behind and to the right of the two pilots. The flight engineer was expected to monitor and operate the vital hydraulic, electrical, fuel, air conditioning, and pressurization systems unsupervised. The design of the short-range 737 was changed radically to provide a flight deck to be operated by a two-person flightcrew.

The airplane systems were first simplified. The fuel system, for instance, has but three tanks: a right wing tank for the right engine, a left wing tank for the left engine, and a center tank to be used by both engines. The fuel boost pump capacity, line sizes, and fuel head were selected to permit fuel from the center tank to be used first, followed by wing fuel, without any crew action following the prestart checklist. When the center fuel was depleted, amber lights annunciated to the crew that the center tank pumps could be turned off. A crossfeed system was provided for nonnormal operations. The simplified system had other benefits besides crew workload reduction.

Multiple sources of power or supply were provided on all systems to have adequate system function when one or more elements failed. Multiple hydraulic pumps were provided, driven in different ways to provide a completely redundant system.

The control system on the Boeing 727 uses cables from the control column to hydraulic actuators for normal operations. However, if the two hydraulic systems fail, a system of cables to the flight controls provides a *manual reversion* method of controlling the aircraft. The 757 is a newer-generation aircraft and takes a different approach to redundancy. It has three hydraulic systems, engine and electrical pump, and a ram air turbine as a backup pump. This replaced the direct cables-to-controls connection. The 747 is similar, with four hydraulic systems and no direct cable-to-control connection. At least one hydraulic system and pump must be operating to move the flight controls. The Airbus A-320 also requires at least one hydraulic system and pump to move flight controls. The A-320's redundancies include three hydraulic systems, two engine-driven pumps, two electric pumps, a power transfer unit (uses one hydraulic system to pressurize another), and a ram air turbine in case ac electric power is lost.

Automatic operation of a system was provided for certain selected equipment failure cases to avoid the necessity of crew intervention at a critical time in flight. The 737 electrical system load-shedding feature is an example. In the event of a single generator failure on the two-generator nonparalleled electrical system, the remaining generator picks up the essential load and nonessential loads are shed. The galley power and other similar loads are shed so that the remaining generator can provide all essential loads without the need for crew attention or intervention. Later, when the crew has time to restore a second generator, the shed loads can be recalled. This same design philosophy has prevailed through the other two-crew designs on the 757, 767, and 747-400 aircraft.

FLIGHT DECK—757/767 AND B-747-400

Flight-deck noise levels are low enough to allow a true "headsets off" environment. While the forward windshields are flat for best optical characteristics, the side windows are curved to prevent turbulent airflow and reduce the associated aerodynamic noise. Wind-tunnel studies showed that the aerodynamic vortex created by the sharp angular change between the flat forward and flat number-two windows contributed to cockpit noise levels at cruise airspeeds. This source of noise has been eliminated in the 757/767 flight deck. The air-conditioning system is designed to further reduce flight-deck noise levels by means of ducting improvements and lower airflow velocities.

Vision characteristics are excellent inside and out the 757/767. Vision through the windshields exceeds Society of Automotive Engineers (SAE) recommendations, resulting in superior collision avoidance capabilities. An extra margin of safety on landing in adverse weather conditions is achieved by improving downward visibility and maintaining a low deck angle on approach.

In this "quiet dark" flight deck, few green or blue lights indicating normal system operation are used. Lighted pushbutton switches that combine the amber malfunction light with the shutoff switch are used to reduce the possibility of incorrect crew action.

The *integrated display system* (IDS) in the B-747-400 consists of six 8-inch square screens. Although all are identical, each performs a different function depending on its location:

- Primary flight display (PFD)
- Navigation display (ND)
- Engine indication and crew-alerting system

The two outboard CRTs, directly in front of the pilots, function as the captain's and first officer's PFDs. Each pilot can use the PFD as the single source for all the primary flight instruments found on a traditional instrument panel. The tape formats used for altitude, airspeed, and heading/track indications were chosen for two reasons: First, they permit sufficient display resolution without disrupting the "basic T" instrument configuration; second, the tape formats more readily accommodate related supplemental information. This information, such as speed bugs and a trend vector, increases pilot situational awareness.

NEW COCKPIT ENHANCEMENTS

The Boeing company has also improved cockpit human machine interface in the B-777 and its new B-787 "Dreamliner" aircraft. The layout of the 777 flight deck is similar to the B-747-400 with the following enhancements:

- Glass cockpit, three axis digital fly-by-wire flight control system with a convention control yoke rather than side-stick controller.

- Flight, navigational and engine information is presented on six large display screens with advanced liquid crystal display (LCD) technology.

- An integrated Airplane Information Management System which provides flight crews information regarding the overall condition of the aircraft, maintenance requirements and key operating functions.

- Ground maneuver camera system with video views of the nose and main landing gear to assist the pilot with ground handling when at the gate area.

The B-787 "Dreamliner" retains operational similarity with the B-777 including fly-by-wire with a control yoke, providing these further enhancements:

- The display screens are twice as large as the 777, providing additional situational awareness to the flight crew. An avionics full-duplex switched Ethernet is used to transmit data between the flight deck and aircraft systems.

- New dual head-up displays (HUDs) which allow the pilots to view critical flight data on a transparent display looking forward through the windscreen instead of down into the cockpit.

- New dual electronic flight bags (EFBs) which are the digital equivalent of the bulky pilot's flight bag. This includes all maps, charts, manuals, and other data the for "paperless cockpit."

- New electrical architecture which replaces bleed air and hydraulic power sources with electrically powered compressors and pumps. This improvement saves weight and enhances efficiency.

Competition between Boeing and Airbus remains high for new aircraft orders. The A380 Flight deck uses similar cockpit layout to other Airbus aircraft with the following enhancements:

- Improved glass cockpit and fly-by-wire flight controls linked to side-sticks as in the past, with improved LCD cockpit displays.

- Two primary flight displays (PFDs) dedicated to critical flight information, and two multi-function displays (MFDs) which display navigation route, moving map, weather, and other information. These MFDs are new to the A-380, providing an easy to use interface with the flight management system.

- Similar to the 787, the A-380 uses an integrated modular avionics system with a full duplex switched Ethernet to transfer data to the aircraft systems.

- The electronic flight bag paperless cockpit is called a network systems server which stores data such as equipment lists, navigation charts, performance calculations, and the aircraft logbook.

The Airbus A-350 is the latest European aircraft following in the wake of the super jumbo A-380 development. Its cockpit will contain the following enhancements:

- Large LCD screens with two central displays, a single primary flight and navigation display with an on-board information screen, plus a head-up display on the windscreen.
- The A-350 will improve the A-380 integrated modular avionics system and will manage many additional critical aircraft functions such as the landing gear, fuel, brakes, pneumatics, cabin pressurization, and fire detections systems.

CREW ALERTING SYSTEMS

The 737 introduced a very simple but elegant and effective crew monitor and alerting system. Almost all the airplane's system controls are located overhead and are outside the normal line of sight of the two crewmembers. When all systems are "on" and operating, no caution lights are observed. In the event of loss of equipment, an amber light annunciates the condition on the overhead panel. The loss is also repeated on master caution lights on the glare shield in front of each crewmember and in small caution panels also on the glare shield. The panel annunciation identifies the system affected and the location of the system controls overhead.

Since this was a new scheme at the time of certification, it got a lot of attention by the certification authorities and designers. To make sure that airplane system operation was not an inordinate burden or substantial workload for the flightcrew, many hours of flight in the simulator and during certification flight testing were devoted to measuring the time spent on airplane subsystem operation. Figure 9-7 shows the typical time a crew spends on various functions. Less than 1 percent of the crew's time from takeoff to landing is spent on system operation, including deliberate equipment malfunction simulation by the certifying agency. This careful attention to system detail design and improvements in monitoring capability has been used for virtually all jet models subsequent to the 737.

GROUND-PROXIMITY WARNING SYSTEM.   Since the advent of powered flight, inadvertent ground or water contact has been a worldwide problem. While much early effort went into avoiding such accidents, no major advance occurred until introduction of the *ground-proximity warning system* (*GPWS*) in the early 1970s. Although there has been a marked reduction in controlled flight into terrain (CFIT) accidents since then, they still occur with distressing frequency and account for close to 75% of worldwide fatalities on commercial transports. For the air carrier fleet, the primary reasons are twofold: Either the airplane did not have GPWS (or it was inoperative), or the crew ignored the GPWS alert. The Flight Safety Foundation (FSF), along with others, has an aggressive program to essentially eliminate CFIT accidents.

The enhanced ground proximity warning system is an advanced terrain warning system used in most modern commercial airline fleets. The EGPWS improves situational awareness and increases warning times using a terrain database with a "look ahead" feature.

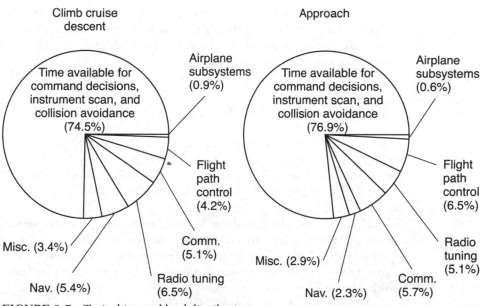

FIGURE 9-7 Typical jet workload distribution.

TRAFFIC COLLISION AVOIDANCE SYSTEM (TCAS). This is an electronic aircraft collision warning system to help prevent dangerous midair collision accidents, which generally involve a great loss of life. Typically, TCAS requires mutual "Mode S" transponder technology which automatically interrogates other aircraft independent of air traffic control systems to display the relative position of other such transponder-equipped aircraft. Upon detecting an intruder aircraft, TCAS issues a traffic advisory or resolution advisory to the flight crew to climb or descend as necessary to avoid the other aircraft. The latest version of the system is TCAS II, Version 7.1, which has been endorsed by ICAO and several leading regulatory agencies. Implementation of this improved TCAS version has been delayed and it is now scheduled for adoption by January 1, 2014 (for forward fit) and January 1, 2017 for retrofitting existing commercial aviation aircraft.

ENGINE-INDICATING AND CREW-ALERTING SYSTEM. The *engine-indicating and crew-alerting system* (EICAS, or ECAM for Airbus aircraft) is a digital computer system which monitors and indicates propulsion and airplane subsystem information for the operation and maintenance of the airplane. EICAS/ECAM interfaces with many airplane components and subsystems. Discrete inputs are implemented in a hierarchy that reflects operational and maintenance requirements:

- Flightcrew alert messages duplicate dedicated subsystem information (usually indicator lights) elsewhere in the flight deck.

- Status and maintenance messages provide lower-priority information on the condition of many subsystem components.

New Airbus and Boeing aircraft models have improved EICAS/ECAM to enhance functionality and reduce display footprint in the cockpit. Each manufacturer has its own strategies for engine indications and crew alerts, but all perform essentially the same basic functions.

AIRCRAFT COMMUNICATIONS ADDRESSING AND REPORTING SYSTEM

The *Aircraft Communications Addressing and Reporting System* (*ACARS*) is a communication datalink system that sends messages, using digital technology, between an airplane and the airline ground base. The operational features of ACARS equipment and the ways in which ACARS is used in service vary widely from airline to airline.

There is nothing new about sending messages between the airplane and the ground. What makes ACARS unique is that messages can be sent, including fuel quantity, subsystem faults, and air traffic clearances, in a fraction of the time it takes using voice communications, in many cases without involving the flightcrew.

ACARS relieves the crew of having to send many of the routine voice radio messages by downlinking preformatted messages at specific times in the flight. These may include the time the airplane left the gate, liftoff time, touchdown time, and time of arrival at the gate. In addition, ACARS can be asked by the airline ground operations base to collect data from airplane systems and downlink the requested information to the ground.

Each ACARS message is compressed and takes about 1 second of air time to transmit. Because of the automatic reporting functions described above, the number of radio frequency changes that flightcrews must make is reduced on ACARS-equipped airplanes. Sending and receiving data over the ACARS network reduces the number of voice contacts required on any one flight, thereby reducing communication workload and costs.

The accurate reporting of event times, engine information, crew identification, and passenger requirements provides for a close control of any particular flight. Airplane system data, such as engine performance reports, can be sent to the ground on a preprogrammed schedule, or personnel on the ground may request data at any time during the flight. This allows ground personnel to observe the engines and systems and can alert them to problems to be investigated.

Improvements in ACARS technology has been brought about by a system known as VHF Data Link (VDL). VDL Mode 2 has been implemented in nearly 2,000 aircraft to transmit ACARS messages and date communications from air traffic controllers known as Controller Pilot Data Link Communications (CPDLC). Voice communications are easily misunderstood, and voice frequencies are easily overloaded. CPDLC was proposed by FAA as a new strategy to cope with the increased demands on air traffic control and has proven very effective and efficient during operations in busy en route airspace.

Flight Management System

The *flight management system* (*FMS*) is an integration of four major systems: the flight management computer system (FMCS), the digital flight control system (DFCS), the autothrottle (A/T), and the inertial reference system (IRS). The basic functions of the FMS are

- Automatic flight control
- Performance management
- Precision navigation
- System monitor

The FMS is designed to allow crew access to the total range of its performance, navigation, and advisory data computation capability at any time and in any flight control mode. For example, when the airplane is under manual control, the pilots, at their option, can get flight optimization data from the flight management computer and appropriate "bugs."

The flight management computer is a major innovation in the FMS design. In addition to navigation, it performs real-time, fully automatic performance optimization and can control the airplane through the flight control system, including the auto throttle. While crews currently determine the most efficient speed and altitude by using the flight operations manuals and calculators, this function is usually performed only periodically. Further, because it is a digital system, the software and programmable databases in the FMS provide the adaptability and growth needed for present and future airline operations, particularly as navigation and air traffic control systems evolve.

Multiple Flight Control Computers

On the newer-generation aircraft, *elevator/aileron computers* (*ELACs*) have primary control of elevators and ailerons and perform the computations for spoilers and yaw damping. One ELAC is active while the other is a backup.

Spoiler/elevator computers (*SECs*) provide primary control of spoilers. If both ELACs fail, the SECs also provide backup control of roll and pitch via spoilers and elevator.

Flight augmentation computers (*FACs*) control rudder for turn coordination and yaw damping. They also compute limits for the flight envelope, wind shear, and speed information displayed on the primary flight display.

Multiple redundant computer flight control systems ensure that no single failure of an electrical, hydraulic, or flight control component will result in a reduction in operational capability.

CENTRAL MAINTENANCE COMPUTER SYSTEM

One technology that is able to assist in maintaining complex systems is the *central maintenance computer system* (CMCS). This system can collect, display, and provide reports of system fault information and test airplane systems. It provides ground maintenance personnel with a centralized location for both testing and access to maintenance data. The net result is a decrease in airplane turnaround time and an increase in dispatch reliability over previous-generation airplanes.

The CMCS was developed to monitor and troubleshoot the complex and integrated systems on the 747-400. It was also created to centralize ground testing, fault information storage, and real-time data monitoring. In previous-generation airplanes, testing, troubleshooting, and fault isolation were confined to a system-by-system approach, and faults that affect multiple systems were often difficult to isolate. Because so many systems on the 747-400 are interdependent, a fault in a single system can ripple through several other systems. Since the CMCS is linked to all major systems, it can provide simultaneous fault monitoring for related systems and can correlate multiple fault indications to a single fault.

Another major problem with previous-generation aircraft was the lack of a common test initiation procedure. Initiated tests went unused when they could have helped to isolate faults. The 747-400 CMCS provides a common design for testing interfacing systems.

The latest Airbus system for real-time aircraft maintenance is known as "AIRMAN" for AIRcraft Maintenance ANalysys, which is an intelligent software system designed to optimize maintenance and minimize aircraft turnaround time on the ground.

FLIGHT PROCEDURES—TAKEOFF, APPROACH, AND LANDING

Procedures to ensure the flightcrew gets the best performance out of the airplane under all conditions are very important in realizing safety. All forms of analysis and simulation, from procedure trainers to all-up, full-function flight simulators, are used to ensure updated procedures. All conditions, functions, and routine and nonroutine takeoff and landing scenarios can be safely tried and perfected.

TAKEOFF

Significant amounts of analysis and study have gone into understanding the physics involved in takeoff, rejected takeoff (RTO), and climbout. On modern jets, takeoff techniques involve accurately setting power, using proper rotation rates at V_r (rotation speed), and achieving and holding target climbout speed and attitude. Correct rotation techniques and procedures are highly stressed in pilot training and are essential to achieving consistent results.

Another problem encountered in early jets, which has improved but still needs further attention, is RTO overruns. Overruns may occur when a decision to stop is made at speeds above V_1 (critical engine failure recognition speed). If the aircraft is near its field-length performance limit, initiation of the first stopping action must occur at or before V_1, and the maximum stopping procedure must be followed to prevent overrun. This procedure involves immediately retarding the throttles and applying full brakes, followed by actuating spoilers. Analysis of 67 rejected-takeoff accidents found that two-thirds of the no-go decisions were made above V_1. Continuing the takeoff from these high speeds would have resulted in successful takeoff and climbout.

A significant amount of analysis and study has taken place over the years to more clearly understand and define takeoff decision speed and its correct use. Information has been developed and distributed to airlines and pilots through meetings and seminars, numerous written articles, and pilot training at manufacturers' schools.

Numerous improvements in aircraft and systems design have led to significantly improving safety during takeoff. The additions of autospoilers and autobrakes discussed earlier provide backup and assurance of achieving full braking performance in an RTO. Brakes and brake antiskid system design have greatly improved stopping capability on wet and dry runway surfaces. On many later models, noncritical warnings and sounds are inhibited during the takeoff roll and initial climb, allowing the crew to focus on the takeoff and reducing the possibility of unnecessary rejected takeoff aborts.

APPROACH AND LANDING

A review of worldwide jet transport accidents indicates that the approach and landing phase of the flight accounts for approximately 50 percent of all flight accidents. Most of these accidents involve failure of the flightcrew to stabilize the aircraft in speed and rate of descent soon enough before touchdown (at least by 500-feet altitude). This lack of stability takes the form of

- Excessive approach height over the threshold
- Floating down the runway
- Incorrect stopping techniques

Emphasis has been placed by the manufacturers on "flying by the numbers" in training aid publications. These numbers include approach speeds, procedures, approach path control, touchdown point, and flying technique. Aircraft design strives to build in approach-speed stability in the lift, drag, and speed relationships of the airplane. Improvements in engine design and engine acceleration characteristics provide better flight path control during descent and allow faster engine spin-up for go-around.

Many new and improved systems are now available at airports and on aircraft for glide slope control to the threshold and provide a consistent touchdown point

1000 to 1500 feet down the runway. Newer aircraft are fitted with autothrottles, automatic landing systems, autospoilers, and autobrakes that allow the aircraft to get on the runway quickly and attain full braking. Again, manufacturer training and education emphasizes consistency in approach and landing procedures to minimize the risk of an accident. For example, the A-350 will have a new overrun warning and protection system to alert the crew to specific aircraft landing performance under differing runway conditions.

MODELING, DESIGN, AND TESTING TOOLS

Airplane design and technological progress seen in this century could not have happened without parallel advancements in computing capability. This was no coincidence, for in many cases it was trying to solve airplane technology problems that gave stimulus for "there must be a better way." Two breakthroughs for the computer (and all avionics for that matter) were the vacuum tube and the transistor. The triode vacuum tube was invented by Lee De Forest in 1906, shortly after the Wright brothers' aeronautical success. It allowed development of an electronic computer.

The computer advancements made to date are what enabled not only all the onboard and support avionics systems but also all the analysis and testing that go into them. It is not possible to cover all this technological progress, as it has been explosive on all fronts. The important part of incorporating any technology is to first make sure it has been analyzed and tested thoroughly. Some of the technologies that have made contributions to safety are discussed next.

COMPUTATIONAL FLUID DYNAMICS (CFD)

Computational fluid dynamics methods have been used extensively in the design of all new-generation aircraft. Advances in supercomputing technology over the years have allowed CFD to affect problems of greater relevance to aircraft design. Use of these methods allowed a more thorough aerodynamic design earlier in the development process, allowing greater concentration on related operational and safety-related features.

WING DESIGN.   The ability of CFD to do the inverse problem (i.e., given the desired resulting flow, what is the shape of the surface geometry?) has revolutionized the transport wing design process. Since the Wright brothers, wing design has been a "cut and try" operation, with the "try" taking place in the wind tunnel or in flight. The advent of sufficiently powerful computational methods allowed some of the try to be shifted to the computer. But the "cut" was still the designer shaping the wing based on experience and intuition. CFD allows a new approach in which the design engineer specifies to the computer the aerodynamic pressures desired on the wing, and the CFD code computes the geometric contouring of the wing surface that will produce those pressures. The engineer does all the design work and initial evaluation with CFD and then picks the best candidates, builds wind tunnel models, and tests them.

This inverse design technique was first used in the development of the B-777 which set a new standard for all future aircraft to follow, allowing achievement of a level of aerodynamic performance not otherwise possible in a time-constrained development program. The CFD methods used for this application were based on nonlinear potential equations coupled with boundary-layer equations to account for viscous effects. Key to this use of CFD was the ability to model enough relevant physics and gain quick turnaround from initial geometry to completed solution. The resulting wing designs are thicker and lighter than their previous counterparts. These characteristics allow wings with greater span or less sweepback. Both features are conducive to better low-speed (landing and takeoff) performance, improved safety margins, and reduced community noise.

CAB DESIGN OF THE B-757. The 757, as originally conceived, was to share the cab of the 727. Only very late in aerodynamic development, when delivery dates were already being quoted to potential customers, was the decision made to develop a new cab. A new cab was desired for two separate but compelling reasons.

Improved downward vision from the 757 flight deck was desired to improve safety over that provided by the 727 configuration. The safety improvement comes from the additional rows of approach lights that can be seen during low-visibility landing. Additionally, it was desired to make the flight decks of the 757 (a standard body) and the 767 (a wide body) essentially identical, with all instruments and controls in the same place, which would make it feasible to obtain a common-type certificate that would allow a given crew to be certified to fly either type of aircraft. The 757 cab design had to accommodate the wide overhead panels and flight-deck instruments that had been designed earlier for the 767. The essential design challenge was to wrap a narrow low-drag aerodynamic shape about these wide components while lowering the number-one window frame to provide improved downward vision. A greater challenge was to produce this design in a very short time. Using computational methods, the 757 cab was carefully designed and tailored with complete success. The resulting elimination of areas of local supersonic flow over the cab at cruise conditions reduced both aerodynamic drag and noise compared to previous designs. Subsequent design of smaller cabs has resulted in more fuel efficient aircraft.

The 737 series of aircraft might not have remained in production beyond the mid-1980s if not for the installation of a modern, large-diameter, high-bypass-ratio turbofan engine. Installation had to be accomplished without incurring an excessive increase in interference drag, weight, or cost. The need, the opportunity, and the technology were available to provide a solution to this challenge. The resultant solution, which allowed a much larger-diameter engine to fit under the wing without increasing the main landing gear length, was made possible by using CFD techniques.

While the linear methods that were used for the above applications can simulate flows about complex configurations, they cannot accurately account for the nonlinear transonic flows that are more prevalent in installation of newer, more efficient, larger-diameter engines close to an aircraft's wing. During the aerodynamic design of a new airplane, risk of significant interference drag due to the engine exhaust was revealed through CFD analysis. Neither the earlier linear-based CFD methods nor conventional wind-tunnel testing techniques, which did not simulate the exhaust, would have detected this problem. Only a very expensive powered-nacelle testing technique would have assessed these interference effects.

Had this problem gone undetected until late in the aircraft's development when powered testing is usually done, any fixes would have been extremely expensive and time-consuming to implement. Again, the need, the opportunity, and the technology were available to provide a solution to this challenge.

Key to success of this application was the ability to model enough relevant physics and to provide solutions quickly enough to support the development schedule. CFD provided the ability to detect problems early in the development process, when fixing or avoiding them was least expensive.

Today, CFD is being used to better address the diverse requirements for nacelle design for cruise, engine-out second-segment climb, and engine-out ETOPS (extended range twin-engine operations). Nacelle design is being optimized to minimize drag at cruise conditions and provide acceptable engine-out drag for second-segment climb and ETOPS. Engine-out drag is not the only criterion of merit. If external flow separation occurs, the separation wake could impinge on other aircraft components, causing premature buffeting of the aircraft, perhaps indicating to the pilot that the aircraft is approaching stall conditions. If it is severe enough, the pilot might be inclined to reduce the angle of attack—a situation not desired during takeoff or climbout. CFD has become a very valuable tool for the designer to provide additional safety benefits that might not otherwise have been feasible.

WIND TUNNEL

Today, *wind-tunnel testing* complements the computational fluid dynamics tools, which focus efforts to a narrowed field of configuration options. During development of the 767 in the late 1970s, Boeing conducted tests in 15 tunnels in four countries for more than 20,000 hours, conducting 100,000 data runs, each consisting of recording an average of 65 independent parameters at approximately 25 different aircraft attitudes. In that 767 program, more than 100 variations of the wing design and untold numbers of different flap and slat positions were tested. Despite intensive use of CFD, the test program for the 777 involved about 20,000 hours of wind-tunnel testing up to the first flight.

The value of wind-tunnel testing goes beyond optimization of wing designs and performance. Safety involving loads determination for structural design, simulator database development, stability and control characteristics for both normal and failed configurations, and aircraft anti-icing control greatly benefits from the wind-tunnel-testing set of tools. The effective and efficient integration of propulsion systems onto the airframe is served, in part, by powered models that provide simulations of installed engines' flow fields.

FLIGHT SIMULATION

The *flight simulator* was recognized very early as a valuable tool in improving aircraft safety. It owes a lot to the pioneering efforts of Ed Link. Early in the initial transition of crews from props to jets, a number of training accidents occurred, particularly during engine-out training. It endangered not only those in the airplane but also those on the ground. The technology that allowed this hazardous training to occur in a simulated environment on the ground was a great benefit to flight safety.

Development of the simulator took two parallel paths, one for engineering evaluation and the other for flight training. Each benefited the other. The engineering simulation allowed the engineer and the test pilot to evaluate design options during design and to define the best operating procedures. It has also been used to evaluate accident and incident data to better understand what might have gone wrong and to find solutions.

At first, simulators were very crude. The engineering simulators were analog and could fly the initialized flight condition on instruments only. Later, a model board was constructed to represent a bird's-eye view of a large ground area using a miniature TV camera and closed-circuit TV that "flew" over the landscape model to provide an accurate view out the cockpit window. This feature greatly improved the evaluations that one could perform, but it had limitations. *Computer-generated imagery (CGI)* was the real breakthrough. It allowed weather and a variety of airports to be evaluated. This technology continues to advance and is providing more and more realism with full flight simulators featuring three axis motion platforms and 360 degree outside world visual projections.

STRUCTURAL TESTS

Structural static tests are paramount to verifying the aircraft design's structural margin of safety prior to in-flight demonstration. Static tests to 150 percent of limit load were to show that the design margin was actually achieved. These tests were in effect before the jet transport era. The B-47 experience showed these structural static tests alone did not ensure full structural integrity. It was then that flight fatigue loads were required to be tested. This requirement produced a much more complicated and involved test but ensured a much safer product. These wing cyclic fatigue tests are performed on all commercial aircraft.

FLIGHT TEST

The ultimate test of an airplane before it enters service comes when it goes through flight testing and is subjected to the real world environment. Flight testing has had to advance in four areas to keep pace with the jet airplane's advanced performance, expanded flight envelope, and the advanced technology incorporated into the airplane systems, such as autoland. These four areas include

1. New flight-test techniques
2. Improved accuracy of instrumentation
3. Improved data-recording systems
4. Real-time monitoring and data analysis

These advances have made it possible to conduct increasingly accurate and thorough tests to ensure that the airplane is ready for service. Typical of such tests is the takeoff minimum unstick speed V_{mu} which is the minimum airspeed at which an airplane can safely lift off the ground and continue takeoff. This test demonstrates the takeoff performance margin existing at unstick (takeoff), ensuring that the airplane will not encounter a hazardous condition such as loss of control or failure to accelerate. This is one of the key tests to ensure that engine thrust is maintained and the control and flap systems function as designed and provide the airplane with the low-speed, high-angle-of-attack controllability needed for safe operation. This test is also used to establish required operational parameters for takeoff field length. Flight tests such as this one have greatly contributed to the safety of the commercial airplane.

ACCIDENT/INCIDENT INVESTIGATION

The industry has always devoted a lot of effort to the technologies involved in finding the causes of incidents and accidents. The *flight data recorder* (FDR) and *cockpit voice recorder* (CVR) are indispensable tools in this task.

Substantial progress was made during 1987 in the NTSB's goal of persuading the FAA to require more and better cockpit voice recorders and flight data recorders aboard aircraft. Efforts by the NTSB to upgrade requirements for the use of FDRs and CVRs in keeping with the times have faced a long, uphill battle. Over the years, the NTSB had difficulty persuading the FAA to improve recorders on larger commercial jets and expand their use on commuter and some corporate aircraft, despite the proven value of recorders as a vital tool in accident investigations.

The FAA had issued a notice of proposed rulemaking to upgrade recorder requirements but failed to act on it for more than 2 years, despite NTSB prodding. On March 25, 1987, however, the Secretary of Transportation announced a requirement for installation of CVRs on newly manufactured turbopropeller commuter aircraft carrying six passengers or more and having two pilots. The secretary also called for installation of digital flight data recorders on older jet aircraft, replacing the foil-type versions currently required. The announcement came after a commuter plane accident at Detroit Metro Airport on March 4, which dramatically helped focus congressional and public attention on the stalled flight recorder rulemaking. Just two weeks after the accident, a congressional panel approved report language directing the FAA to "take immediate steps to require the installation of CVR and FDR devices on all commuter aircraft in line with NTSB recommendations."

The number of FDR data channels and the crashworthiness of these vitally important records have increased significantly. The early generation of FDRs only recorded time, altitude, airspeed, vertical acceleration, heading, and radio transmission event marker. The number of parameters has increased so that today, with all-digital avionics systems onboard the new airplanes, a more detailed record can be made and stored for a longer time. This area is another way in which digital avionics contributes to aircraft safety.

The Boeing 787 Dreamliner will be the first aircraft equipped with enhanced airborne flight recorders (EAFR); a combined cockpit voice and flight data recorder (CVR/FDR) with crash protected memory and the capability to record datalink messages and cockpit imagery. In 2002, the NTSB added crash protected image recorders to its Aviation Most Wanted Safety Improvement List of items needing urgent attention by the FAA; therefore, development of cockpit image recorders will be a future issue in the area of accident investigation.

FUTURE AIRCRAFT TECHNOLOGIES

The technological opportunities for continued improvement in flight safety are as good as ever. The analytical tools, computing capability, simulation, and testing capability all far exceed those of the past. And new technologies are continuing to unfold. These opportunities are challenging to all in the aviation community.

Of all the technologies available for improving safety while providing substantial gains for future growth, system capacity, and efficiency, electronics technology appears to have the most immediate promise. Accident statistics continue to bear out the fact that the manufacturers need to assist flightcrews in doing their tasks while improving their capacity. Chip-level redundancy combined with fault-tolerant design will provide far greater reliability levels for essential and critical functions. Airplane safety, capability, and total system capacity and efficiency will be enhanced.

Boeing and Airbus are using a number of these features on their next generation aircraft.

WEATHER DETECTION

Weather remains a major cause of delays and accidents. Real-time, accurate knowledge of the weather ahead is required. Aircraft must be able to detect weather conditions, alert the flightcrew to avoid dangerous situations, and effectively cope with turbulence, precipitation, and wind shear.

Today's airborne weather radar needs improvement. In the current air carrier fleet, we are limited to a simple view, and we need a real time "big picture" integrated display. The future will likely bring advanced airborne weather radar with vertical as well as plan cross-sectional views. Ground and satellite weather detection improvements made available in the flight deck will combine to enhance flight planning, performance, and safety.

Onboard system enhancements to aid the pilot in unexpected wind shear encounters are being incorporated on current aircraft. These wind shear alert and guidance systems are being combined with crew training and improved ground-based alerting and weather radar systems. Emphasis is on detection and avoidance of hazards occurring along the flight path. If wind shear is unavoidable, the wind shear system quickly alerts the crew and provides pitch attitude guidance. Improvements are also planned to the current ground-based, low-level wind shear alert system. Terminal Doppler weather radars will be installed at critical airports. More information on next generation weather automation systems is included in Chapter 8 on air traffic system technologies.

COMMUNICATION AND NAVIGATION SYSTEMS

As discussed at length in Chapter 8 on air traffic system technologies, aviation is transitioning toward a new communications, navigation, and surveillance/air traffic management (CNS/ATM) environment. This will inevitably change the way pilots communicate and navigate. For starters, most ATC messages would be sent by datalink to and from the airplane rather than by voice; navigation would be based more on satellites than on land-based navigation aids; and traffic surveillance would be achieved by networking tactical information among all aircraft. The benefits of these changes, particularly to the airlines, are potentially enormous. A global navigation and communications system will allow for higher traffic density and improved safety, while at the same time enabling airplanes to fly at optimum altitudes and on more direct routes.

As envisioned, the transition from today's airspace environment, in which routes are rigidly assigned by controllers on the ground, will move first to a system in which flight plans are originated in the airplane and approved by controllers on the ground, a NextGen system, where flight plan changes will be initiated in the airplane in real time and simply monitored from the ground.

But this does not mean that the conventional terrestrial navigation systems will disappear overnight. In the United States, the FAA plans to keep very high frequency omnidirectional range (VORs) and distance-measuring equipment (DME) in operation for at least another decade. During this time, pilots' dependency on them will decrease as more flight operations use satellites to navigate. But the susceptibility of GPS to interference probably means VORs will be around for a long time to come, most observers believe.

Cockpits of the future will bring to the flight deck increased capabilities, some of which will include digital communications to accommodate high-speed data to and from the cockpit, WAAS/GBAS approaches and landings, GPS-based en route navigation, and automatic dependent surveillance, allowing new sophistication in flight tracking and traffic avoidance. Also during this time, today's VHF voice communications will slowly give way to increased digital communications; and surveillance systems will become more sophisticated, moving away from today's Mode S transponders, air traffic control (ATC) surveillance radar, and Traffic and Collision Avoidance System (TCAS) to ADS-B (automatic dependent surveillance—broadcast). The result will be a better tactical picture of traffic and therefore far less need for voice callouts of other aircraft by air traffic controllers.

There is a tremendous need to provide data to the cockpit from the ground. In the future, digital communications are expected to make today's VHF communication radios obsolete, even those that are now undergoing updates for 8.33Khz requirements in Europe. The new infrastructure will require radios that can transmit in both voice and high-speed digital formats. The VHF Datalink (VDL)

Mode-2 system is already in operation, and the FAA has already proven the efficiency of the CPDLC system in high density air traffic areas. ADS-B avionics under development include a *cockpit display of traffic information (CDTI)* feature that enables the pilot to "see and avoid" other aircraft electronically. Using ADS-B technology, the aircraft emits signals that tell its position to other similarly equipped aircraft in the area. This information is depicted visually on a display in the aircraft cockpit and will greatly enhance a pilot's awareness of the position of other ADS-B equipped aircraft and lead to safer operations. The United Parcel Service (UPS) efficient use of CDTI at its hub in Louisville has proven that substantial fuel savings can be gained by use of ADS-B technology.

By reducing dependency on voice transmissions, pilot workload, presumably, will be greatly reduced. For en route applications, communications via datalink will be implemented first. Most observers agree it will probably be at least a decade before we see heavy use of digital communications in the terminal area, however. The transition from voice to digital communications will evolve fairly slowly, observers believe, in part because pilots need to become more comfortable with the idea of sending and receiving information in this format and in part because thousands of older airplanes will require new avionics to receive and send data via digital link.

Avionics makers must develop new radios that can be updated through software changes rather than expensive hardware upgrades. If we have learned anything from the personal computer industry, it is that the new technology constantly being developed can quickly make yesterday's equipment obsolete. The challenge for aviation will lie in developing avionics systems that have reasonably long, useful lives at a reasonable cost.

The potential for change and innovation in the navigation and communications infrastructure is limitless; and when it finally evolves, the modern cockpit will be a markedly different place from what it is today.

DISPLAYS

Most observers did not imagine that the liquid crystal display (LCD) would replace the cathode-ray tube as quickly as it did. Advances in commercial LCD production, however, helped bring costs down, while avionics makers enhanced off-the-shelf products, for unmatched levels of fidelity. Today, all new avionics suites for jets are being designed around large-format LCDs. Just how big the displays will grow depends largely on what happens in the market for commercial LCDs.

Color flat-panel displays or three-dimensional holographic displays have the potential to eliminate pilot errors by providing virtual reality for the crew. As is the case in most of the commercial transport category airplanes and in virtually all general-aviation planes, the pilot must use mental gymnastics to locate him- or herself in space when flying in instrument meteorological conditions (IMC). Display technology coupled with GPS accuracy can be the means for safety improvement in all weather conditions for all airspace users.

The future vision of avionics makers is to someday use displays that spread across the entire cockpit. Some are looking at new technologies such as gas-plasma displays or projection displays to meet this goal. For now, however, LCD is the most cost-effective way to bring the information needed for tomorrow's flight environment to the cockpit.

Using LCDs, avionics makers plan to integrate terrain, traffic, and weather information and display it on top of charts and approach plates, which will provide a seamless source for information, layered in such a way that pilots can better interpret the overall flight situation. Tomorrow's flight deck will be more dependent on sophisticated computer software that will greatly enhance the capabilities of the avionics, while at the same time adding to the complexity of systems. To help pilots better manage the technology, cursor-control devices, similar to a computer mouse, will be used, at first to scroll through simple menus and later to plan entire flights.

HEAD-UP DISPLAYS (HUDS)

HUDs will become much more commonplace in tomorrow's cockpits, experts believe, particularly if manufacturers can bring costs down. A number of HUD makers may shift their engineering focuses and begin looking at the use of smaller lenses and direct-projection HUDs, such as those used by the military in the C-17. This would eliminate the need for the large optics projectors and image collimators used today, making the systems far simpler to install and maintain.

HUD presents vital aircraft performance and navigation information to the pilot's forward field of view. The information on the HUD is collimated so that symbols appear to the human eye to be at infinity and overlaid on the actual outside scene. The use of HUD is gaining broad acceptance in aviation as flightcrews realize the tremendous benefits to situational awareness it can provide.

Also, use of the new enhanced flight vision system (EFVS), a display that allows pilots to see through low-visibility weather, could become commonplace in the cockpit of the future. Forward looking infrared (FLIR) thermal cameras have been developed to display a real-world visual image to the cockpit allowing the pilot to "see" in very low visibility conditions. Additionally, HUD systems are also being developed to display new synthetic vision system (SVS) images. In contrast to the EFVS, SVS produces a computer generated artificial image using moving map and terrain databases to provide a more realistic view of the outside environment. Recently, display researchers have even combined EFVS with SVS to improve the visual picture. Combining the real work reliability of the EFVS FLIR with the terrain and situational awareness of SVS could dramatically reduce issues associated with low visibility approaches in the future.

ELECTRONIC FLIGHT BAGS (EFBS)

In the near future, EFBs will begin to show up with greater regularity in the cockpit. A number of avionics makers have already introduced one or are developing such devices, which pilots would use to review charts and approach plates, perform weight and balance calculations, and plan flights, among other functions.

Eventually, with the inclusion of chart databases that can be viewed on the MFD and EFBs, paper may vanish from the cockpit. Of course, the transition from a paper chart that the pilot can hold in his or her hands to electronic-only format will be slow, especially for airlines in third world countries without financial resources or technology.

NEXT GENERATION FLIGHT OPERATIONS

As discussed in Chapter 8 on air traffic system technologies, in a NextGen environment, pilots would be released from the rigid discipline of being spaced in nose-to-tail time blocks, along less-than-optimum routes, and often at inefficient altitudes. Extended to its ultimate application, NextGen would embrace the airplane's complete operation, from start-up at the originating airport to shutdown at the destination after having flown a direct, nondeviating course, using performance figures straight out of the airframe manufacturer's operating handbook.

To make this new era possible, operators will need to install new avionics, the most critical being based upon ADS-B technology. The current Mode-S transponders will eventually take on a much more sophisticated role as ADS-B is implemented. Also required for NextGen will be datalink radios, which would be used to pass as much information as practical over discrete controller-pilot frequencies. Voice mode will still be essential for emergencies and other nonstandard communication, but for the most part, tomorrow's skies will be much quieter.

Once the transition to a NextGen-based traffic regime begins, observers believe airlines will begin updating the current fleet fairly rapidly. The technological challenges involved are huge, but not insurmountable, which means the future may be nearer than we realize. From a human factors point of view, major changes in NextGen flightdecks provide new opportunities for human error as well as significant training challenges.

KEY TERMS

Fly-by-wire

Extended-range twin-engine operations (ETOPS)

High-lift systems

Antiskid system

Engineered Materials Arrestor System (EMAS)

Aging Aircraft Safety Rule (AASR)

Widespread fatigue damage (WFD)

Primary flight display (PFD)

Multifunction display (MFD)

Liquid crystal display (LCD)

Electronic flight bag (EFB)

VHF Data Link (VDL)

Controller Pilot Data Link Communications (CPDLC)

Cockpit display of traffic information (CDTI)

Traffic Collision Avoidance System (TCAS)

Wing spoilers (speed brakes)

Thrust reversers

Fail-safe or damage tolerance design

Multiple-site damage (MSD)

High Altitude Clear Air Turbulence (HICAT)

Wind shear

Ground-proximity warning system (GPWS)

Engine-indicating and crew-alerting system (EICAS)

Aircraft Communications Addressing and Reporting System (ACARS)

Flight management system (FMS)

Central maintenance computer system (CMCS)

Computational fluid dynamics (CFD)

Wind-tunnel testing

Flight simulator

AIRcraft Maintenance ANalysys (AIRMAN)

Enhanced Airborne Flight Recorder (EAFR)

Enhanced Flight Vision System (EFVS)

Forward Looking Infrared (FLIR)

Synthetic Vision System (SVS)

Heads-Up Display (HUD)

Flight data recorder (FDR)

Cockpit voice recorder (CVR)

REVIEW QUESTIONS

1. Discuss some of the early developments in jet engine technology that were included in the Boeing B-47 and Dash 80 aircraft.

2. What was the purpose of pod-mounted engine installation?

3. What were some of the challenges to safety resulting from such radical airframe designs as highly swept wings, high wing loading, increased speeds, and long-duration flights at high altitudes?

4. Describe some technological improvements in the following stopping systems.
    a. Antiskid system   *BRAKES*
    b. Fuse plugs in the wheels   *OVERHEATING/SHRAPNEL*
    c. Engineered Materials Arrestor System (EMAS)
    d. Wing spoilers   *LIFT/LOW SPEED*
    e. Thrust reversers   *BRAKES*
5. What is meant by a fail-safe design?   *"BACK UP" STRUCTURE*
6. List and discuss the parameters that define the age of an aircraft.
7. What are some of the technological advances that enhanced the flight-handling characteristics of today's generation of aircraft?
8. Who are the three major participants in the structural safety process? Describe their individual roles.
9. What are some of the approach solutions to the problems of turbulence, winds, wind shear, ice and precipitation, and volcanic ash?
10. How do ice and precipitation affect airplane operation?   *DECREASE LIFT*   Describe several technologies designed to address this problem.   *ANTI ICE BOOTS, HEAT,*
11. List and briefly describe at least five flight-deck technology changes that have made a significant contribution to improving safety.
12. What is the purpose of GPWS, TCAS, and EICAS?   *ENGINE INDICATING AND CREW ALERTING SYSTEM*
13. What are the basic functions of the flight management system (FMS)?
14. What is computational fluid dynamics (CFD)?
15. What is the purpose of the engineering simulation?
16. What data are recorded on the FDR and CVR?
17. Describe some of the problems and improvements needed to gather better weather information.
18. Describe at least three major features likely to be present in the flight deck of the future.
19. Discuss some of the technological advances in communications, navigation, and displays including HUDs and EFBs.
20. Describe the functions of EFVS and SVS.

REFERENCES

Boeing Airliner Magazine. Various dates. Selected articles. Published bimonthly. Seattle, Wash.

Curtic, Michael T., Jentsch, Florian, and Wise, John A 2010. Aviation Displays article in *Human Factors in Aviation*, 2nd ed., pp. 439–478.

Hall, John, and Ulf G. Goranson. 1992. "Structural Damage Tolerance of Commercial Jet Transports." FSF 45th IASS and IFA 22d International Conference. Long Beach, Calif.

Hennigs, N. E. 1990. "Aging Airplanes." *Boeing Airliner*. July/September: 17–20.

Henzler, K. O. 1991. "Aging Airplanes." *Boeing Airliner*. January/March: 21–24.

Krois, Paul; Piccione, Dino; and McCloy, Tom 2010. Commentary on NextGen and Aviation Human Factors article in *Human Factors in Aviation*, 2nd ed., pp. 701–710.

Mosier, Kathleen L. 2010. The Human in Flight: From Kinesthetic Sense to Cognitive Sensibility article in *Human Factors in Aviation*, 2nd ed., pp.147–173.

Norris, Guy. 1996. *Boeing 777*. Osceola, Wis.: Motor Books.

Salas, Eduardo and Maurino, Dan 2010. *Human Factors in Aviation,* Second Edition. Burlington, Mass. Academic Press.

Schairer, George. 1992. "The Engineering Revolution Leading to the Boeing 707." FSF 45th IASS and IFA 22d International Conference. Long Beach, Calif.

Shaw, Robbie. 1996. *Boeing Jetliners*. Osceola, Wis: Motor Books.

Soekkha, Hans M. (ed.). 1997. "Aviation Safety." Selected papers on flight operations. *Proceedings of the IASC-97*. Rotterdam, The Netherlands: VSP BV.

Taylor, Laurie. 1997. *Air Travel—How Safe Is It?* 2d ed. London, England: Blackwell Science, Ltd.

WEB REFERENCES

www.airbus.com
www.boeing.com
www.geaviation.com
www.faa.gov/news
www.flightglobal.com
www.flightsafety.org
www.usatoday.com

AIRPORT SAFETY

LEARNING OBJECTIVES

After completing this chapter, you should be able to

- Discuss the regulatory chronology of airport certification.
- Compare and contrast the four airport certification classes.
- Discuss the airport certification process.
- Identify and differentiate between the two types of airport certifications.
- Explain what the Airport Certification Manual is and discuss its significance.
- Identify the elements of an Airport Certification Manual.
- Discuss airport inspection and compliance with the provisions of FAR Part 139.
- List and explain the required elements of an Airport Emergency Plan.

- Identify the sources of safety hazards and the applicable regulations that govern the following:

  Airport terminal buildings

  Hangars and maintenance shops

  Ramp operations

  Aviation fuel handling

  Aircraft rescue and firefighting (ARFF)

  Deicing

- Describe the airport surface operational environment.

- List the individual groups involved with airport surface operations.

- Define and discuss the following terms relating to runway safety:, *runway incursion, operational error, pilot deviation,* and *vehicle/pedestrian deviation.*

- Discuss runway incursions, trends, and statistics with respect to rates and severity.

- List and discuss control strategies and initiatives that are being proposed to minimize the occurrence of runway incursions.

INTRODUCTION

A commercial *airport,* by definition, is a tract of land (or water) that provides facilities for landing, takeoff, shelter, supply, and repair of aircraft and has a passenger terminal. This results in a diverse collection of complex operations each with its own set of safety issues. This chapter provides an overview of the certification process, the certification manual, and safety issues concerning airports. Extended discussions are conducted on some of the more hazardous operations such as fuel handling and runway incursions. Highlights of Part 139 and other relevant regulations are also reviewed.

AIRPORT CERTIFICATION

The Federal Aviation Act of 1958 was broadened in 1970 to authorize the FAA Administrator to issue operating certificates to certain categories of airports serving air carrier aircraft. This authority was established as a result of the Airport and Airway Development Act of 1970 that added a new section (612) to the 1958 Act. The 1970 Act also added to the 1958 Act a clause that barred any person from operating an airport or any air carrier from operating in an airport that did not possess a certificate if it was required to or if the airport was in violation of the terms of an issued certificate. The intent of the certification was to establish minimum safety standards for the operation of airports to ensure the safety of the flying public. To be certified by the FAA, airports are required to meet certain standards for airport design, construction, maintenance, and operations as well as firefighting and rescue equipment, runway and taxiway guidance signs, control of vehicles, management of wildlife hazards, and record keeping. The FAA works with certificated airports to help them meet these standards through periodic consultations and site inspections.

In 2004, the FAA issued a final rule which made major revisions to its regulations pertaining to airport certification. These regulations, found at 14 CFR Part 139, established new certification requirements for airports serving scheduled air carrier operations in aircraft with more than nine passenger seats. In addition, under the changed certification process, airports are now reclassified into four new classes based upon the type of air carrier operations.

AIRPORT CERTIFICATION CLASSES

- **Class I**—These airports must comply with all Part 139 requirements and may serve any air carrier operations covered under FAR Part 139. Some of the provisions of the new rule for these larger airports are:
  - New personnel training and hazardous material (HAZMAT) response standards
  - Airport Emergency Plan retains its triennial exercise requirement with new requirements to plan for fuel storage fires, HAZMAT and security incidents, and water rescue situations
  - New wildlife hazard management plan standards

- **Class II**—These airports serve scheduled operations of small air carrier aircraft (10–30 seats) and unscheduled operations of large air carrier aircraft (30+ seats). Class II airports are *not* permitted to serve scheduled large air carrier operations. Some of the provisions of the new rule for these airports are:
  - Similar recordkeeping, personnel training and HAZMAT response standards as Class I
  - New requirements for snow and ice control plan, self-inspection standard, protection of NAVAIDS, and wildlife hazard management
  - New requirement for Airport Emergency Plan but no triennial exercise required

- **Class III**—These airports serve *only* scheduled operations of small air carrier aircraft (10–30 seats), but are not applicable to certain airports in the State of Alaska. Some of the rules for these airports:
  - Similar requirements for Class II airports
  - Aircraft rescue and firefighting (ARFF), some alternative compliance measures are allowed from strict Class I or II standards
  - Airport Emergency Plan—no triennial exercise required

- **Class IV**—These airports serve only unscheduled operations of large air carrier aircraft (30+ seats). Air carrier operations are so infrequent at these airports that FAA previously only required them to comply with certain Part 139 requirements. This continues to be the case under the new rules with include the following:
  - New recordkeeping and personnel training standards similar to Classes I, II and III
  - New requirements to mark and light airport obstructions
  - New requirements for Airport Emergency Plan but no triennial exercise required

In the United States, there are approximately 551 Part 139 airports as of March 2011, broken down as follows:

Class I 381 airports

Class II 49 airports

Class III 35 airports

Class IV 86 airports

An airport that meets FAR Part 139 criteria is issued an Airport Operating Certificate (AOC). To obtain a certificate, an airport must maintain operational and safety standards under categories mentioned earlier. The FAA provides guidance on meeting the requirements of Part 139 certification through documents called *Advisory Circulars* (*ACs*). The FAA may also issue *CertAlerts* through the Airport Safety and Operations Division of the Office of Airport Safety and Standards that serve as a quick means of providing both the public and FAA staff with important certification-related guidance.

Fully certificated airports must maintain an *Airport Certification Manual* (ACM) that details operating procedures, facilities, and equipment and other appropriate information as listed below.

AIRPORT CERTIFICATION MANUAL

Every certificated airport that serves air carriers is required to have an ACM in accordance with Part 139. The ACM is a working document that outlines the means and procedures used to comply with the requirements of Part 139. While each airport has its own unique features and operational requirements, its Airport Certification Manual will contain basic elements appropriate to its class, which include the following:

1. Lines of succession of airport operations responsibility.

2. Each current exemption issued to the airport from the requirements of Part 139.

3. Any limitations imposed by the FAA Administrator.

4. A grid map or other means of identifying locations and terrain features on and around the airport that are significant to emergency operations.

5. The location of each obstruction required to be lighted or marked within the airport's area of authority.

6. A description of each movement area available for air carriers and its safety areas, and each road that serves it.

7. Procedures for avoidance of interruption or failure during construction work of utilities serving facilities or NAVAIDS that support air carrier operations.

8. A description of the system for maintaining records.

9. A description of personnel training.

10. Procedures for maintaining the paved areas of the airport.

11. Procedures for maintaining the unpaved areas of the airport.

12. Procedures for maintaining the safety areas of the airport.

13. A plan showing the runway and taxiway identification system, including the location and inscription of signs, runway markings, and holding position markings.

14. A description of, and procedures for maintaining, the marking, signs and lighting systems.

15. A snow and ice control plan.

16. A description of the facilities, equipment, personnel, and procedures for meeting the aircraft rescue and firefighting requirements.

17. A description of any approved exemption to aircraft rescue and firefighting requirements.

18. Procedures for protecting persons and property during the storing, dispensing, and handling of fuel and other hazardous substances and materials.

19. A description of, and procedures for maintaining, the traffic and wind direction indicators.

20. An Airport Emergency Plan (AEP).

21. Procedures for conducting the airport self-inspection program.

22. Procedures for controlling pedestrians and ground vehicles in movement areas and safety areas.

23. Procedures for obstruction removal, marking, or lighting.

24. Procedures for protection of NAVAIDS.

25. A description of public protection at the airport.

26. Procedures for wildlife hazard management.

27. Procedures for airport condition reporting.

28. Procedures for identifying, marking, and lighting construction and other unserviceable areas.

29. Any other item that the FAA Administrator finds is necessary to ensure safety in air transportation.

(*Source: FAA.gov*)

The Airport Emergency Plan is an integral part of the ACM although it may be published and distributed separately to each user or airport tenant. The plan must be extremely detailed and provide sufficient guidance on response to all emergencies and abnormal conditions that the airport is likely to encounter. Emergencies include aircraft accidents; bomb threats; sabotage; hijackings; major fires; natural disasters such as floods, tornadoes, and earthquakes; and power failures. Emergency response plans should spell out procedures for coordinating internally with airport and firefighting personnel and the control tower. In addition, procedures should be established for coordination with external agencies such as law enforcement, rescue squads, fire departments, ambulances and emergency transportation, and hospital and medical facilities. The Airport Emergency Plan must be reviewed at least annually, and for Class I airports there must be a full-scale exercise of the plan every 3 years.

AIRPORT INSPECTION AND COMPLIANCE WITH FAR PART 139

The FAA has a team of nearly 35 Airport Certification Safety Inspectors to ensure that airports with AOCs are meeting the requirements of FAR Part 139. These inspections are usually performed annually, but may also be conducted on an unannounced basis at any time. The process used by these FAA inspectors includes the following steps: (FAA.gov)

- Pre-inspection review of office airport files and the specific *Airport Certification Manual.*
- In-briefing with airport management.
- Administrative inspection of airport files and paperwork including self-inspection, the Airport Master Record, and Notices to Airmen (NOTAMs).
- Movement area inspection of runway areas, markings, lighting, signs, and similar items including the safety of ground vehicle operations and wildlife considerations.
- Aircraft rescue and firefighting inspection including timed-response drills, emergency medical training, equipment and protective clothing.
- Fueling facilities inspection of fuel farm and mobile trucks, including review of tenants and fire safety training procedures.
- Night inspection of runway/taxiway and apron lighting and signage including the airport beacon, wind cone, and obstruction lighting for compliance with Part 139 requirements.
- Post-inspection (exit briefing) with airport management to discuss findings; a letter of correction noting violations, if any, and safety recommendations may also be issued.

If the FAA determines that an airport is not meeting its obligations, the FAA can impose a daily cumulative fine for each day the airport continues to violate any requirement of Part 139. In extreme cases, the FAA could revoke the airport's certificate or limit the areas of an airport where air carriers can land or take off.

OPERATIONAL SAFETY

As mentioned earlier, airport operations are complex and diverse, with hazards and their severity varying by the type of operation. For the purposes of our discussion, airport safety issues will be discussed under the following headings:

- Airport terminal buildings
- Hangars and maintenance shops
- Ramp operations
- Specialized services
  - Aviation fuel handling
  - Aircraft rescue and firefighting (ARFF)
  - Deicing
- Runway incursions

AIRPORT TERMINAL BUILDINGS

These facilities are laid out to minimize passenger delays and maximize throughput of travelers from the airport terminal entry point to the aircraft. Inadequate layout and facility size leads to overcrowding which can cause dangerous conditions that lead to injuries. Due consideration to safety should be given during construction and modification of the facilities so as to minimize hazards that cause falls and to protect people from harmful contact with machinery. Most of the design standards and requirements for building materials, fire protection, and building egress are covered by building codes (e.g., the *International Building Code*); the *National Fire Protection Association* (NFPA); the *American National Standards Institute* (ANSI); and other local, state, and federal agencies. For operational issues, OSHA standards (which incorporate many of the standards from the above-mentioned organizations) should be adhered to. A summary of important safety considerations for airport terminal facilities follows:

- Emergency evacuation and egress routes, signs, and other marking should conform to the Life Safety Code, NFPA 101, and OSHA regulations 29 CFR Part 1910, Subpart E, "Exit Routes, Emergency Action Plans, and Fire Prevention Plans." Some additional notes are that
  - Exits should be well lighted and marked.
  - Panic hardware should be installed and properly working on the inside of exterior doors used for emergency exits.
  - Exits doors should be kept unobstructed and unlocked in the direction of egress.
  - Emergency evacuation plans should be posted in all buildings.
- All areas must be accessible to persons with disabilities (more information on the Americans with Disabilities Act of 1990 and this topic can be found in Advisory Circular 150/5360-14).

- All slip and trip hazards must be eliminated, and fall prevention strategies should be implemented on stairways, escalators, and moving walkways.

- Stairs, ladders, and ramps should be kept clean and free of obstacles or slippery substances and should be maintained in good condition. Both interior and exterior stairways should be sufficiently lighted so that all treads and landings are clearly visible. Prevention of falls, slips, and trips on any walking and working surface is regulated under 29 CFR Part 1910 Subpart D.

- There must be safe access to heating, ventilation, and air-conditioning systems for maintenance to permit safe and convenient replacement of heavy equipment (e.g., pumps and motors). In this connection a safe means for maintaining luminaries and windows must be provided.

- Skylight and roof openings must be provided with protective screens to prevent individuals from falling through.

- Doors should open automatically to eliminate inconvenience to people with baggage and to the handicapped. Also, doors should open away from the person going through.

- Escalators, elevators, and moving walks should comply with ANSI Standard A17.2-2010. Speeds for moving walkways and escalators should be set at the lower limit of the acceptable range.

- Moving walkways should have emergency stop buttons at both ends and, wherever feasible, every 50 meters apart. Attention-grabbing techniques should be installed to warn passengers that they are approaching the end of a moving walkway.

- Ticket counters should be ergonomically designed to permit safe body postures during baggage handling and data entry. Weight scales and postweigh-in baggage conveyors should be installed as close to floor level as practical to minimize lifting, and entrances to the scales should be left free so as to permit baggage sliding as opposed to lifting. OSHA regulates ergonomic hazards (or any other recognized hazard that does not have a specific regulation) under the *General Duty Clause* listed in Section 5(a) of the OSHA Act.

- All electrical systems and power service should be installed, maintained, and inspected according to the National Electrical Code, NFPA 70. OSHA regulates electrical practices under 29 CFR Part 1910 Subpart S.

- Fire extinguishers and sprinkler systems should be installed and maintained in accordance with NFPA 10 and NFPA 13. OSHA addresses fire prevention under 29 CFR Part 1910 Subpart L.

- Maintenance and other operational issues:
  - Roofs should be inspected regularly, especially after violent storms.
  - Maximum allowable loading per square foot of floors should be prominently displayed in all storage areas to avoid damage or collapse.
  - Floors and stairways should not be waxed if it creates a slipping hazard.

- Planning and design guidance for airport terminals is included in Advisory Circular (AC) 150/5360-13 and AC 150/5360-9 for nonhub locations.

HANGARS AND MAINTENANCE SHOPS

Construction of hangars is covered by federal, state, and local building codes as well as many National Fire Protection Association standards. Operational issues are mostly covered by OSHA and EPA requirements. Some of the highlights include the following:

WALKWAYS AND WORKING SURFACES

- Adequate aisle space and access to fire extinguishers, hoses, sprinkler control valves, and fire alarm stations should be maintained.
- If work is performed at elevations in excess of 4 feet, the worker must be protected from falling by providing guardrails 42 inches high with an intermediate rail approximately halfway between the top rail and the standing surface. If guardrails are not feasible, an equivalent effective means of fall protection (safety harnesses, nets, etc.) must be provided. Regardless of height, if a work platform is adjacent to a hazard (e.g., dangerous equipment), then the worker must be protected from falling. Toe boards should be provided for the work platform on the open sides to prevent objects from falling on persons passing beneath or if there is moving machinery or if there is equipment with which falling material could create a hazard.

ELECTRICAL

- Large metal complexes such as aircraft docks should be permanently bonded and grounded and secured to a permanent structure such as a hangar floor. Primary sources of electric power to hangars should be approved for the location according to the National Electrical Code, NFPA 70, and aircraft hangars, NFPA 409.
- All other electrical operational issues such as grounding, wiring, receptacles, and circuit breakers are covered by 29 CFR Part 1910 Subpart S.

COMPRESSED GAS, FLAMMABLES, AND HAZARDOUS AND TOXIC SUBSTANCES. Storing and handling of compressed gases, and flammable and combustible liquids are addressed under 29 CFR 1910 Subpart H, Hazardous Chemicals. Many of these standards reference or incorporate those developed by the NFPA and the Compressed Gas Association. Toxic and hazardous substances are addressed under 29 CFR Part 1910 Subpart Z.

- 29 CFR 1910.101(b) addresses the in-facility handling, storage, and utilization of all compressed gas cylinders by referencing procedures specified in the Compressed Gas Association Pamphlet P-1-1965. Compressed gas cylinders should be secured and stored upright when not in use and be stored in carts designed for that purpose when in use. Gas cylinders should have their caps on except when in use. Cylinders of compatible gases should be stored together, and oxygen cylinders should be stored separately from flammable gas cylinders and flammable or combustible substances.

- General-purpose compressed air systems should not exceed 30 pounds per square inch unless specifically required (e.g., for tire inflation and special-purpose pneumatic tools). The practice of using compressed air to clean personal clothing should be avoided.

- All flammable and combustible materials (paints, hydraulic fluids, greases, solvents, cleaning agents, etc.) should be stored and used in accordance with NFPA 30. Information on the use and handling of specific materials and processes can be found in 29 CFR Part 1910 Subpart H and 29 CFR Part 1910 Subpart Z, Toxic and Hazardous Substances.

- Storage of flammables should be limited to quantities that are permitted for immediate short-duration uses in the shop areas. These chemicals should be stored in labeled cabinets designed and approved for flammable storage. Bulk storage of flammables should be in separate buildings designed for that propose that include design features such as explosion-proof electrical systems, spill containment, and automatic fire protection. Storage quantity requirements for shop and bulk storage can be found in 29 CFR 1910.106.

- Painting and stripping are hazardous activities that use toxic chemicals and therefore must be performed in well-ventilated areas that are separate from other maintenance activities and are designed for that purpose (e.g., paint booths). In addition, painting and stripping generate toxic wastes that must be contained and either recycled or disposed of appropriately. Therefore, these operations are governed by both OSHA and EPA regulations.

- Miscellaneous on-site use of all chemicals is covered under OSHA's Hazard Communication Standard, 29 CFR Part 1910 Subpart 1200. This standard comprehensively addresses the issues of chemical hazard evaluation, employee hazard communications, and employee personal protective equipment and procedures. In summary, this subpart requires:
  - Developing and maintaining a written hazard communication program for the workplace, including lists of hazardous chemicals present on site
  - Labeling of containers of chemicals in the workplace and containers of chemicals being shipped to other workplaces
  - Preparing and distributing material safety data sheets (MSDS) to employees
  - Developing and implementing employee training programs that detail the hazards of chemicals on site and what employees must do to protect themselves

MISCELLANEOUS

- The control of hazardous energy (lockout and tag-out) standard (29 CFR 1910.147) covers the servicing and maintenance of machines and equipment in which the unexpected energization or start-up of machines or equipment, or release of stored energy (electrical, hydraulic, pneumatic, or gravity), could cause injury to employees. This standard requires maintenance personnel, through written equipment-specific procedures, to deenergize stored energy and positively block off all potential energy sources to systems that are being worked on by using personalized locks and tags.

- Material handling equipment such as overhead and gantry cranes that are used to move large aircraft sections are regulated under 29 CFR 1910.179. Slings, chains, and hoists used to move or lift smaller aircraft parts and sections (e.g., aircraft jacking) are regulated under 29 CFR 1910.184.

- The use of powered material handling equipment, such as fork trucks and hand trucks, and the use and maintenance of batteries used to power these and other equipment are regulated under 29 CFR 1910.178.

- Hoist units such as those used to reach large aircraft tails should comply with safety requirements listed under 29 CFR Part 1910 Subpart F, Powered Platforms, Manlifts, and Vehicle-Mounted Work Platforms.

- Heating and ventilation systems in hangars should be in accordance with NFPA 409 or the equivalent.

- Welding, cutting, brazing, and other hot work that has the potential to start fires is regulated under 29 CFR Part 1910 Subpart Q.

- Noise evaluation and hearing protection for processes such as riveting, sand blasting, grinding, polishing, and other noisy operations that exceed 85 decibels (time-weighted A scale for 8 hours) are regulated under 29 CFR 1910.95.

- The safe use of pedestal grinders and other grinding wheels is regulated under 29 CFR 1910.215, Abrasive Wheel Machinery.

RAMP OPERATIONS

The ramp area is generally designed for aircraft, not the vehicles that service and/or operate in the proximity of the aircraft. Most of the signs and markings are for aircraft. The ramp area sees a diverse collection of high-paced activities that involve aircraft, vehicles, and individuals working in close proximity to one another. Some of these activities include

- Aircraft ground handling that may include taxiing, towing, chocking, parking, or tie-down
- Aircraft refueling
- Aircraft servicing—catering, cleaning, food service, etc.
- Baggage and cargo handling
- Power supply
- Routine checks and maintenance

Individuals involved with the above activities are exposed to several of the occupational hazards (and hence regulated by similar regulations) mentioned earlier, including

- Cuts (from antennas, pitot tubes, static discharge wicks).
- Slips, trips, and falls (on the ground and from elevations).
- Strains and sprains (from baggage handling).
- Exposure to hazardous materials (fuel).
- Contact with moving parts (propellers) and bumps (undersurface of fuselage).
- Electrical hazards (tools, motor, generators)—electrical wiring should conform to National Electric Code, NFPA 70, for class I, Group D, Division I hazardous locations.
- Biohazards (bloodborne exposures during cleaning, covered under 29 CFR 1910.1030).
- High-pressure air and other fluid exposures from pressurized systems.
- Noise from engines and other equipment. Hearing protection is a significant health and safety issue during ramp operations.

In addition, there is a potential for significant injury from exposure to jet blast, moving aircraft, and moving vehicles. Since most of the operations take place outdoors, weather conditions (heat, cold, snow, rain, ice, and wind) can increase the risk of injury. Hazards due to aircraft refueling are discussed in the next section.

SPECIALIZED SERVICES

These comprise

- Aviation fuel handling
- Aircraft rescue and firefighting (ARFF)
- Deicing

AVIATION FUEL HANDLING.   Fuel handling is an important safety issue not only to the fuelers, but also to other airport personnel, the traveling public, and the operation of the aircraft. Failure to adhere to safe operating procedures when fueling aircraft and/or transporting fuel from one location to another on the airport can result in major disasters. This potential for disaster is well recognized as millions of gallons of aircraft fuel are handled each year without major incidents for the most part. The basics types of aviation fuel are aviation gasoline (avgas), jet A or jet B (commercial jet uses) fuels, and JP series (military jet uses) fuels. Guidance on fuel handling in an airport environment is covered under 14 CFR Part 139.321(b) and the FAA advisory circulars in the 150 series (e.g., AC 150/5230-4A), all of which reflect NFPA standards (Aircraft Fuel Servicing, NFPA 407, 2007, and Aircraft Fueling Ramp Drainage, NFPA 415, 2008). It is also necessary that fueling operations be conducted in accordance with procurement document specifications and all other applicable local, state, and federal regulations. Special attention must be paid to federal, state, and local hazardous materials regulations and to agency-specific fuel spill avoidance requirements. The major concerns with fuel handling are

- Health hazards to fuelers
- Fuel contamination
- Explosions and fires during fueling or fuel transfer
- Explosions and fires during fuel tank repair
- Hazards from spills

HEALTH HAZARDS TO FUELERS   Repeated contact with aviation fuels can cause skin irritation and dermatitis. Skin area exposures to fuels should be washed immediately with soap and water. Eyewash stations should be conveniently located for immediate flushing in case the eyes or face comes into contact with fuel. Fuels such as avgas contain additives such as benzene. Hence, fuels can be toxic if inhaled or swallowed. Fuels can also affect the central nervous system by causing narcosis, which leaves an individual in a state of stupor or unconsciousness. Threshold limits for fuel vapors in the breathing zones are established by the *American Conference of Governmental Industrial Hygienists (ACGIH)* and 29 CFR 1910 Subpart Z (which reflects ACGIH standards).

FUEL CONTAMINATION   Fuel contamination is a major safety issue as it can affect the operation of the aircraft. Contamination can occur by refueling an aircraft with contaminated fuel or the wrong fuel grade or type. Fuel is considered contaminated if it contains any material other than what is called for in the specifications. The two most common contaminants found in fuels are rust and water (free or emulsified). Other contaminants found include microbial growths, paint, metal, rubber, and lint. Commingling, or the mixing of different fuel grades and types, is also a serious issue. For example, mixing avgas with jet fuels can reduce the volatility and antiknock features of fuels needed for reciprocating engines, which can result in engine failure. This could also lead to lead deposits in turbine engines. To prevent commingling, Advisory Circular 150/5230-4A requires the use of separate pumps, lines, standard couplings, and color codes for each type of fuel. Also, all fuel-dispensing systems should be clearly marked and labeled to indicate the grade and type of fuel they contain.

EXPLOSIONS AND FIRES DURING FUELING OR FUEL TRANSFER   There are three essential requirements for a fuel to burn. First, the fuel must be in its vapor form; second, it must develop the right fuel-to-air ratio with the surrounding atmosphere; and third, there must be a source of ignition to start the fire. In fuel handling, the first two conditions invariably exist, and there is very little one can do to completely eliminate these hazardous states. Hence, the primary objective during fuel handling is to control or eliminate ignition sources. Several ignition sources and their control are discussed below.

STATIC ELECTRICITY   Static charge discharge is one of the most important sources of ignition and is a constant threat to safe fueling. When two dissimilar materials come in contact with each other, a static charge of electricity develops by the exchange of positive and negative charges across the contact surfaces. Static charges build up when fuel gets pumped through the fuel lines. The quantity of static electricity produced is directly proportional to the fuel flow rate. Also, due to their lower volatility, jet fuels are generally more susceptible to ignition from static discharges than are aviation fuels. Static electricity also gets generated when fuel is allowed to free-fall through the air, as when a tank is filled through a spout. An electrically charged atmosphere can build up charges on an aircraft, as can rain, snow, or dust blowing across the aircraft.

While completely eliminating static charge is impractical, several steps can be taken to minimize its buildup. Some of these steps are as follows:

- Follow correct *grounding* and *bonding* procedures, as listed in AC 00-34A, "Aircraft Ground Handling and Servicing."
- Avoid pumping contaminated fuels since dirt and water particles in the fuel will add to the static charge accumulation.
- Reduce fuel flow rates to decrease turbulence, where feasible. Reduced flow rates allow longer periods for the static charge to dissipate.
- Prevent fuel from free falling through the air to the bottom of the tank.

SPARKS   Arcing may occur when electrical connections are made or from faulty equipment and improperly maintained ramp vehicles. To avoid sparks, the following precautions should be taken:

- Aircraft batteries should not be installed or removed, nor should battery chargers be operated during fueling operations.
- Aircraft ground power units should be located as far away as possible from the fueling points.
- Battery-powered tools as opposed to electrically powered tools should be used near aircraft during fueling.
- Aircraft radios and radars should be shut off. Also, fueling should not be performed within 100 feet of energized airborne radar.
- Flashings used near the fueling area should be UL-approved for use in such locations.
- Workers near the fueling area should be advised against wearing footwear with metal nails or cleats or metal plates on the heels.

MISCELLANEOUS   Other issues to be considered for minimizing ignition sources include these:

- No-smoking rules should be strictly enforced.
- Employees should not be allowed to carry matches, lighters, and other smoking paraphernalia when engaged in fueling operations.
- Welding, cutting, or other hot work should not be conducted within at least 35 feet (preferably 50 feet) of fueling. Similar distance restrictions should be maintained for open-flame lights and exposed-flame heaters.
- Fuel pits should be located at least 50 feet away from a terminal building or concourse.

EXPLOSIONS AND FIRES DURING FUEL TANK REPAIR   Maintenance work on aircraft fuel cells and tanks, and bulk fuel storage tanks present similar inhalation and fire hazards as described above. It is extremely important that cells and tanks be drained and thoroughly vented before any maintenance work is attempted, especially hot work such as welding, brazing, and cutting. If tanks are large enough for an employee to enter and perform work, then confined-space testing, entry, and rescue procedures should be adhered to as required by 29 CFR 1910.146. Positive pressure should be employed for venting and should be undertaken with an approved blower (with nonsparking blades and an isolated nonsparking blower motor) using a grounded hose.

HAZARDS FROM SPILLS   Fuel spills present varying degrees of safety issues depending on the size of the spill and the kind of response required. While simple flushing may be appropriate for small spills, larger spills may require the use of absorbents, adsorbents, and chemical fixants. Leaks from underground storage tanks (USTs) present a different set of safety issues. Spills and USTs are regulated by EPA (as mentioned in earlier chapters) and local and state jurisdictions. Also, depending on the response planned by the facility, OSHA regulations 29 CFR 1910.120 could apply. Some of the issues to consider in connection with fuel spills include

- Leak/spill prevention
- Emergency response procedures
- Reporting and notifications
- Spill control/containment
- Cleanup procedures
- Employee training

AIRCRAFT RESCUE AND FIREFIGHTING (ARFF).   Responding to aircraft crashes is a high-risk job leaving individuals exposed to a wide variety of safety hazards such as dealing with fire, smoke, chemicals, blood, and other body fluids and lifting in awkward postures. AARF training, therefore, should include varying levels of OSHA training depending on the level of responder involvement. To develop an understanding of the safety issues involved with aircraft rescue and firefighting, a review of ARRF is in order.

As mentioned earlier, all certificated airports must have a plan to respond to all conceivable emergencies. The airport emergency plan is an integral part of the Airport Certification Manual. A major emergency that an airport has to plan for is aircraft crashes, which are handled under the aircraft rescue and firefighting plan. This plan is based on the largest aircraft that is likely to be serviced by the airport. Detailed requirements and procedures for ARFF equipment and agents are contained in

- Part 139.315 (Aircraft rescue and firefighting: Index determination)
- Part 139.317 (Aircraft rescue and firefighting: Equipment and agents)
- Part 139.319 (Aircraft rescue and firefighting: Operational requirements)
- FAA Advisory Circulars in the 150 series

Aircraft rescue services and firefighting equipment are based on an index as determined by the length of the air carrier aircraft and the frequency of its departure. If there are five or more average daily departures of air carrier aircraft in a single index group serving a given airport, the longest index group with an average of five or more daily departures will be the established index for the airport. If there are less than five average daily departures of air carrier aircraft in a single index group serving that airport, the next-lower index from the longest index group with air carrier aircraft in it will be the index established for the airport. The minimum designated index for a certificated airport is Index A. Highlights of index classifications and other ARFF requirements are summarized below.

INDEX CLASSIFICATIONS

INDEX A

Aircraft length: Less than 90 feet

Requirements: One vehicle carrying at least

1. 500 pounds of sodium-based dry chemical or halon 1211; OR
2. 450 pounds of potassium-based dry chemical and water with a commensurate quantity of aqueous film-forming foam (AFFF) to total 100 gallons, for simultaneous dry chemical and AFFF foam application

INDEX B

Aircraft length: At least 90 feet but less than 126 feet

Requirements: Either of the following:

1. One vehicle carrying at least 500 pounds of sodium-based dry chemical or halon 1211, and 1500 gallons of water, and the commensurate quantity of AFFF for foam production; OR

2. Two vehicles:
   a. One vehicle carrying the extinguishing agents as listed above in Index A(1) or Index A(2); AND
   b. One vehicle carrying an amount of water and the commensurate quantity of AFFF so that the total quantity of water for foam production carried by both vehicles is at least 1500 gallons.

INDEX C

Aircraft length: At least 126 feet but less than 159 feet

Requirements: Either of the following:

1. Three vehicles:
   a. One vehicle carrying the extinguishing agents as listed above in Index A(1) or A(2); AND
   b. Two vehicles carrying an amount of water and the commensurate quantity of AFFF so that the total quantity of water for foam production carried by all three vehicles is at least 3000 gallons; OR
2. Two vehicles:
   a. One vehicle carrying the extinguishing agents as listed above in Index B(1); AND
   b. One vehicle carrying water and the commensurate quantity of AFFF so that the total quantity of water for foam production carried by both vehicles is at least 3000 gallons.

INDEX D

Aircraft length: At least 159 feet but less than 200 feet

Requirements: Three vehicles:

1. One vehicle carrying the extinguishing agents as listed above in Index A(1) or Index A(2); AND
2. Two vehicles carrying an amount of water and the commensurate quantity of AFFF so that the total quantity of water for foam production carried by all three vehicles is at least 4000 gallons.

INDEX E

Aircraft length: At least 200 feet

Requirements: Three vehicles:

1. One vehicle carrying the extinguishing agents as listed above in Index A(1) or Index A(2); AND

2. Two vehicles carrying an amount of water and the commensurate quantity of AFFF so that the total quantity of water for foam production carried by all three vehicles is at least 6000 gallons.

ADDITIONAL AIRCRAFT RESCUE AND FIREFIGHTING (ARFF) REQUIREMENTS Further operational requirements listed in Part 139.319 include details regarding the following areas:

- Rescue and firefighting capability
- Procedures for reduction in ARFF capacity if necessary
- Hazardous materials guidance
- Emergency access road requirements
- Vehicle communications
- Vehicle marking and lighting
- Vehicle readiness
- Response requirements
- Personnel selection and qualification requirements

DEICING.   Ice and snow on control, airfoil, and sensor surfaces can have serious repercussions on the operation of the aircraft. Hence, when freezing or near-freezing conditions are likely to be present, the aircraft should not be allowed to take off before being sprayed with deicing fluid. Deicing presents five distinct hazards:

- Damage to the aircraft by the deicing equipment
- Application of deicing fluid to areas where it should not be applied, such as static ports, pitot heads, angle-of-attack sensors, the engine, and other inlets
- Hazards (inhalation and ingestion) to the deicing crew and to the environment, since deicing fluids are considered hazardous materials and are moderately toxic
- Hazards to the crew and passengers if the cabin air intakes are not shut off during deicing
- Falls, slips, and other safety hazards during the application process

Deicing crews should have training as required by OSHA's Hazard Communication standard (29 CFR 1910.1200). Also, Material Safety Data Sheets (MSDS), which list health hazards and protection information on the deicing fluids being used, should be made readily available. Vapors or aerosols of deicing fluids can cause nose and throat irritation, headaches, vomiting, and dizziness. For high-volume applications it is recommended that personnel be fitted with respirators in accordance with 29 CFR 1910.134. Ingestion of deicing fluids can affect the kidneys and the central nervous system. However, these fluids are so highly diluted with water that large quantities would need to be consumed to be lethal. Application crews usually work on elevated platforms (baskets at the end of booms) and would therefore have to be adequately protected from falling through safety harnesses and guardrails. Finally, deicing spent fluid is a regulated waste and therefore should be used, collected, and recovered/disposed of in accordance with EPA and other federal and local regulations. The FAA has recently adopted the European deicing standard called the Standard International Aircraft Ground Deice Program. Advisory Circular 120-60B is the primary FAA guidance document for the development of an air carrier's ground deicing program.

RUNWAY INCURSIONS

BACKGROUND

The prospect of two planes colliding is horrifying. It conjures an image of twisted, flaming wreckage falling from the sky. And yet such catastrophic collisions can easily occur on the ground when the safe, coordinated use of runways breaks down. The worst aviation disaster on record was, in fact, the result of two 747s colliding on a runway at Tenerife Airport in the Spanish Canary Islands. It was a foggy night in 1977 when the captain of one of the 747s failed to obey a "hold in position" instruction from air traffic control and started his takeoff roll before the other 747 cleared the runway. This catastrophic incident resulted in the loss of 583 lives. In the United States, the deadliest runway incursion accident occurred in August 2006 in Lexington, Kentucky, when Comair Flight 5191, a regional jet, crashed after taking off from the wrong runway, killing 49 of the 50 people on board (NTSB.gov). Improving runway safety has been on the NTSB list of "Most Wanted Transportation Safety Improvements" for nearly two decades.

In recent years, the FAA has joined forces with ICAO member countries and industry groups to take an integrated approach to the serious problem of runway incursions. A Global Runway Safety Symposium at ICAO in May 2011 recommended a multidisciplinary approach to improving runway safety outcomes. Terminology and definitions used have been harmonized to provide a better understanding of the problem of runway incursions.

DEFINITIONS

Both ICAO and FAA have jointly defined a runway incursion as follows:

Any occurrence at an aerodrome involving the incorrect presence of an aircraft, vehicle, or person on the protected area of a surface designated for the landing and take-off of aircraft.

In order to promote global harmonization and effective data sharing, both ICAO and FAA have also established a standard classification scheme to measure the severity of runway incursions (Table 10-1).

TABLE 10-1   Severity Classification Scheme (ICAO and FAA)

Severity Classification	Description
A	A serious incident in which a collision is narrowly avoided.
B	An incident in which separation decreases and there is significant potential for collision, which may result in a time-critical corrective/evasive response to avoid a collision.
C	An incident characterized by ample time and/or distance to avoid a collision.
D	An incident that meets the definition of runway incursion such as the incorrect presence of a single vehicle, person, or aircraft on the protected area of a surface designated for the landing and takeoff of aircraft but with no immediate safety consequences.
E	Insufficient information or inconclusive or conflicting evidence precludes a severity assessment.

SOURCE: ICAO Manual on the Prevention of Runway Incursions, Doc 9870, 2007. www.icao.int

According to the latest FAA National Runway Safety Plan (2009–2011), the rate of runway incursions has remained steady during the 4-year period ending in September 2008. The sheer scope of the potential problem is staggering. During this period, there were nearly 250 million operations in the United States (approximately 170,000 takeoffs and landings per day) at FAA towered airports, resulting in 1,353 runway incursions. The majority (92 percent) were categories C and D events involving little or no risk of collision, but the remainder (8 percent) were categories A and B, serious events.

It is clear that the majority of runway incursions are caused by pilot deviation, generally a violation of an ATC direction or other Federal aviation regulations. In the year ending September 2009, 63 percent of runway incursions were caused by the pilot, while 21 percent were caused by vehicle/pedestrian deviations and 16 percent by operational errors caused by air traffic control. In the next 9-month period, ending July 2010, the results were similar: of 787 runway incursions during this period, the findings are as follows (see Fig. 10.1)

- Pilot deviations 66 percent
- Vehicle/pedestrian deviations 18 percent
- Operational errors by ATC 16 percent

The reasons for these significant problems are varied and complex but can be boiled down to human error in a challenging and confusing environment.

Airport Surface Environment

The typical airport surface environment is a complex system of markings, lighting, and signage coupled with layouts that vary by airport. Under adverse weather conditions, the complexity of the environment increases as visibility diminishes and

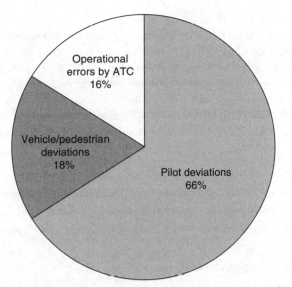

FIGURE 10-1   Runway Incursions by type, October 1, 2009–July 30, 2010. [*Source: FAA National Runway Safety Plan (2009–2011) www.faa.gov*].

visual cues get concealed. Cryptic signs and surface markings together with inexperienced pilots pose additional difficulties as does nighttime flying where airport lights blend with background city lights. It is within this environment that large numbers of individuals with varying levels of experience, training, and language proficiency must concentrate on performing their tasks against a backdrop of intense radio communications for coordinated actions and procedures that are needed for smooth and safe operations—all this while interfacing with large numbers of airplanes and ground vehicles of differing makes and models that are in close proximity to one another. Given the complexity and intensity of the operation, it is easy to understand why even well-trained, highly conscientious individuals remain vulnerable to error, especially when faced with unusual or unexpected situations. Runway incursions are invariably the result of human error. With expected growth in air traffic over the next several years likely to exceed planned capacity requirements, even if the incursion rate is reduced or held steady, the absolute number of these events could increase over time.

Since runway incursions result from human error, it would be beneficial to review some of the jobs and tasks involved with conducting airport operations so that one can get a better understanding of which task elements have the greatest potential to cause errors. Airport ground operations are managed by air traffic controllers, pilots, and airport personnel. These groups jointly manage aircraft and vehicular traffic, and monitor and maintain conditions of runways, taxiways, aprons, signs, markings, and lighting. Controllers serve as the hub for the entire operation. Issuing clearances, instructions, and advisories to pilots and other ground vehicle operators, controllers guide all traffic that operates on the aprons, taxiways, and runways. Controllers monitor the surface conditions together with the identity, location, and movement of aircraft and other vehicles by looking out through the tower windows. Pilots for their part use airport layout maps, and surface taxiway/runway signs, markings, and lighting that they observe from the cockpit window, to guide them from the runways to the gates and back. Pilots and controllers rely primarily on visual feedback to maintain situational awareness and separation. Vehicle operators manage baggage, fuel, catering, and miscellaneous ground handling functions. They are required to follow controllers' instructions to maneuver, especially if they have to cross runways. A sampling of vehicles required at most airports includes

- Fuel trucks for aircraft refueling
- FAA vehicles for monitoring and maintaining navigation equipment, radars, visibility-measuring equipment, and certain lighting systems
- Airport authority vehicles to check for foreign objects and monitor runway conditions
- Maintenance vehicles for maintenance tasks
- Snowplows and mowers for maintaining runways and surrounding areas
- Baggage, catering, and utility trucks for flight service
- Emergency vehicles for crash and fire rescue
- Construction vehicles for maintaining airport surfaces or new construction

As can be seen from the preceding discussion, there is ample opportunity for error given the significant combination of tasks, procedures, and equipment that must be coordinated to permit the operation of the system.

CONTROL STRATEGIES AND FUTURE INITIATIVES

The FAA's Office of Runway Safety has sponsored several initiatives to improve this critical safety area. In its document entitled *National Runway Safety Plan 2009–2011*, FAA outlines several programs, which should further the progress of increasing runway safety over the next several years:

SAFETY MANAGEMENT SYSTEMS (SMS) IMPLEMENTATION.   In 2005, ICAO directed its member nations to establish an SMS program at all certificated international airports. FAA has taken the following steps to implement SMS externally and internally at it more than 560 airports certificated under FAR Part 139:

- The FAA's notice of proposed rulemaking was issued in October of 2010. FAA will review existing Part 139 requirements including all public and industry comments before issuing a final rule.
- FAA has issued Advisory Circular 150/5200-37 to introduce SMS to U.S. airports. Runway safety will remain a top priority using the SMS process.

- Internally, SMS will eventually encompass all FAA runway safety processes. Working through the Runway Safety Council, a working group has been formed to examine the root causes of runway incursions.

TRAINING AND EDUCATIONAL OUTREACH PROGRAMS

- Quarterly air traffic controller refresher training at every control tower.
- Pilot training and instruction during flight instructor review clinics and during flight reviews to increase situational awareness.
- Utilization of the FAA Safety Team and the AOPA Air Safety Foundation to promote educational programs including runway safety videos at both the national and regional levels. These outreach programs will train general aviation pilots in runway safety principles.
- Runway Safety Action Teams have been established to develop action plans for specific local airports.
- FAA promotion of industry association and ICAO runway safety seminars such as the Global Runway Safety Symposium in Montreal, Canada (May 2011).

TECHNOLOGY DEVELOPMENT TO CONTROL RUNWAY INCURSIONS

- Airport Movement Area Safety System (AMASS) has been installed at the nation's top 34 airports to provide an automatic visual and audio alert to controllers when the system detects potential collisions.
- Airport Surface Detection Equipment, Model X (ASDE-X) is being deployed at 35 of the busiest U.S. airports to provide precise surface detection automatically using a combination of radar, transponder, and ADS-B technology.
- Runway Status Lights (RWSL) are being deployed at 22 large airports. RWSL are red lights embedded in the pavement to signal potentially unsafe situations.
- Final Approach Runway Occupancy Signal (FAROS) is designed to provide a visual alert of runway status to pilots in flight automatically flashing the Precision Approach Path Indicator (PAPI) lights whenever a runway is occupied.
- Electronic Flight Bag (EFB) with Moving Map Displays. Using this new technology, pilots are able to see exactly where their aircraft is on the airfield.
- Low Cost Ground Surveillance (LCGS) Systems for use at certain small and medium-sized airports which provide ASDE-X–like situational awareness.

LOCAL AIRPORT SOLUTIONS   As previously discussed, FAA encourages development of runway safety initiatives at the local level to solve problems that are unique to a particular airport. In addition, participation in information sharing at the regional, national, and global levels is encouraged so that successes and lessons learned in one locale can be applied with changes, where appropriate, to other regions. Local airports and regions are encouraged to maintain and update a central information source of newly implemented corrective actions and their effectiveness in preventing runway incursions. To promote safer ground operations, the FAA offers all airports the ability to create their own web sites on which they can post airport-specific information and advisories. Informative booklets on airport basics, operations at towered and nontowered airports, phraseology, and emergency procedures are also readily available from the FAA.

CONCLUSIONS

Despite causing civil aviation's worst disaster at Tenerife in 1977, runway incursions are very infrequent events, and collisions that result in deaths are even more infrequent. The FAA's report on incursion severity for the period of FY2004 to FY2007 documents a total of 1353 runway incursions spread over nearly 250 million hours of operation. This approximates 5.4 runway incursions per million operations. Of these 1353 incursions, 92 percent were minor in nature. While this is a good record by any measure, there is serious concern that an increasing number of airplanes, vehicles, and people are mistakenly coming dangerously close to each other. With predicted increases in air traffic volume in a system that is already constrained, the number of incursions would most certainly increase. What's more, even if the incursion rates stay the same, there will be an increase in the absolute number of incursions if nothing is done about it, which of course would increase the probability of catastrophic incursions. It is for this reason that the FAA places a very high priority on controlling runway incursions. Despite the fact that pilot errors result in more runway incursions than any other cause, minimizing runway incursions requires a joint effort from all groups involved in the management of the airport operations system, i.e., the pilots, controllers, vehicle operators, and other ground staff.

KEY TERMS

Airport

Airport Operating Certificate

FAA Advisory Circulars

CertAlerts

Airport Certification Manual

Airport Emergency Plan

National Fire Protection Association (NFPA)

American National Standards Institute (ANSI)

American Conference of Governmental Industrial Hygienists (ACGIH)

Bonding

Grounding

Aircraft rescue and firefighting (ARFF)

Runway incursion

Final approach runway occupancy signal (FAROS)

Electronic Flight Bag (EFB)

Low Cost Ground Surveillance (LCGS)

Operational error

Pilot deviation

Vehicle/pedestrian deviation

Runway incursion incident rates

Runway incursion severity categories (A, B, C, and D)

Airport surface detection equipment (ASDE)—X series

Airport movement area safety system (AMASS)

Runway status lights (RWSL)

International Civil Aviation Organization (ICAO)

REVIEW QUESTIONS

1. Which act was responsible for initiating the airport certification process and what were its salient features?

2. What part number regulates airport certification? What are its highlights, and when was it first promulgated?

3. Discuss the highlights of the airport certification process and their significance.

4. What are the four classes of airport certifications and what are their differences?

5. What is the Airport Certification Manual and why is it so significant?

6. List at least 10 elements required in the Airport Certification Manual.

7. Where would one locate the requirements of the Airport Certification Manual?

8. List and explain the required elements of an Airport Emergency Plan.

9. List five to seven safety hazards associated with each of the following:
   a. Airport terminal buildings
   b. Hangars and maintenance shops
   c. Ramp operations
   d. Aviation fuel handling
   e. Aircraft rescue and firefighting (ARFF)
   f. Deicing

Biggest Risk / Ways to get in trouble

10. What is a runway incursion?

11. List and explain the different categories of runway incursions.

12. Describe the airport surface operational environment.

13. List individual groups together with their functions and the equipment they use for airport surface operations.

14. Define and discuss the following terms: *runway incursion, pilot deviation, operational error and vehicle/pedestrian deviation.*

15. Discuss runway incursion trends with respect to rates and severity.

16. List and explain several strategies to minimize the occurrence of runway incursions.

17. List and explain the technologies being proposed to eliminate runway incursions.

18. What is being done in training, and outreach programs to prevent runway incursions?

REFERENCES

Barnett, A., G. Paul, and J. Ladelucs. 2001. *Fatal US Runway Collisions Over the Next Twenty Years*, FAA Grant Report.

Flight International. 2000. *Airline Safety Review of the 1990's*. January 25 issue. Reed Business Information.

FAA 2010, *Runway Safety Program*. FAA Order 7050.1A Sep 16, 2010

———— 2009, *FAA Annual Runway Safety Report 2009*

———— 2009, *FAA National Runway Safety Plan, 2009–2011*

———— 2011, *FAA Runway Safety News/Current Events*

Landsberg, B. 1995. "Stop Look and Listen." *AOPA Pilot*, vol. 38, no. 8, pp. 156–159.

National Safety Council. 2000. *Aviation Ground Operation—Safety Handbook*. 5th ed. Itasca, Ill.: National Safety Council Press.

Rodrigues, C. C. 2001. "Safety Analysis of Runway Incursions." *Proceedings of the International Conference on Industrial Engineering—Theory, Applications and Practice*. November 18–20. San Francisco, Calif.

Wenham, T. P. 1996. *Airliner Crashes*. Stephens Limited, Sparkford, Somerset, UK.

Wood, R. H. 1997. *Aviation Safety Programs—A Management Handbook*. 2d ed. Englewood, Colo.: Jeppesen Sanderson, Inc.

WEB REFERENCES

AOPA. 2011. *Air Safety Foundation Runway Safety Program*. http://www.aopa.org/

FAR Part 139 http://www.access.gpo.gov

Flight Safety Foundation May 2009, *Reducing the Risk of Runway Excursions, Report on the Runway Safety Initiative*. www.flightsafety.org

NTSB 2011, Most Wanted Transportation Safety Improvements, Aviation, Improve Runway Safety www.ntsb.gov

AVIATION SECURITY

LEARNING OBJECTIVES

After completing this chapter, you should be able to

- Distinguish between security and safety.
- List the categories of attacks on civil aviation.
- Discuss the salient features of:
  - ICAO Annex 17 and ICAO Security Manual
  - Security Management Systems (SeMS)
  - The 9/11 Commission and its recommendations
  - The Homeland Security Act
- Explain the role of the Transportation Security Administration (TSA).

- Discuss the role of intelligence gathering and analysis regarding the security threat.
- Explain the mission of the National Counterterrorism Center (NCTC).
- Discuss the importance of the Intelligence Reform and Terrorism Prevention Act of 2004 (IRTPA).
- Discuss the theory and concepts behind the following security strategies:
  - Imaging technologies
  - Explosive trace detection technology
  - Explosive detection systems (EDS)
  - Metal detectors
  - Biometrics and future checkpoint systems
  - Strengthening of aircraft and baggage containers
  - Cockpit door reinforcement
  - Computer-assisted passenger screening

INTRODUCTION

Security and safety are often considered synonymous topics, and the discussion of one invariably invokes references to the other. However, it is important to realize that there is a fundamental difference between the two. *Safety* usually refers to measures taken against the threat of an accident, whereas *security* refers to protection from threats motivated by hostility or malice. While the end objective of safety and security is to minimize risk by preventing injuries and loss of lives and property, their core safety practices are designed to prevent *unintentional* acts whereas security practices are designed to avert *intentional* acts. The intentional acts addressed in this chapter are the attacks against commercial aviation. While attacks on civil aviation can be carried out against airliners, airports, and airline offices, airliners have been the most common target for attacks. The spectrum of attacks on civil aviation includes

- Bombings, shooting, and miscellaneous attacks at airports
- Bombings, shooting, hijackings, and miscellaneous attacks (e.g., commandeering) on aircraft on the ground or in flight
- General-aviation and charter aviation aircraft attacks
- Off-airport facility attacks
- Shootings (from the ground) at aircraft during takeoff and landing

This chapter reviews the history of attacks against civil aviation together with the regulations and security measures that evolved to minimize the occurrence and severity of these attacks.

REVIEW OF ATTACKS ON CIVIL AVIATION

The first attack on civil aviation dates back to 1930 when Peruvian revolutionaries seized a Fokker F-7 aircraft in South America. The incident received little attention and never resulted in any international effort to combat potential threats to international aviation. At the time of ICAO's formation in Chicago 1944, no one foresaw aviation safety as a serious threat; hence it was not addressed. Therefore, until the mid-1960s, airlines and airports gave security matters little attention. Low-technology security applications, such as airport fences, were intended as a safety measure to separate aircraft from wildlife rather than terrorists. However, attacks against civil aviation rose rapidly in the 1967–1976 decade. While there were only 32 worldwide hijackings from 1961 through 1967, there were 290 hijacking attempts (successful and unsuccessful hijackings) worldwide during the following 4 years after 1968. In the United States, from 1930 to 1967, only 12 commercial aircraft hijackings were attempted, of which only seven were semisuccessful (Dorey, 1983). However, in 1968 there was a surge of hijackings in the United States in which several flights traveling to the Caribbean were repeatedly detoured to Havana, Cuba. Of the 22 hijackings in 1968, 19 of these flights were redirected to Cuba. The climax occurred in 1969 when 33 regularly scheduled U.S. airliners were hijacked (seven of these attempts failed). By 1969, the number of U.S. passengers and crewmembers who were being detoured to Cuba totaled 1359 (Moore, 1991). This led to a great deal of money being spent on security by the air carriers and the FAA on x-ray systems, magnetometers, training programs for screening personnel, and air marshals, after which the relative level of security immediately rose. Consequently, the threat of hijacking in the United States as well as the actual incidences of hijackings diminished over time, giving way to more frequent occurrences of both threats and actual assaults against U.S. flag carriers in Europe, Asia, and Africa, peaking with the bombing of Pan Am Flight 103 over Lockerbie, Scotland, in December 1988. The decades that followed saw a decline in attacks against aviation. As can be seen from Fig. 11-1, all three major forms of attack

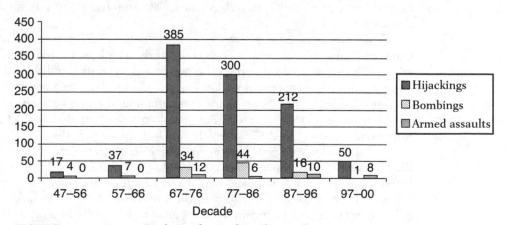

FIGURE 11-1   Frequency of main forms of attack on airlines. (*Source: http://www.gwu.edu/*)

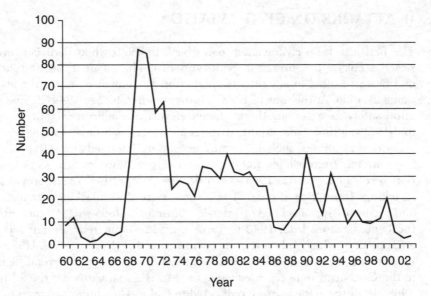

FIGURE 11-2   Worldwide air carrier hijackings.

on airliners display almost the same trend over time, rising steeply in the 1967–1976 decade and then declining in the following two decades. Also, note the 1969 peak in the 40-year worldwide hijacking trend in Fig. 11-2.

From 1947 to 1996 there were 1098 total incidents of attacks on airliners, compared with 129 attacks on airports and 249 attacks on airline offices. Unlike attacks on airliners, attacks on airports (bombing and armed assaults) peaked in the 1977–1986 decade. With regard to the modes of attack, hijackings were by far the most common form of attack on commercial aviation. Hijackings constituted 87 percent (959 of 1098) of all incidents on airliners during this 50-year period. The other significant forms of attack on airliners were bombings (on the ground or in midflight) and armed assault (shooting at aircraft on the ground or in flight, throwing hand grenades, etc.).

While the number of attacks was on the decline, the severity of these attacks demonstrated a different trend. The decade of the 1980s was a disastrous one for aviation. This period confirmed the existence of a dangerous trend toward greater violence against air transportation. Overall, 25 planes were sabotaged by explosives, causing 1237 casualties. Some of the more infamous incidents included the Air India bombings in June 1985 in which 329 people perished off the coast of Ireland, the Korean Air disaster in 1987 in which 115 lost their lives, the midair explosion in 1989 of the French UTA flight over the Niger desert in which 171 were killed, and the bombing of Colombian Avianca in 1989 near Bogotá in which 107 people died. In addition to these sabotages there were 17 other documented failed attempts between 1982 and 1987, as noted in the President's 1989 Commission Report on Aviation Security and Terrorism. By comparison, there were 650 deaths in the 1970s and 286 deaths in the 1960s. Things cooled off in the 1990s only to give way to the events of September 11, 2001, when 3247 (including onboard passengers, crew, and people on the ground) were killed in four separate terrorists acts on the same day.

The increases in severity of aviation attacks were the results of terrorists changing their tactics and philosophies and making use of new technologies. They gained access to more sophisticated and lethal technologies, such as automatic weapons and deadly plastic explosives. They attacked airports by using pistols and bazookas, and they developed numerous ingenious ways to turn innocuous-looking suitcases and radios into lethal bombs. The character of airline hijackings also changed from the lone hijacker of the early 1960s who was making a personal or political point to its use by the 1970s as an organized terrorist tactic. In the era leading up to September 11, 2001, hijacking for the most part did not explicitly threaten the lives of hostages. The operating philosophy was to negotiate with the hijackers until an amicable settlement was reached, thereby minimizing casualties. However, after the September 2001 attacks, where hijackers rammed airplanes into buildings, murdering passengers and people on ground and killing themselves, things have definitely changed in anti-hijacking tactics. The last aircraft hijacked on September 11, 2001 was United Airlines Flight 93. On that flight, after learning of the World Trace Center attacks, the passengers and crew were able to thwart the suicide mission of the terrorists by mounting an assault against the hijackers. The Boeing 757 crashed in Somerset County, Pennsylvania, during the struggle and was the only hijacked aircraft that failed to reach its intended target (the U.S. Capitol) on that fateful day.

During the following decade, hijackings have been largely unsuccessful and aircraft defenses have significantly improved. Flight attendants and pilots receive extensive anti-hijacking training. Passengers have also helped prevent terrorists from igniting explosives (shoe bombing attempt on December 22, 2001 and underwear bombing attempt December 25, 2009), and hijackers have been arrested and brought to justice in nearly every incident since September 11, 2001.

THE REGULATORY MOVEMENT

The terrorist events of September 11, 2001 have indeed changed the Aviation Security (AVSEC) landscape on a global basis. There have been significant regulatory developments in the international arena and in U.S. policy along the way. The international developments will be discussed briefly, followed by a historical discussion of the evolution of aviation security in the United States.

INTERNATIONAL RESPONSE TO TERRORISM

As stated above, no one considered aviation security as a major issue at the time of the formation of ICAO at the Chicago Convention in 1944. The focus at that time was the planning and development of air transport to ensure safety of flight in international air navigation. Gradually, through a series of ICAO meetings and conferences, the international community began its focus upon crimes committed on board aircraft. Some major milestones of International Aviation Security include:

- *Tokyo Convention*—on offenses and certain other acts committed on board aircraft (1963), recognized the issues and authorized the airline captain to take appropriate action to restrain persons interfering with safety of flight.
- *Hague Convention*—for the suppression of unlawful seizure of aircraft (1970), adopted 14 articles relating to hijacking and provided guidelines to governments dealing with this problem.
- *Annex 17*—SARP entitled: "Security-Safeguarding International Civil Aviation against Acts of Unlawful Interference" (1974). This SARP sets forth the basis for the ICAO Civil Aviation Security program and outlines the minimum standards for aviation security worldwide. It has been a living document, evolving with security problems around the world. The 12th amendment to Annex 17 has been approved by the ICAO council for adoption in 2011.
- *ICAO Security Manual*—(Doc 8973—restricted) is the primary document providing member States with guidance material to assist with the implementation of international security measures. The seventh edition of Doc 8973 (2010) consists of five volumes:
  - Volume I: National Organization and Administration
  - Volume II: Recruitment, Selection and Training
  - Volume III: Airport Security, Organization, Programme and Design Requirements
  - Volume IV: Preventive Security Measures
  - Volume V: Crisis Management and Response to Acts of Unlawful Interference

In recent years, ICAO has been deeply involved in strengthening worldwide aviation security. In light of the attempted sabotage of Northwest Airlines Flight 253 on December 25, 2009 (the underwear bomber), the 37th session of the ICAO assembly issued a strong joint Declaration on Aviation Security in Montreal in October, 2010, urging member States to take action. The declaration seeks to enhance international cooperation to counter threats to civil aviation using an eight point program, strengthening, among other things:

- Focus on ICAO Annex 17—Security, to address current and emerging threats.
- Security screening procedures using modern technology.
- New security measures to protect airport facilities and improve in-flight security.
- Harmonized measures and best practices for air cargo security.
- Enhanced travel document security (especially machine readable travel documents).
- Correction of deficiencies identified under the Universal Security Audit Programme (USAP).

- Increased information exchange and early detection and dissemination of information on security threats.

- Shared best practices in key security areas.

A "Security Roadmap on Aviation" was recently adopted at the ICAO Regional Conference in New Delhi, India on February 18, 2011. The roadmap was applauded by the U.S. TSA Administrator and world officials as a means to enhance the ICAO Joint Declaration on Aviation Security to address ever-changing and emerging security threats in the global arena. The International Air Transport Association (IATA) also advocates a business-like approach to security that builds on ICAO Annex 17 standards under the banner of "Security Management Systems" (SeMS). As stated in the IATA fact sheet, similar to SMS, a Security Management System is integrated into a company's overall business to make security one of the company's core values. In this manner, a security culture is formed with risk management processes that require continuous assessment of security threats, risks, and vulnerabilities. IATA has identified six core elements in an air carrier's SeMS:

- Senior Management and Corporate Commitment
- Resource Management
- Threat Assessment and Risk Management
- Management of Emergencies and Incidents (resilience)
- Quality Control and Quality Assurance
- A documented aviation security program

IATA is a strong proponent of the SeMS concept which could lead to continuous security improvements on a global basis. The foundation and development of this program is on the agenda of the ICAO Aviation Security Panel during its meetings in 2011.

EVOLUTION OF AVIATION SECURITY IN THE UNITED STATES

Similar to ICAO, the FAA response to aviation security has slowly evolved. This section will discuss the history of aviation security in the United States and the changes brought about by the events of September 11, 2001.

THE HIJACKING ERA (1968–1987).   In response to the major increase in hijackings during this era, FAA began to beef up the Federal Aviation Regulations. Initially, airport security matters were governed under FAR Part 139, Subpart 335 (Public Protection) and Subpart 337 (Wildlife Hazard Management) as discussed in Chapter 10. These sections were intended to prevent inadvertent entry of unauthorized persons or vehicles to the aircraft movement area and to prevent damaging collisions with wildlife other than birds through fencing or other means. Subpart 325 (Airport Emergency Plan) required each certificate holder to develop and maintain an airport emergency plan designed to minimize the possibility and severity of personal injury and property damage on the airport in an emergency. This concentrated more on managing the emergency rather than on preventing it. Part 139 was mainly concerned with deterring *mistaken* entry of humans and animals into air operations areas. However, to stem the increase of hijackings during the 1970s, there was a need to specifically deter access to air operations areas by individuals and ground vehicles that were *unauthorized* to do so. FAR Part 107 (entitled Airport Security) was therefore issued in 1972, which required the airport operator to immediately adopt and put into use facilities and procedures designed to prevent persons and vehicles from unauthorized access to air operations areas. Under Part 107, airports were required to

- Improve or establish protection against unauthorized access to air operations areas.
- Establish authorized access to air operations areas through a suitable identification system.
- Identify vehicles operating in air operations areas.

As an example, FAR 107.14 mandated the upgrading of access control identification and security systems at all major commercial airports in the United States. Such systems were required to ensure that only authorized personnel could enter restricted areas. Additionally, passenger screenings as well as carry-on item checks were required for high-density flights. By the end of 1972 several hijacking acts, including the unenviable distinction of a California man being hijacked twice in a single trip, led the government to issue two emergency regulations. One gave carriers 30 days in which to institute a 100 percent search of all passengers and carry-on items, and the other gave airport operators 60 days in which to station at least one law enforcement officer at each passenger checkpoint during boarding and reboarding. These resolutions caused several problems for carriers and airport operators. At most airports, carriers had been sharing a single metal detector (magnetometer). This had been sufficient because they were required to screen only "selected" individuals. Now every passenger was required to be screened, and there were not enough magnetometers in existence to meet the new requirements. Carriers did not have enough time to recruit, hire, and train the required number of additional employees.

Airports' funding problems were equally severe. Funds were not budgeted for full-time law enforcement officer coverage, and the airports and their city or joint authority were hard pressed to find the necessary additional funds in such a short time. Despite all these problems on January 5, 1973, the 100 percent screening of air passengers and their carry-on items was instituted. The results, as expected, were long lines of travelers who were quite tolerant of these procedures that were in place to stop hijackings. The significant difference between this rule and previous antihijacking measures was the universality of the new regulation. Previously, FAA had required air carriers to conduct a weapons scan of only those passengers who fitted a hijacker profile, about 1 percent of the 500,000 passengers boarding airliners daily. As a result of this program, not a single airliner was hijacked in the United States in 1973.

AIR CARRIER SECURITY.   On January 31, 1972, Part 121.538 was issued to cover the air carrier community. Within this rule, air carriers were required to adopt and implement a screening system that would detect weapons and explosives in carry-on baggage or on the person of passengers. An order was issued requiring each carrier to submit its screening program to the FAA Administrator no later than June 5, 1972. The carrier's security program was required to

- Prevent or deter unauthorized access to its aircraft.
- Ensure that a responsible agent or representative of the certificate holder would check in baggage.
- Prevent cargo and checked baggage from being loaded aboard its aircraft unless they were handled in accordance with the certificate holder's security procedures.

In response to a Presidential statement ordering renewed emphasis on air security, Part 121.538 was amended again on March 9, 1972. The screening requirement was to be put into immediate effect, and the new deadline became May 8, 1972. Further refinements took place on September 11, 1981 when Part 121.538 was rewritten as Part 108 of the FARs. Both parts were first amended as a result of the President's 1972 Commission Report on Aviation Security and Terrorism.

FAR Part 108 (entitled Aircraft Operator Security) based security requirements upon aircraft complexity rather than upon certification. It categorized airplanes into three groups according to configured seating capacities: more than 60 seats, 31 to 60 seats, and less than 31 seats. The rationale used by the FAA was that larger airplanes would be more attractive to the hijacker. Part 108 required the adoption of a comprehensive security program for operations with 31 through 60 seats. The program was to be comparable to that required for operations with airplanes having more than 60 seats. The smaller operator, however, would only have to implement those portions of the program that required the following:

- Procedures for contacting a law enforcement agency and arranging for a response to an incident when needed.
- Instructions for all crewmembers and internal employees in the appropriate procedures. Each operator must also be prepared to implement its full security program upon notification of specific threats by the FAA.

For operation of smaller aircraft of 1 to 30 seats, no security program was required unless passengers had uncontrolled access to a sterile area. In those cases where passengers had uncontrolled access, or where passengers were discharged into a sterile area, provisions were required to properly screen the passengers. Carriers were required to control access to the sterile area through surveillance and escort procedures or through screening procedures of another carrier.

In summary, by virtue of federal regulations in Part 107, airports were responsible for preventing unauthorized access to the air operations area and for providing law enforcement support at passenger-screening stations. Part 108 required airlines to institute programs that would prevent firearms and explosives aboard their aircraft through passenger and baggage screening and other methods. While Parts 107 and 108 addressed criminal acts and aircraft sabotage, Part 139 addressed operational safety issues of separating the public from air operations.

ANTI-HIJACKING OR AIR TRANSPORTATION SECURITY ACT OF 1974.   This Act (Public Law 93-366), promulgated in August 5, 1974, contained two titles:

- Title I, Anti-hijacking Act (punitive provisions for hijackers and security standards for foreign air transportation and services)
- Title II, Air Transportation Security Act (security regulations for U.S. airports and carriers)

This law provided the statutory basis for the December 5, 1972, rules requiring carriers to institute 100 percent screening of passenger and carry-on items, and for airport operators to station at least one law enforcement officer at each passenger checkpoint during boarding and preboarding. Additionally, as part of its obligation under this Act, the FAA began a research and development program that emphasized the development of devices to protect air travelers against acts of criminal violence and aircraft piracy. Application of the above-mentioned regulations varied greatly across the airlines. Regulatory requirements were not implemented consistently across the industry. Airlines had complete discretion on how they implemented the requirements. They also had the right to express their concerns and suggest changes based on trial and error. This resulted in inconsistent application of regulations across the airline industry, and in early 1975 the Air Transportation Association sought to work out a single standard security program. Their effort produced the FAA's Air Carrier Standard Security Program, which attempted to bring some structure to the diverse interpretations of the new rules. Today, the Transportation Security Administration (TSA) has continued this program now known as the Aircraft Operations Standard Security Program (AOSSP).

FAR Part 109 regulations that were established in 1979 govern indirect air carrier security and provide additional protection against criminal activity. This part prescribes aviation security rules governing each air carrier, including air freight forwarders, food service, and cooperative shipping associations engaged indirectly in air transportation of goods. Each indirect air carrier is required to have a security program designed to prevent or deter the unauthorized introduction of explosives or incendiary devices into any package cargo intended for transportation by air.

AIR MARSHAL PROGRAM ESTABLISHED. The increase in the rate of hijackings during the late 1960s and early 1970s caused the public to exert pressure on the U.S. government to implement security procedures at airports and to mandate security requirements for U.S. air carriers. This resulted in the establishment of the Anti-Hijacking Program of the Federal Aviation Administration. One element of this program was the federal Air Marshal Program, which began as the *Sky Marshal Program* in 1968 and continued through the 1970s and part of the 1980s as a program that was initially instituted to stop hijackings to and from Cuba. The Air Marshal Program was criticized by airlines and passengers who considered it more an airborne risk than a deterrent for hijackers. This program gained in importance, however, after the hijacking of TWA 847 in June 1985. During this incident, two Lebanese Shiite Moslems hijacked a Boeing 727 departing from Athens and diverted it to Beirut, where they were joined by additional hijackers. During a 2-week confrontation, the hijackers demanded the release of Shiite prisoners held by Israel and murdered Robert Stethem, a U.S. Navy diver who was a passenger on board the plane. In response to this hostage nightmare and the rapid surge in Middle East terrorism, then-President Ronald Reagan directed the Secretary of Transportation, in cooperation with the Secretary of State and the Attorney General, to explore immediately an expansion of the FAA's armed Federal Air Marshal Program aboard international flights for U.S. air carriers. On August 8, 1985, Congress enacted *Public Law 99-83*, the *International Security and Development Cooperation Act*, which established the explicit statutory basis for the FAA-Federal Air Marshal Program and allowed for assessment of security at foreign airports and approval of foreign air carrier security programs. This statute authorized Federal Air Marshals to carry firearms on board and to make arrests without warrant for any offense against the United States committed in their presence, if they had reasonable basis to believe that the person to be arrested had committed or was committing a felony. Three weeks after the TWA 847 hijacking, the FAA imposed new regulations, requiring that all scheduled carriers and public charter operators carry Federal Air Marshals on a priority basis, without charge, and that they provide seating selected by the marshals, even though it might mean bumping full-fare passengers. Today TSA has assumed responsibility for the Federal Air Marshal Service whose mission is to detect, deter, and defeat hostile acts targeting U.S. air carriers, airports, passengers and crews.

THE BOMBING ERA (1988–2000). On December 21, 1988, Pan Am flight 103 exploded over the village of Lockerbie, Scotland, killing the 259 people aboard and 11 people on the ground as well as damaging several residential buildings. Investigations by a flight commission found that the crash was due to an explosion of a bomb placed in the luggage compartment of the aircraft. It was asserted that the luggage was coming from direct passengers boarding in Frankfurt, as well as from some possible suitcases transferred from Air Malta flight 180 to Pan Am 103 at the Frankfurt Airport.

Prior to the bombing of Pan Am 103, the U.S. civil aviation security system was not sophisticated enough to detect the level of device used to destroy the plane. The system, created in the early 1970s, was designed to prevent hijackings. In the mid-1980s, it was converted to an antisabotage system at high-threat international locations. Accordingly, the 1988 Pan Am 103 tragedy can be regarded as an aberration because the U.S. antisabotage security measures in the FAA's Air Carrier Standard Security Program (ACSSP) were designed to prevent such a tragedy.

The major issue in the Pan Am 103 incident is the reconciliation of baggage and passengers. U.S. carriers were required to conduct a positive baggage–passenger reconciliation in 1988 at designated international locations. According to investigation findings, the Pan Am tragedy occurred because the airline was x-raying all interline bags at certain international high-threat locations instead of conducting a reconciliation process and physically searching all unaccompanied bags. Because Pan Am did not identify and physically search all the unaccompanied interline bags, flight 103 left Frankfurt with several extra bags, one of which contained a bomb.

This failure was duly noted in the 1990 report of the Presidential Commission on Aviation Security and Terrorism, established in the aftermath of the disaster. The commission's mandate was to comprehensively study and evaluate the practices and policy options with respect to preventing terrorist acts involving aviation. The commission made a number of recommendations to prevent the recurrence of such a tragedy. The *Aviation Security Improvement Act of 1990* (Public Law 101-604), passed on November 16, 1990, implemented many of the recommendations of the commission. This Act has been described in the 1992 FAA Annual Report to Congress as perhaps the most comprehensive, far-reaching legislative initiative designed to improve all aspects of aviation security. It mandated many regulatory actions affecting several agencies, required new reports, created new organizations and staffing requirements, and empowered the FAA to promote and strengthen aviation security through an expedited, more focused research and development (R&D) program.

The Aviation Security Improvement Act of 1990 generated many changes, especially within the FAA. It created a new high-level position of assistant administrator for civil aviation security to head the office of Civil Aviation Security. This office now dealt with all aspects of security, including drug and narcotics interdiction. In addition, a new office of Intelligence and Security was established within the U.S. DOT to coordinate transportation security activities. It is headed by a director of intelligence and security, who reported directly to the Secretary of Transportation.

The 1990 Act also charged the FAA with enhancing its oversight of airport security and appointed FAA security managers at major airports.

WHITE HOUSE COMMISSION ON AVIATION SAFETY AND SECURITY, 1996. The late 1990s witnessed significant changes in the direction and emphasis of aviation security in the United States. The principal triggering event for this new emphasis and importance was the catastrophic loss of TWA 800 off Long Island, New York, in July 1996. The early model Boeing 747 was carrying 230 passengers and crew when it exploded minutes after departing John F. Kennedy International Airport, bound for Paris. The terrific force of the explosion had torn the aircraft apart, and the disturbing recovery images, along with vivid eyewitness accounts, riveted the attention of a shocked U.S. public for many weeks; but the irony was that the FBI eventually ruled that TWA 800 was not the result of a terrorist act.

Nevertheless, the traveling public was frightened, and the media questioned the perceived safety and security of domestic airline operations. Within weeks, President Clinton announced the creation of the *White House Commission on Aviation Safety and Security* (the Gore Commission) which outlined sweeping changes calling for regulatory reform and additional research directed toward new, safer technologies. Special attention was given to an action plan to deploy new high-technology machines to detect the most sophisticated explosives. The report included 57 recommendations, 31 dealing with improvements in security for the traveling public. The commission recommended that the federal government consider aviation security a national security concern. It created within the FAA a new high-level position of assistant administration for civil aviation security to head the office of Civil Aviation Security.

Several other events had come together at about the same time to enable this significant change to occur. First, several major terrorist events within the United States, beginning in 1993, made it clear to the U.S. public that the existence of two large oceans no longer guaranteed the absence of major international terrorist acts on our territory. These attacks were well known at that time and included the World Trade Center garage bombing and the murders at the headquarters of the Central Intelligence Agency near Washington, D.C.

The existence of a serious terrorist threats within the United States impelled the FAA to convene an outside advisory panel composed of representatives from other government agencies, air carriers and airport authorities, and various citizens and professional groups with the purpose of recommending improvements in baseline aviation security measures. The working group's recommendations were passed on to the *White House Commission on Aviation Safety and Security*, and had a major impact on its first and final reports. Many of the Gore Commission's recommendations (over 30 dealing with security issues) were given the force of law and financing by the passage of the *Aviation Security and Anti-terrorism Act of 1996.*

A second piece of the groundwork for the major change in aviation security was the emergence of successful new security technology, both in explosives detection and in other areas such as human factors and aircraft container hardening. The existence of an approved explosives detection system, made it conceivable that effective technical measures could be taken to block the introduction of explosives aboard aircraft. Further, the rapid development and improvement of trace explosives detectors raised the possibility of redundant technical measures to check baggage for explosives, based on a totally different technical approach.

Combined with the apparently successful bombing of TWA 800, practically within sight of New York City, the situation in July 1996 made the social and political pressure to institute significant improvements in baseline security measures irresistible. Within 3 months, congressional legislation appropriated federal funds for a large-scale purchase of expensive security equipment.

Legislation authorized other security enhancements, such as background checks on security screeners, vulnerability assessments at airports, and the increased use of dogs for detecting explosives. While security implementation strategies continued throughout the late 1990s, they lacked the urgency and thoroughness that would be needed to thwart the events of September 11, 2001. Improvements in aviation security developed complications because government and industry often found themselves at odds, unable to resolve disputes over financing, effectiveness, technology, and potential impacts on operations and passengers. To make a point of this, the inspector generals (IGs) of the FAA conducted approximately 173 tests at eight U.S. airports between December 1998 and May 1999. Much to their dismay, the inspector generals were able to gain a total of 117 unauthorized accesses to restricted areas during the testing period. The IGs were able to gain access by following authorized employees into restricted areas, riding unguarded elevators, going through unlocked gates, and walking through cargo areas. The investigators concluded that problems that plagued airport operators in the past still existed, and that airport operators and air carriers had major deficiencies in the areas of unauthorized access control procedures, employee training, and FAA oversight programs (Anderson, 2000).

THE GLOBAL WAR ON TERROR (2001 TO DATE).   The terrorist events of 2001 changed the face of aviation forever while fundamentally modifying the thinking and approach to security in the United States. This catastrophic event called for marked new trends and measures in security not only in aviation but for all modes of transportation. On September 11, 2001, four passenger aircraft were hijacked and crashed by terrorists. The first airplane was American Airlines 767 (Flight 11) which was flying from Boston to Los Angeles when it was hijacked and flown into one of the World Trade Center towers. All 11 crewmembers, 76 passengers, and 5 hijackers were killed. The second jet that was crashed into the second World Trade Center tower was United Airlines 767 (Flight 175). All 9 crewmembers, 51 passengers, and 5 hijackers were killed. The impacts against the towers, together with the heat generated from the explosion of the aircraft, caused both towers to collapse. The third aircraft was American Airlines 757 (Flight 77) on a flight from Dulles to Los Angeles. It was hijacked and flown into the Pentagon, collapsing part of the structure. All 6 crewmembers, 53 passengers, and 5 hijackers were killed. The fourth aircraft was United Airlines 757 (Flight 93) on a flight from Newark to San Francisco. It was hijacked and crashed into a field near Pittsburgh. All 7 crewmembers, 34 passengers, and 4 hijackers were killed. The total death toll from this event was 3274. Next, the U.S. response to these serious aviation security problems will be discussed.

THE 9/11 COMMISSION (www.9-11commission.gov)

The National Commission on Terrorist Attacks Upon the United States ("The 9/11 Commission") was an independent bipartisan commission created by Congress and the President to prepare a full and complete account of the circumstances surrounding the terrorist attacks, and to provide recommendations designed to guard against future attacks. The full comprehensive report of the commission was issued on July 22, 2004, consisting of 585 pages. The executive summary of the report, which is available on the 9/11 Commission's Web site listed above, outlines sweeping general and specific findings, plus numerous recommendations, (source: www.9-11commission.gov)

General findings:

- *Imagination*—The most important failure was one of imagination. The Commission did not believe national leaders understood the gravity of the terrorist threat.
- *Policy*—Terrorism was not the overriding national security concern for the U.S. government.
- *Capabilities*—The United States tried to solve the al Qaeda problem with old, insufficient, Cold War capabilities.
- *Management*—The United States missed opportunities to thwart the 9/11 plot, and could not find a way to pool intelligence information.

Specific weakness and adverse findings were made by the 9/11 Commission in the following areas:

- Unsuccessful diplomacy
- Lack of military options
- Problems within the intelligence community
- Problems in the FBI
- Permeable borders and immigration controls
- Permeable aviation security system
- al Qaeda financing was not detected
- The United States had an improvised homeland defense and communication was poor at senior government levels
- Emergency response was determined and saved lives but effective decision making in New York was hampered by poor communication and weak command and control
- Congress and the executive branch responded slowly to the rise of the transnational terrorism threat
- The Commission stated that at the time of the report (2004), the United States was safer than at 9-11-2001, but was not safe yet; therefore, recommendations were issued as follows.

RECOMMENDATIONS OF THE 9/11 COMMISSION.   The U.S. should take steps to:

1. Attack terrorists and their organizations
2. Prevent the continued growth of Islamic terrorism
3. Protect against and prepare for future terrorist attacks

The 9/11 Commission specifically recommended:

1. Creation of a National Counterterrorism Center
2. Creation of a new National Intelligence Director
3. Effective sharing of information across U.S. government agency boundaries
4. Congressional action to improve oversight of homeland security
5. Clarification of roles, missions and authority between DoD and DHS officials among other agencies

TRANSPORTATION SECURITY ADMINISTRATION

On November 19, 2001, Congress enacted the *Aviation and Transportation Security Act (ATSA)*, Public Law 107-71, 115 Stat. 597, which established the Transportation Security Administration (TSA) as an operating administration within the Department of Transportation (DOT). A year later, on November 28, 2002, Congress enacted *The Homeland Security Act* Public Law 107-296, 117 Stat. 745, which created the Department of Homeland Security (DHS) and is headed by a cabinet-level Secretary. The TSA subsequently moved from the DOT to the DHS in March 2003 as part of a massive federal government integration of all agencies whose mission it was to prepare for, respond to, and protect the United States from domestic emergencies, especially terrorism.

According to its Web site, the mission of TSA is to protect the nation's transportation systems to ensure freedom of movement for people and commerce. Security screening is at the heart of TSA, which now numbers approximately 50,000 employees. TSA conducts 100 percent passenger screening of over 600 million people who fly each year, including inspection of passenger baggage at over 700 security checkpoints and nearly 7000 baggage screening areas each day. The multiple layers of TSA aviation security include, among other things:

• An airport document checker
• Behavior detection officers
• Secure flight—the behind-the-scenes watch list which matches passenger names to government lists of suspected terrorists
• Federal Air Marshals who fly aboard selected U.S. commercial aircraft
• Federal flight deck officer who are pilots armed and trained by the Federal Air Marshal Service on the use of firearms and other tactics
• Crew Member self-defense training program available at over 20 sites across the country

- Mobile dog explosive detection teams
- Air cargo programs and initiatives such as the Certified Cargo Screening program to allow the prescreening of cargo prior to arrival at the airport. This program is part of the Improving America's Security Act of 2007, Public Law 110-53, which implements the ambitious 9/11 Commission recommendation mandating 100 percent inspection of all air and sea cargo entering the United States.

TSA REGULATIONS.   The TSA issues Transportation Security Regulations (TSRs) which are codified in 49 CFR Chapter XII, Parts 1500 through 1699. TSA regulations are broken out into subchapters. Subchapter A contains administrative and procedural rules. Subchapter B contains rules that apply to many modes of transportation. Rules for civil aviation security are contained in Subchapter C.

Subchapter A, 49 CFR 1500, outlines and defines the terms used in the TSRs.

- *Part 1520* addresses the protection of sensitive security information. This part outlines the type of information that may not be released under the Freedom of Information Act. Contained in this section is the duty to protect any information that is given to a person in performance of his or her duties and the responsibility to report to TSA, DHS or DOT when he or she becomes aware that sensitive security information has been released to unauthorized individuals.
- *Part 1540* outlines the Civil Aviation Security General Rules. Part 1540 also contains prohibitions regarding making fraudulent or intentionally false statements or entries in compliance reports. Also prohibited by Part 1540 is the interference with screening personnel while they are performing their duties and the carriage of weapons, explosives, or incendiaries by individuals into specified areas at airports. Other requirements outlined in Part 1540:
  - The security responsibilities of employees and persons who access the airport.
  - The responsibilities of persons who wish to enter any area that requires screening.
  - The responsibility of airmen to present certain certifications to TSA for inspection when so requested.
- *Part 1542* contains the regulations for airport security. These airport requirements primarily address access control, law enforcement support and fingerprint-based criminal-history records checks, among other things.
- *Part 1544* addresses airport operator security for air carriers and commercial operators. Part 1544 requires that the aircraft operator not permit persons to have unauthorized explosives, incendiaries, or weapons when on board an aircraft.
- *Part 1546* provides the rules for foreign air carrier security including the screening of individuals and property, access to cargo, and bomb or air piracy threats.

- *Part 1548* provides the rules for indirect air carriers that operate within the United States including the acceptance of cargo and security threat assessments.

- *Part 1550* was created to require security programs for both passenger and all-cargo operations using aircraft with a maximum certified takeoff weight of 12,500 pounds or more.

ROLE OF INTELLIGENCE

The comprehensive 9/11 Commission Report details how al Qaeda was allowed to become a real danger to the United States in the years preceding 2001. The attack was driven by Osama Bin Laden, who built a dynamic and lethal terrorist organization over the course of a decade. The Report surprisingly concludes that "the 9/11 attacks were a shock, but they should not have come as a surprise." The extremists had given plenty of warning that they meant to kill Americans in great numbers. With 20/20 hindsight, it is clear that the 9/11 attacks were able to succeed due to a serious failure of the U.S. intelligence community.

NATIONAL COUNTERTERRORISM CENTER

A major recommendation of the 9/11 Commission was to build a "Unity of Effort" combining all strategic intelligence and operational planning under one office. The National Counterterrorism Center (NCTC) was established by Presidential Executive Order 13354 in August 2004. This order was codified into law by the *Intelligence Reform and Terrorism Prevention Act of 2004 (IRTPA)*. The establishment of NCTC implements a key recommendation of the 9/11 Commission as follows:

> Breaking the older mold of national government organization, this NCTC should be a center for joint operational planning and joint intelligence, staffed by personnel from the various agencies. (9/11 Commission Report pg. 403)

Today, the NCTC is manned on a 24/7 basis in a modern operations center colocated with the Central Intelligence Agency (CIA) and Federal Bureau of Investigation (FBI) operations centers. It is staffed by more than 500 personnel from over 16 different U.S. departments and agencies. The Director of NCTC has a unique, dual line of reporting: (1) to the President for counterterrorism planning and (2) to the Director of National Intelligence (DNI) regarding intelligence matters. The DNI began operations on April 22, 2005 as another major recommendation of the 9/11 Commission and the IRTPA legislation was implemented on that date. Today, the Director of National Intelligence serves as the single head of the intelligence community, overseeing such major intelligence offices as:

- The Central Intelligence Agency
- The Defense Intelligence Agency
- The National Security Agency
- The National Reconnaissance Office

SCOPE OF THE TERRORISM THREAT.   The latest NCTC Annual Report on Terrorism contains detailed statistical data which indicates that the scope of worldwide terrorism is very large indeed. Some interesting NCTC observations related to terrorist incidents include the following statistics from 2009:

- Approximately 11,000 terrorist attacks occurred in 83 countries, resulting in over 58,000 victims, including nearly 15,000 fatalities.
- Sunni extremists groups such as al Qaeda and the Taliban conducted about one half of the attacks.
- Most attacks were perpetrated by terrorists applying conventional fighting methods such as armed attacks, bombings, and kidnapping. Figure 11-3 illustrates the number of deaths by method used.

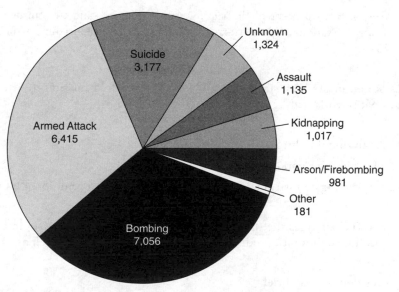

FIGURE 11-3   Worldwide terrorism deaths by method used, 2009. (*Source: http://www.nctc.gov/*)

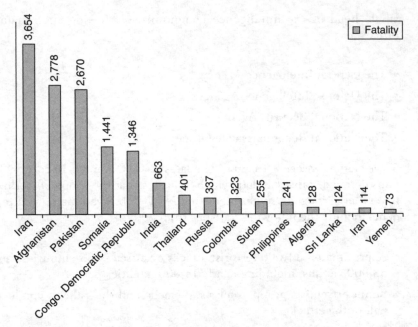

FIGURE 11-4 Worldwide terrorism deaths by country, 2009. (*Source: http://www.nctc.gov/*)

- Attacks in Iraq, Afghanistan, and Pakistan accounted for about 60 percent of all terrorist attacks. Figure 11-4 illustrates the number of deaths by country.

- Well over 50 percent of the victims were Muslims, and two-thirds were civilians. (Source: National Counterterrorism Center 2009 Report on Terrorism, April 30, 2010)

As stated on its Web site, the NCTC has a three-fold mission to lead the effort to combat terrorism at home and abroad by

1. Analyzing the threat
2. Sharing information with intelligence partners
3. Integrating all instruments of national power to ensure unity of effort.

NCTC analyzes the threat by integrating all counterterrorism intelligence from across the community, producing detailed analytic products such as

- President's Daily Brief
- National Terrorism Bulletin
- Information on the chemical, biological, radiological, and nuclear threats

NCTC shares information with U.S. intelligence agencies by conducting analysis of more than 30 dedicated intelligence networks located under one roof. It routinely makes these counterterrorism products available to the following:

- Intelligence users in 75 U.S. government agencies through the NCTC Interagency Threat Analysis and Coordination Group.
- Intelligence users also obtain information by means of a daily secure video teleconference system on a 24/7, 365 days a year basis.

NCTC integrates all instruments of national power in the Global War on Terror through joint planning and strategic plans such as

- The National Implementation Plan for the War on Terror
- The National Strategy for Aviation Security in close coordination with the Department of Homeland Security under the National Security Presidential Directive 47/Homeland Security Presidential Directive 16 of March 26, 2007

DEPARTMENT OF HOMELAND SECURITY (www.dhs.gov)

The National Strategy for Aviation Security cited above details the comprehensive U.S. Government strategy in this area. The Department of Homeland Security coordinates the operational implementation of this strategy, which is an over-arching national plan to optimize aviation security integration on a government-wide basis. Six supporting plans have been issued to ensure collaborative interagency effort. (Source: www.dhs.gov)

1. Aviation Transportation System Security Plan (DHS and DOT)
2. Aviation Operational Threat Response Plan (DoD and DoJ)
3. Aviation Transportation System Recovery Plan (DHS and DOT)
4. Air Domain Surveillance and Intelligence Integration Plan (DoD and DNI)
5. Domestic Outreach Plan (DHS and DOT)
6. International Outreach Plan (DOS)

REVIEW OF SECURITY TECHNOLOGIES

Commercial aviation can be protected from the threat of explosives in two ways: by preventing explosives from reaching the aircraft (e.g., by using explosives detection technologies) or by mitigating the effects of an explosive by protecting the aircraft from an onboard explosion (e.g., via aircraft hardening and hardened containers).

While each of these approaches will be discussed individually, a combination of these two approaches may provide the best protection of commercial aviation.

The first stage of the deployment of new, advanced security equipment began in 1997 when Congress appropriated $144 million to this end. After the events of September 11, 2001, the implementation plan was sped up, and all airports were required to install baggage and passenger screening equipment by December 2002. TSA began deploying state-of-the-art Advanced Imaging Technology (AIT) in 2007. In March 2010, TSA began deploying AIT units which were purchased under the $1 billion appropriation to TSA, authorized by the American Recovery and Reinvestment Act jobs program brought about by the worldwide recession of 2009–2010. Currently, there are nearly 500 imaging technology units at 78 airports nationwide, with additional units planned for deployment during 2011. (www.tsa.gov)

IMAGING TECHNOLOGIES

Imaging technologies work either by sensing the natural radiation emitted by the human body (millimeter wave) or by exposing subjects to a specific type of radiation and then measuring the radiation reflected by the body (backscatter). These systems can detect metallic weapons or plastic explosives by sensing the differences in reflected radiation between the human body and the weapons or explosives. The screening systems then generate televisionlike digital images. Operators are trained to identify potential threat objects in these images, often with the assistance of image-enhancing software that highlights unusual features.

FULL BODY SCAN.   TSA uses two types of advanced imaging technology today, millimeter wave and backscatter x-ray technology. Millimeter wave scanners come in two varieties: active and passive. Active scanners predominate the market; passive technology is still under development. Active scanners work much like radar, they direct millimeter wave energy at the passenger and then interpret the reflected energy. An example of an active millimeter wave scanner is the L3 Communications ProVision machine (Fig. 11-5). Since these modern scanners can display images that reveal private parts of the human anatomy, some privacy objections were raised by the public. To alleviate the privacy concerns, in February 2011, the TSA began testing a new software system on its AIT machines which eliminates passenger-specific images and instead "auto-detects" potential threat items and indicates their location on a generic outline of a person. The generic outline is identical for all passengers. If no potential threat items are detected, an "OK" will appear on the monitor with no outline of the body.

Backscatter x-ray imaging is also used for full body scanning by detecting radiation reflected from the passenger. X-ray units used for inspecting carry-on luggage and people employ low-dosage, low-energy radiation. Higher dosage units are used for checked baggage. Older generation units produced images of low quality and poor contrast. Most of the older analog systems have given way to modern digital video amplification and processing systems that produce clearer images and increased contrast ratios. On its Web site, TSA assures the public that current

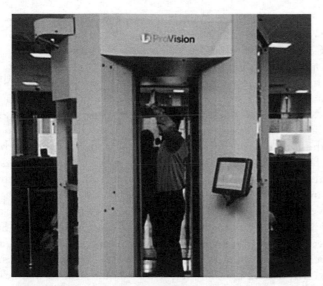

FIGURE 11-5   Advanced Imaging Technology, L3 Communications ProVision Active Millimeter Wave Scanner. (*Source:* www.tsa.gov)

technologies are safe and meet national health and safety standards. TSA states that the energy projected by millimeter wave technology is thousands of times less than a cell phone transmission. A single scan using only backscatter technology produces exposure equivalent to two minutes of flying on a plane.

EXPLOSIVE TRACE DETECTION TECHNOLOGY

Explosive trace detection technologies are based on the direct chemical identification of either particles of explosive material or vapor-containing explosive material. Thus, the presence of a threat object or bomb is inferred from the presence of particulate matter or vapor. The main difference between trace detection and electromagnetic imaging is that in trace detection, a sample of the explosive material must be transported to the instrument in concentrations that exceed the detection limit. Trace detection technologies cannot be used to detect the presence of metallic weapons. All trace detection equipment is regulated by TSR Part 1544.213. The two distinct steps in trace detection are sample collection and chemical identification. Trace detection practices are more commonly used for baggage screening (as opposed to people screening) in aviation security. Some basics of this technology are discussed next.

SAMPLE COLLECTION.   Explosive substances can be detected by instrumentation in their vapor or solid form. Initial efforts in the development of trace detection technology were focused on collecting vapor around the person or baggage. However, because many modern explosives do not readily give off vapor at room temperature, the focus has expanded to include detection of particulates of explosive materials on the skin and other surfaces.

If traces of explosive material are to be detected, they must be concentrated from an air sample (vapor technologies) or dislodged from a substrate (particulate technologies). In vapor detection, large amounts of air must be collected, from which small amounts of the substances of interest are extracted. In particle detection, pieces of explosive material must be removed from the surface to which they are adhering. Both trace detection approaches have strengths and weaknesses depending on the type of explosive material being sought. Vapor technologies are more effective for detecting explosive materials with high vapor pressures, while particulate technologies are more appropriate for detecting explosive materials with low vapor pressure such as military plastic explosives.

Samples can be taken either by having the subject walk through a portal (the puffer machine approach) or by passing a hand-wand device over the subject or piece of baggage (the swab approach). Either method may be implemented as a *contact* or *noncontact* technique. In contact portal sampling, passengers walk through a portal by pushing open a door or by rubbing against paddles or brushes. In a noncontact system, an airstream passes over the passengers as they walk through the portal. Hand-wand devices may be used to sample air around the person or to make physical contact. In general, contact methods focus on gathering particulates of explosive material from the hands or clothing of the subject. Noncontact methods may use the airstream to dislodge particles, or they may distill a sample of explosive vapor from the airstream.

Although using a hand-wand device is a potentially efficient sample collection technique, it is more labor-intensive and more time-consuming than collecting samples by using an automated portal. The optimum solution may be to attach a hand-wand device to a portal-based trace detection system as a higher-level surveillance accessory. This is a common technique used with metal detection portals. In trace detection, a single chemical-identification instrument could be served by both the portal and the hand-wand device sample collection mechanisms.

TSA is expanding its use of explosive trace detection, especially using the swab approach. To ensure the health of TSA workers, swabs are disposed of after each use. When tested, the swab is placed inside a fixed or portable machine which analyzes the content for potential explosive residue. American Recovery and Reinvestment Act funding is also being used to purchase additional explosive trace detection equipment.

EXPLOSIVE DETECTION SYSTEMS (EDSs)

Bulk *explosive detection systems* includes any device or system that remotely senses a physical or chemical property (or combination thereof) of an object in an attempt to detect the presence of an explosive concealed in a container (e.g., passenger baggage). Today TSA screens 100 percent of all baggage placed on an airplane, covering two million passengers processed each day. This equipment must function without causing unreasonable delays so reliability and throughput are very important.

There have been major improvements in EDS baggage screening technology in recent years. TSA has recognized this trend and has fostered technology improvement through priority funding provided by the American Recovery and Reinvestment Act. In fact, a new standard has evolved in this area of Checked Baggage Inspection Systems (CBIS). At present, average in-line, medium speed equipment can inspect around 600 bags per hour. TSA standards for high volume screening is throughput of at least 900 bags per hour with a low false alarm rate.

Companies are meeting this challenge by providing improved image quality and on-screen resolution using three-dimensional (3D) images very similar to the magnetic resonance imaging (MRI) technology used in medical radiology offices today. In-line checked baggage systems of the future will be extremely efficient, able to achieve a throughput of over 1000 bags per hour, and will be easier to maintain with built-in diagnostics tools. These high-speed screening systems will also be upgradable and allow airports to increase baggage handling capability without major capital investments.

METAL DETECTORS

Screening procedures currently used in U.S. airports, at least during routine operations, involve metal detection portals for screening passengers and x-ray imaging systems for screening hand-carried baggage. Metal detection devices impose a time-varying magnetic field in the space within the portal that induces eddy currents in metallic or ferromagnetic objects passing through that space. Various methods are used to detect these eddy currents; and when they exceed a preset level, an audible or visual alarm goes off, and an operator intervenes to ascertain the presence or absence of a dangerous object or weapon. The effectiveness of this security screening system depends not only on the performance of the equipment, but also on the performance of the personnel operating the equipment and resolving the alarms.

Metal detectors vary from portals to handheld systems depending on the application and size of the airport. Most passengers are required to pass through a portal-type metal detector, but a more scrutinized search could be conducted with a handheld or portable metal detector. Metal detectors are not as effective as other systems, and their deficiency lies in their inability to catch all forms of dangerous weapons. Their greatest weakness is that they do not detect metals incapable of being magnetized. Newer-generation metal detectors can search for ferrous or nonferrous objects. Metal detectors are used broadly around all the airports and are one of the most important sources of security.

An important advance in metal detectors is the multizone concept. Through a multizone approach for portal-type metal detectors, sensitivity can be assigned to specific areas of the body zones that the metal detector would scan. Each zone has its own sensitivity adjustment. This permits the detection response of each zone to be independently adjusted. The horizontal zone corresponding to the foot region of the body can be made more or less sensitive than the waist region or visa versa. Use of metal detectors in airport security is regulated by TSA under its Part 1544.209.

BIOMETRICS AND FUTURE CHECKPOINT SYSTEMS

The term *biometric* is used to refer to the emerging field of technology devoted to identifying individuals by using biological traits such as those based on retinal or iris scanning, fingerprints, or face recognition. *Biometric technology* is defined as the automated use of physiological or behavioral characteristics to determine or verify identity. This technology has traditionally been used in place of passwords in top-secret government areas but is increasingly finding uses in a wide range of security applications. One such application is to prevent unauthorized employees from gaining access to certain areas and assets. Another is to quickly identify low-risk users, such as prescreened airport passengers, so that security personnel can focus on the much smaller category of "high-risk" passengers.

Biometric systems are basically of two types: verification and recognition. By measuring a physical feature or repeatable action of the individual, we establish a biometric identification. Verification systems require that the individual seeking access have some sort of identification, such as a smart, e-card, that is then matched with some physical characteristic of that person to make the verification. A reference measurement of the biometric is obtained when the individual is programmed into the device.

TSA conducted a biometric pilot program called Registered Traveler from 2006 to 2008. The pilot program concluded in July 2008, at which time it became solely a market-driven venture. The leading biometric contractor had over 200,000 subscribers at 20 of the busiest airports in the United States at the height of its program.

In addition to using biometrics contractors, TSA has chosen to focus on the 9/11 Commission recommendation that the government take over the No Fly Watch List system. This is the careful computer screening of passenger information by comparing names to the federal government intelligence watch lists. The Intelligence Reform and Terrorist Prevention Act of 2004 (IRTPA) codified this recommendation so that as of November 1, 2010, the new TSA Secure Flight Program requires airlines to collect and transmit to TSA the following Secure Flight passenger data:

- Name as it appears on government issued I.D. when traveling
- Date of birth
- Gender
- Redress number (this is a special number given passengers who have resolved denied boarding problems with the TSA, to prevent future delays as a misidentified passenger)

In the future, TSA is considering asking airline passengers to provide more information, both biometric and biographical, to allow the U.S. government to begin an enhanced screening process. Passengers who comply with this proposed new program could see benefits such as quicker clearance through security check-points. On the international front, ICAO has been promoting the use of machine readable travel documents (MRTDs), which are used to quickly process passengers across borders. The checkpoint of the future would move passengers into low, medium and high risk categories and screen them accordingly.

Strengthening Aircraft and Baggage Containers

Another approach to aviation security is to try to strengthen aircraft frames and to plan redundancies in vital systems such as controls, electrical systems, and hydraulics, to mitigate the effects of bomb blasts in flight. A further alternative is to use hardened baggage containers that can control the effects of bomb blasts in checked baggage.

Retrofitting is difficult. It is easier and more practical to incorporate such design measures during the design stage of the aircraft. The FAA has engaged in extensive studies with military experts and airframe manufacturers to learn how aircraft fail due to explosions in flight and to discover measures to increase chances of aircraft survival. Explosives tests have been carried out to check calculations, both in the United States and in the United Kingdom. The best known of these efforts was the explosives testing on a Boeing 747 in Bruntingthorpe in England in May 1997. Several simultaneous experiments were run with four independent bombs. The experiments tested the effects of various protective measures to different parts of the interior cargo hold of the aircraft and also tested a model of a hardened baggage container.

The aircraft hardening experiments in this case were run by experts from the United Kingdom, not the United States. They appeared to indicate some promise for the future, in which the application of material of relatively small weight may contribute significantly to the resistance of aircraft to bombs at certain locations in their cargo holds. However, due to the size of the current U.S. fleet, hardening an aircraft is neither economically feasible nor reachable. Therefore, hardened containers are being investigated by the FAA and other international airworthiness authorities as an alternative near-term solution.

Using HULDs (hardened unit-loading devices) to protect aircraft from explosive attacks is not a new concept. Airlines that operate in high-risk areas of the world have been using custom-built containers. Because these containers are much too heavy for general use, only one or two are used per aircraft for carrying select items. However, new advances in technology and the need for increased security after the terrorist attacks of 2001 has resulted in greater use of these containers.

Hardened cargo containers are also being considered to mitigate the threat of improvised explosive devices (IEDs). The IRTPA legislation required TSA to carry out a pilot study to evaluate the use of blast-resistant containers, and the act authorized $2 million for this program. In 2009, DHS issued a solicitation for a lighter weight HULD (Reduced Threat) to be tested at the FAA Technical facility near Atlantic City.

The future success of hardened containers could radically change the detection capability requirements for explosives detection equipment for checked baggage screening. Of course, for any given container, a large enough bomb can be constructed to overcome it. However, a larger bomb is more susceptible to detection, and increasing the mass of explosives that need to be detected would relax the requirements on the detection equipment.

COCKPIT DOOR REINFORCEMENT

On January 10, 2002, the FAA published new standards to protect cockpits from intrusion and small-arms fire or fragmentation devices, such as grenades. The rule required U.S. operators of as many as 7000 airplanes to install reinforced doors by April 9, 2003. Congress appropriated $97 million to help the airlines defray the cost of cockpit door replacement. Current cockpit door regulations and security standards are located at 14 CFR Section 25.795.

CONCLUSIONS

High-technology detection methods can yield results, but it is clear that these results are expensive and are not perfect. The small added yield in security is arrived at only at some very great expense. Further, the cost is not just one of acquisition. There is also the matter of operations and maintenance and the opportunity costs to passengers standing in line, among others.

Although technology has its place in the overall security scheme, it is no panacea for air transportation security issues. Greater attention needs to be paid to security personnel. They play a critical role and are the ones who interpret the results of the technological analyses.

Until the TSA was established and took over security screening, most security personnel were low-paid workers who lacked the required training. The profession was plagued by questions of morale, turnover, stress, and ineffectiveness. FAA insiders had criticized security personnel as being inadequate. The ATA responded by developing standards that called for better selection, improved work conditions, and rewards for effective security personnel. In addition, many airlines upgraded their personnel screening. Nevertheless, despite new training requirements, the process remained uneven. The length of training was too short, and the subject matter was very loosely defined. However, when the TSA took over this function in 2001, conditions improved for security personnel in terms of pay, training, and other benefits. This has led to improved levels of security at airports.

A related problem is that of airport workers and staff. Airports are like cities; their personnel engage in a variety of activities and represent vocations ranging from salespeople to mechanics to cleaners. Because this diversity of workers provides terrorists with numerous opportunities, the FAA requires background checks of previous employment records and the wearing of badges. Despite these efforts, however, these measures do not guarantee security. Terrorists can forge badges, although forging computerized badges is a far more difficult task, requiring insider collaboration or access to the computerized database and badge process. Terrorists can follow a potentially riskier path, such as using threats to family members to pressure employees or simply bribing them to gain access.

There are alternative technological and tactical approaches to airport security. One might be able to divert some level of resources from hardware applications toward improved intelligence gathering to intercept terrorists long before they arrive at the airport. This suggestion, of course, raises the question of how intelligence gathering has been improved by implementation of the recommendations of the 9/11 Commission, and how much more can be done in this critical area. From an economic point of view, the use of resources to gather intelligence is more attractive than elaborate security technology. Although expensive hardware guarding the front door is very effective, better intelligence can be applied across the board to transportation, government, and business toward the improved security of them all, particularly since even the most effective technology remains vulnerable to changes in terrorist tactics. If we become experts in weapons and explosives detection, the threat could become biological (germs in the water supply) chemical or radiological (dirty bombs).

In addition, technology suffers from problems defined by two related axioms. The first is that security is very site-specific. What works well and inexpensively in one place is not necessarily a sound approach in some other place. The second axiom is political: Regulators seldom recognize the real-world needs of security. No matter how well conceived the legislation or regulation may be, it is not likely to have anticipated the kinds of problems described.

An important concluding point is that, in spite of all that has been accomplished to ensure the safety and security of the traveling public, the terrorist, in one sense, always has the upper hand. While we must protect every element of the transportation system at all times (not just aviation), the terrorist has the luxury of being the only one who knows the time, the place, and the method of the next attack.

KEY TERMS

Security versus safety

ICAO Security Manual (Doc 8973)

Security Management Systems (SeMS)

The 9/11 Commission

The Homeland Security Act

Intelligence Reform and Terrorism Prevention Act of 2004 (IRTPA)

Department of Homeland Security (DHS)

National Counterterrorism Center (NCTC)

Improving America's Security Act of 2007

International Civil Aviation organization (ICAO) Annex 17

The International Security and Development Cooperation Act of 1985

Anti-hijacking or Air Transportation Security Act of 1974

The Air Marshal Program

Aviation Security Improvement Act of 1990

Aviation Security and Antiterrorism Act of 1996

White House Commission on Aviation Safety and Security, 1996

Aviation and Transportation Security Act of 2001

Transportation Security Administration (TSA)

X-ray imaging technologies

Machine Readable Travel Documents (MRTDs)

Checked Baggage Inspection System (CBIS)

Hardened Unit Loading Device (HULD)

Improvised Explosive Devices (IEDs)

Active millimeter-wave imaging

Explosive Trace detection technology

Explosives detection systems (EDSs)

Metal detectors

Biometrics

REVIEW QUESTIONS

1. What is the difference between security and safety?

2. Give three examples of attack categories on civil aviation.

3. Describe the international response to terrorism by ICAO and the world community.

4. What is the Air Marshal Program and why was it instituted?

5. Discuss the hijacking era. When did it end?

6. Discuss the role and functions of the Transportation Security Administration (TSA).

7. What was the reason for the 9/11 Commission? What impact did its recommendations have on the U.S. intelligence community?

8. Explain the operation of the National Counterterrorism Center (NCTC).

9. Discuss some of the problems encountered in the gathering and analysis of intelligence for security purposes.

10. Give an example of four different security devices used to prevent weapons and explosives from entering an aircraft, and explain the workings of these devices.

REFERENCES

Anderson, Theresa. 2000. "Airport Security." *Security Management*, pp. 73–74. February.

FAR 107, Airport Security.

FAR 108, Airplane Operator Security.

FAR 109, Indirect Air Carrier Security.

ICAO Annex 17. *International Standards and Recommended Practices—Security, Air Carrier Standard Security Program.*

ICAO Security Manual (Doc 8973)

National Commission on Terrorist Attacks upon the United States (The 9-11 Commission Report) www.9-11commission.gov

National Counterterrorism Center, 2009 Report on Terrorism (April 30, 2010.) www.nctc.gov

National Strategy for Aviation Security. (March 26, 2007) www.nctc.gov

Ott, James. 2011 "Airport Checkpoints of the Future." Aviation Week, (February 4, 2011).

Price, Jeffrey C. and Forrest, Jeffrey S. 2008, *Practical Aviation Security: Predicting and Preventing Future Threats*, Burlington, MA Butterworth-Heineman.

Security Management Systems (SeMS) Fact Sheet. International Air Transport Association www.iata.org (March 2010)

Sweet, K. M. 2002. *Terrorism and Airport Security*. Symposium Series, vol. 8. The Edwin Mellen Press, Lewistown, N.Y.

WEB REFERENCES

http://www.airsafe.com/index.html

http://www.faa.gov

www.dhs.gov

www.l-3com.com

www.nctc.gov

www.dni.gov

www.iata.org

www.tsa.gov

www.9-11commission.gov

http://www.icao.int

AIRLINE SAFETY

LEARNING OBJECTIVES

After completing this chapter, you should be able to

- Discuss the international development aspects of airline Safety Management Systems (SMS).
- Define and discuss the features of the Four Pillars of SMS.
- Explain the development of airline SMS in the United States.
- Discuss the role of accident investigation and auditing in airline safety matters.
- Explain the importance of communications to an effective airline safety program.
- Describe the role of ALPA in airline safety.
- Discuss the features of the SMS infrastructure at a major U.S. airline.
- Explain the phased SMS implementation process.

INTRODUCTION

A commercial airline, like any other major business, has to integrate and manage several different functions/departments while delivering products and/or services to the customer and (hopefully) turn a profit. Safety is a major function, and it is important that it permeates (or should permeate) through all departments. Safety must be built into every product, service, or operation and must be an integral part of every employee's job responsibility. The airline industry in particular has always placed great emphasis on safety and has moved aggressively to identify and control problems that cause accidents. Airlines have learned, sometimes the hard way, that active management of risk is an absolute requirement for a healthy company. This chapter brings together many of the concepts discussed in the previous chapters. This chapter will discuss the fundamental components and elements of a modern airline Safety Management System (SMS).

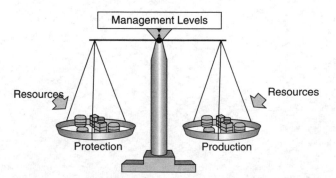

FIGURE 12-1   The balance between safety (protection) and airline efficiency (production) of services. (*Source: ICAO Safety Management Manual, 2nd edition 2009. www.icao.int*)

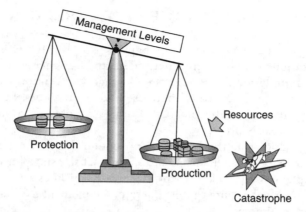

FIGURE 12-2 Improper allocation of resources, safety yields to production with catastrophic results. (*Source: ICAO Safety Management Manual, 2nd edition 2009. www.icao.int*)

An ideal airline SMS would have unlimited resources to manage the safety (protection) demands of the business. As indicated in the ICAO Safety Management Manual (Doc 9859, 2nd Edition, 2009), business resources are limited and therein lies the apparent management dilemma. Safety (protection) and efficiency of airline services (production) should never be in competition but should be in balance in management's decision-making process as indicated in Fig. 12-1. If an airline "cuts corners" on safety processes, seeking to maximize its profits, the result could be a catastrophic accident (Fig. 12-2). On the other hand, if profits are entirely disregarded and safety protection is the only focus, the business may be headed for bankruptcy (Fig. 12-3). The better approach, of course, is a balanced allocation of resources where safety management is a core business function, closely intertwined with business objectives and not in competition with of the aspects of the business.

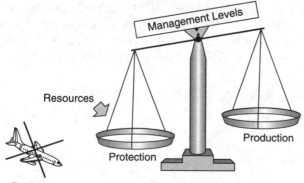

FIGURE 12-3 Resources out of balance with too much protection could lead to bankruptcy. (*Source: ICAO Safety Management Manual, 2nd edition 2009. www.icao.int*)

INTERNATIONAL DEVELOPMENT OF AIRLINE SMS

As previously discussed in Chapters 1 and 2, ICAO is the United Nations organization that issues international standards related to worldwide aviation. Since its foundation in 1944 at the Chicago Convention, ICAO has been in the forefront of airline safety matters, and its member States (190 countries) have agreed to abide by the ICAO annexes know as Standards and Recommended Practices (SARPs).

Annex 6 to the Chicago Convention deals with International Commercial Air Transport. Amendment 33 of Annex 6 set the stage for international awareness and implementation of SMS by the world's airlines. It provides that ICAO member States shall require that its operators implement a Safety Management System that as a minimum:

(a) Identifies safety hazards

(b) Ensures that remedial action necessary to maintain an acceptable level of safety is implemented

(c) Provides for continuous monitoring and regular assessment of the safety level achieved

(d) Aims to make continuous improvement to the overall level of safety

THE ICAO SMS FRAMEWORK OF SAFETY COMPONENTS AND PROGRAM ELEMENTS—THE FOUR PILLARS

The ICAO's definition of SMS is

"an organized approach to managing safety, including the necessary organizational structures, accountabilities, policies and procedures."

The details of SMS are contained in the ICAO Safety Management Manual, 2nd Edition, 2009 (Doc 9859). The ICAO SMS framework consists of four components and 12 elements, which are in the process of implementation at airlines worldwide, including in the United States. The four components (pillars) of a SMS are as follows:

I. Safety Policy and Objectives

II. Safety Risk Management

III. Safety Assurance

IV. Safety Promotion

The international airlines are leading the implementation of SMS principles, and all successful ICAO approved programs will contain these four components.

Additional details of the SMS process will be provided in Chapter 13. This chapter will discuss the specific requirements of an organization's SMS program under each of the four pillars mentioned earlier. (ICAO Safety Management Manual, 2nd Edition, Chapter 8, Appendix 1 and FAA Advisory Circular No. 120-92A, Appendix 1 dated 8/12/2010.)

SAFETY POLICY AND OBJECTIVES

The first Pillar of SMS has five key elements, which will be discussed based upon ICAO and the voluntary FAA guidance provided in anticipation of a final SMS rule.

MANAGEMENT COMMITMENT AND RESPONSIBILITY (ELEMENT 1.1)   It is appropriate that this element is placed first on the list for without it, the other elements are nearly impossible to achieve. ICAO explains the requirements of element 1.1 as follows:

> *The (organization) shall define the organization's safety policy which shall be in accordance with international and national requirements, and which shall be signed by the Accountable Executive of the organization. The safety policy shall reflect organizational commitments regarding safety; shall include a clear statement about the provision of the necessary resources for the implementation of the safety policy; and shall be communicated with visible endorsement, throughout the organization. The safety policy shall include the safety reporting procedures; shall clearly indicate which types of operational behaviors or unacceptable; and shall include the conditions under which disciplinary action would not apply. The safety policy shall be periodically reviewed to ensure it remains relevant and appropriate to the organization.*

This element is critical because senior management is obviously in control of the safe and efficient operations of the aviation organization. It is through management that employees are hired, trained, and dismissed if necessary. The starting point for this element is identification of the "Accountable Executive," who is a single, identifiable person having final responsibility for the organization's SMS. A safety policy should be developed by senior management and be clearly communicated to all employees. It is extremely important that management leadership be strong, demonstrable, and visible if the airline's SMS program is to succeed.

SAFETY ACCOUNTABILITIES (ELEMENT 1.2)   This element reflects the changing nature of the safety department in an SMS environment. ICAO explains the roles and responsibilities as follows:

> *The (organization) shall identify the Accountable Executive who, irrespective of other functions, shall have ultimate responsibility and accountability, on behalf of the (organization), for the implementation and maintenance of the SMS. The (organization) shall also identify the accountabilities of all members of management,*

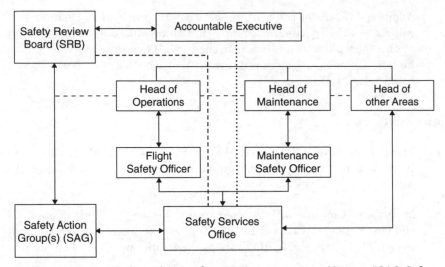

FIGURE 12-4   Functional view of an SMS organization. (*Source: ICAO Safety Management Manual, 2nd edition, 2009 page 8–7. www.icao.int*)

irrespective of other functions, as well as of employees, with respect to the safety performance of the SMS. Safety responsibilities, accountabilities and authorities shall be documented and communicated throughout the organization, and shall include a definition of the levels of management with authority to make decisions regarding safety risk tolerability.

Under this element, safety becomes the responsibility of all members of management, not just the safety department. Figure 12-4 depicts the role of a *"Safety Services Office"* and its functional relationship to the Accountable Executive, and other management officials responsible and accountable for safety matters in their departments.

Note the addition of a Safety Action Group (SAG) and Safety Review Board (SRB) to assist management officials accomplish safety responsibilities.

APPOINTMENT OF KEY SAFETY PERSONNEL (ELEMENT 1.3)   In an SMS environment, the safety manager will be the person to whom the Accountable Executive has assigned the day-to-day management of the SMS. ICAO points this out as follows:

The (organization) shall identify a safety manager to be the responsible individual and focal point for the implementation and maintenance of an effective SMS.

Under normal circumstances, the safety manager communicates to the Accountable Executive through either the Safety Action Group or the Safety Review Board (Fig. 12-4). The SRB is a very high level strategic direction group chaired by the Accountable Executive, which includes senior line functional managers. Once strategic direction has been developed by the SRB, it directs the Safety Action Group to implement the tactical decision reached. In this typical SMS organization, the organization safety manager serves as the secretary of the Safety Action Group to record its actions and keep score of corrective actions taken.

COORDINATION OF EMERGENCY RESPONSE PLANNING (ERP) (ELEMENT 1.4)   This is an area of classical responsibility of an airline safety department. ICAO describes this function as follows:

> The (organization) shall ensure that an emergency response plan that provides for the orderly and efficient transition from normal to emergency operations and the return to normal operations is properly coordinated with the emergency response plans of those organizations it must interface with during the provision of its services.

Airlines prepare an ERP document in writing to describe what actions should be taken following an accident and who is responsible for each action. An annual emergency response drill is usually required to ensure that the plan is adequate and that designated personnel are appropriately trained.

SMS DOCUMENTATION (ELEMENT 1.5)   Careful documentation is an essential element of an SMS organization. ICAO describes this last element of Safety Policy in this fashion:

> The (organization) shall development an SMS implementation plan, endorsed by senior management of the organization, that defines the organization's approach to the management of safety in a manner that meets the organization's safety objectives. The (organization) shall develop and maintain SMS documentation describing the safety policy and objectives, the SMS requirements, the SMS processes and procedures, the accountabilities, responsibilities and authorities for processes and procedures, and the SMS outputs. Also as part of the SMS documentation, the (organization) shall develop and maintain a Safety Management Systems Manual (SMSM), to communicate its approach to the management of safety throughout the organization.

The development of an SMS manual is very important to a proper documentation system. For organizational ease of reference, the manual should cover each of the SMS four pillars and each of the 12 elements of an effective SMS program.

SAFETY RISK MANAGEMENT

The second pillar of SMS has two key elements, which will also be discussed in detail in Chapter 13.

HAZARD IDENTIFICATION (ELEMENT 2.1)   Safety risk management (SRM) begins with a clear understanding of an organization's functional systems, which are analyzed by experienced operational and technical personnel to detect the presence of hazards. ICAO describes the hazard identification process in this manner:

> *The (organization) shall develop and maintain a formal process that ensures that hazards in operations are identified. Hazard identification shall be based on a combination of reactive, proactive and predictive methods of safety data collection.*

How hazards are identified will depend on the resources, culture, and complexity of the organization. It must be a formal process and is a routine, daily activity in a mature SMS airline organization. A further discussion of hazards will be provided in the Chapter 13.

RISK ASSESSMENT AND MITIGATION (ELEMENT 2.2)   After the safety hazards have been identified, they must be analyzed to determine their consequences to the organization. ICAO describes this step as follows:

> *The organization shall develop and maintain a formal process that ensures analysis, assessment and control of the safety risks in operations.*

The conventional method to analyze risk is to break it into two components, the probability of occurrence and the severity of the event should it occur. Typically, one tool often used is called a "risk tolerability matrix," which visually displays a risk to determine if it is acceptable or needs further mitigation. The goal of such mitigation is to reduce the risk to as low as reasonable practicable. Chapter 13 will discuss the SRM and risk mitigation process in some detail.

SAFETY ASSURANCE

The primary task of safety assurance is control. This third pillar of SMS has three key elements, which will also be discussed in Chapter 13. (Element 3.1)

SAFETY PERFORMANCE MONITORING AND MEASUREMENT (ELEMENT 3.1)   After the hazards are identified and the risks are assessed and properly mitigated in Pillar II, safety performance must be monitored and measured in order to be effective. This is an internal process of the SMS team, focusing on data and information regarding the performance of the safety system. ICAO describes this element as follows:

> *The (organization) shall develop and maintain the means to verify the safety performance of the organization and to validate the effectiveness of safety risk controls. The safety performance of the organization shall be verified in reference to the safety performance indicators and safety performance targets of the SMS.*

An airline using the SMS process could use a variety of tools to measure safety performance including, but not limited to:

- Safety studies to determine system deficiencies
- Internal or external safety audits to ensure the integrity of SMS processes
- Internal safety investigations of minor matters that are not investigated by state authorities

THE MANAGEMENT OF CHANGE (ELEMENT 3.2)   It has been stated as a matter of truth that the only thing constant in aviation is that things will surely change. A formal process for managing these frequent changes is necessary for an efficient airline SMS program. ICAO outlines the requirement this way:

> The (organization) shall develop and maintain a formal process to identify changes within the organization which may affect established processes and services; to describe the arrangements to ensure safety performance before implementing changes; and to eliminate or modify safety risk controls that are no longer needed or effective due to changes in the operational environment.

The formal change management system should consider how critical the system is to the airline, and whether the change introduces new, unforeseen hazards or risks. Efficient management of change is a good indicator of a mature SMS process at the airline.

CONTINUOUS IMPROVEMENT OF THE SMS (ELEMENT 3.3)   Rather than maintain the safety status quo, the SMS process seeks continuous improvement using tools such as internal evaluations and independent audits to obtain information. ICAO focuses on substandard performance as follows:

> The (organization) shall develop and maintain a formal process to identify the causes of substandard performance of the SMS, determine the implications of substandard performance of the SMS in operations, and eliminate or mitigate such causes.

Continuous improvement can be enhanced by a strong safety culture and committed management officials. Proactive evaluation of facilities, equipment and procedures is necessary, and vigilance is a must for SMS effectiveness.

SAFETY PROMOTION

The final pillar of SMS calls for the airline to continuously promote safety as a core value of the organization.

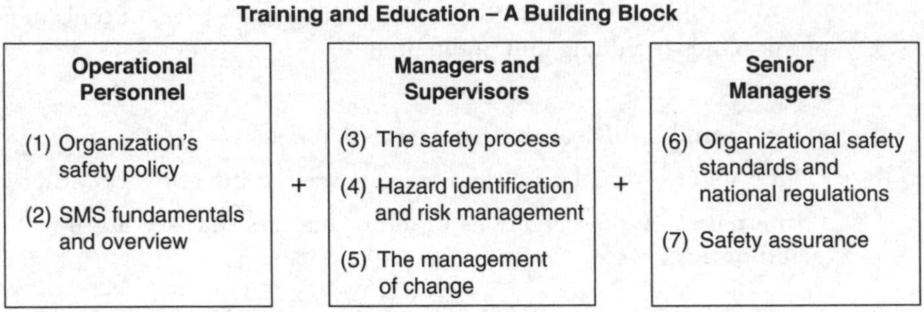

Training and Education – A Building Block

Operational Personnel		Managers and Supervisors		Senior Managers
(1) Organization's safety policy (2) SMS fundamentals and overview	+	(3) The safety process (4) Hazard identification and risk management (5) The management of change	+	(6) Organizational safety standards and national regulations (7) Safety assurance

FIGURE 12-5   Safety training and education fundamentals. (*Source: ICAO Safety Management Manual, 2nd edition, 2009 pp. 9–17*)

TRAINING AND EDUCATION (ELEMENT 4.1)   Safety training should be a formal, recurrent process using a "building block" approach suitable for each employee (Fig. 12-5). ICAO states the following regarding training and education:

> *The (organization) shall develop and maintain a safety training program that ensures that personnel are trained and competent to perform the SMS duties. The scope of the safety training shall be appropriate to each individual's involvement in the SMS.*

As indicated in Fig. 12-5, the training program should be ideally tailored to an employee's role in the organization. Additionally, ICAO recommends that the Accountable Executive should also receive special SMS training on roles and responsibilities, safety policy, SRM, and safety assurance.

SAFETY COMMUNICATION (ELEMENT 4.2)   SMS culture encourages a learning environment where employees communicate openly, up and down the management chain without fear of reprisal. ICAO puts it this way:

> *The (organization) shall develop and maintain formal means for safety communication that ensures that all personnel are fully aware of the SMS, conveys safety-critical information, and explains why particular safety actions are taken and why safety procedures are introduced or changed.*

Examples of ICAO recommended safety communication methods include, but are not limited to the following:

- Employee access to the airline SMS media, safety process, and procedures
- Safety newsletters, notices, and bulletins
- Use of Web sites or email to keep employees in the loop

DEVELOPMENT OF AIRLINE SMS
IN THE UNITED STATES

Following ICAO's introduction of SMS into worldwide aviation, the FAA initiated voluntary SMS pilot projects and voluntary implementation of Safety Management Systems in the United States in 2007. The objectives of the SMS pilot projects were threefold:

1. Develop SMS implementation strategies
2. Develop FAA oversight interfaces
3. Gain SMS experience for FAA and service providers

Major U.S. airlines are participating in these voluntary programs, and SMS implementation is also underway in the international airline community. The SMS infrastructure of a major U.S. airline will be discussed later in this chapter.

As discussed in Chapter 1, the crash of Colgan Air Flight 3407 near Buffalo, New York, on February 12, 2009 has had a profound effect on commercial airline safety through the passage of the Airline Safety and Federal Aviation Administration Extension Act of 2010 (Public Law 111-216 of August 1, 2010). This legislation has accelerated implementation of SMS in the United States. Section 215 of this act required the FAA to move forward with SMS rulemaking on an expedited basis.

STATUS OF FAA SMS RULEMAKING

By FAA Order 1110.152, the FAA established the Safety Management System Aviation Rulemaking Committee (ARC), which was chartered to provide recommendations to the FAA on the development and implementation of SMS regulations and guidance. The detailed recommendations of this ARC are contained in its final report. (see FAA-2009-0671-0094; dated March 31, 2010.)

Essentially, the SMS ARC recommended a detailed blueprint following ICAO international guidelines. The ARC recommended FAA SMS regulations and guidance be closely aligned and consistent with the same ICAO SMS framework of the four pillars and twelve elements discussed earlier. Accordingly, the FAA issued its Notice of Proposed Rulemaking entitled, "Safety Management Systems for Part 121 Certificate Holders" on November 5, 2010. (see Federal Register Volume 75, Number 214, page 68224–68245.)

CHARACTERISTICS OF PART 121 SAFETY MANAGEMENT SYSTEMS.   Public Law 111-216 requires a final FAA SMS rule by August 1, 2012. Although changes will certainly be proposed by airlines and other service providers before that date, it is clear that the final United States version of SMS will be based upon the ICAO system and exhibit fundamental characteristics as recommended by the ARC. The features of the FAR Part 121 SMS will include the following:

- Alignment with ICAO SMS framework and international acceptability
- Phrased promulgation of regulations and implementation of SMS requirements by FAA
- Scalability and flexibility of the regulations to accommodate a broad range of organizations
- SMS processes should not change existing regulatory standards for FAR Part 121
- Consistent with Public Law 111-216, FAA recognition of existing voluntary safety systems as part of the airline's SMS such as the following:
  - Aviation Safety Action Program (ASAP)
  - Flight Operational Quality Assurance Program (FOQA)
  - Line Operations Safety Audit (LOSA)
  - Advanced Qualification Program (AQP)

FAA Guidance on the SMS Program. *Advisory Circular 120-92A, Subject: Safety Management Systems for Aviation Providers* (August 12, 2010) is the primary FAA guidance document, which evolved from international and domestic SMS activities in recent years. It is based upon the ICAO Chicago Convention Annex 6 materials and the ICAO Safety Management Manual. As stated on the FAA Web site, the essential idea of SMS is to provide a systematic approach to achieving acceptable levels of safety risk. The four SMS pillars are depicted on the FAA Web site as shown in Fig. 12-6.

Additional information regarding the FAA SMS-phased implementation process will be provided later in this chapter.

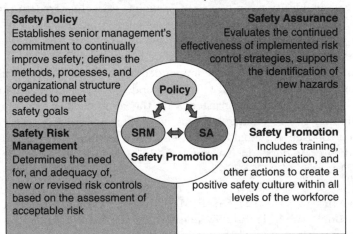

The Four SMS Components

Safety Policy
Establishes senior management's commitment to continually improve safety; defines the methods, processes, and organizational structure needed to meet safety goals

Safety Assurance
Evaluates the continued effectiveness of implemented risk control strategies, supports the identification of new hazards

Safety Risk Management
Determines the need for, and adequacy of, new or revised risk controls based on the assessment of acceptable risk

Safety Promotion
Includes training, communication, and other actions to create a positive safety culture within all levels of the workforce

Policy

SRM ⟷ SA

Safety Promotion

FIGURE 12-6 The four SMS pillars. (*Source: http://www.faa.gov/*)

ACCIDENT INVESTIGATION AND AUDITING

ACCIDENT INVESTIGATION

An important component of the safety effort at any organization is an incident and accident investigation system. By investigating accidents, organizations learn how to prevent future accidents. Some progressive companies investigate all unintended events, irrespective of whether these events led to injuries, illnesses, property damage, or equipment damage. Most companies, however, require some criteria to be met before an investigation is triggered. Investigations are initiated, for example, if accidents

- Require reporting on the OSHA 300 log
- Generate workers' compensation claims
- Cause lost workdays
- Meet NTSB definitions of accidents/incidents
- Result in an environmental spill
- Cause property and/or equipment damage over a preset dollar value

Once an investigation is triggered, the processes used to investigate accidents, whether they are FAA/NTSB, OSHA, or EPA related, are essentially the same. A formal notification system should be in place to provide timely notice of such events to the safety department. A typical investigation process is now described.

When the safety department is informed of an event, an immediate determination is made about whether the event meets the company's criteria for an investigation. When an investigation begins, depending on the circumstances, one or more individuals may be assigned to conduct the investigation, especially when it is clear that there may be more than one operating department involved, such as for incidents occurring during pushback. The manner in which the investigation is conducted is a function of the type of event and the circumstances under which it occurred. Typically, all key personnel involved in an incident or accident are asked to submit written statements describing the facts and circumstances as they saw them. In addition, provisions are made immediately for conducting interviews with the key individuals involved in the incident or accident. These interviews can be done in person or over the telephone, depending on the specific circumstances. In person interviews are preferred. Similarly, interviews can be done individually or with a part or all of an entire crew. The people being interviewed are told that the purpose of the interview is safety only, that the information will not be used for any disciplinary or other purpose, and that it will remain confidential, to the extent permitted by law. Information learned from these interviews, along with written statements, forms an important element of these safety investigations.

In addition to the interviews and personnel statements, other records and documents may be obtained and reviewed. Such material might include training records, training manuals and syllabi, aircraft and procedures manuals, information bulletins, and similar material. Another important source of information is the air-craft flight data recorder, which is read out as part of the investigation of many in-flight events. Other relevant information is identified and obtained as necessary, including accident reports, technical reports, and any other documentation that may contain information relevant to the incident or that can provide useful information for formulating recommended practices and corrective actions.

Following the collection of basic information, the investigator assembles and issues an accident report. These are brief, synoptic reports that describe the basic facts and circumstances of the event under investigation, including a history of the flight, a summary of damage and injuries, an analysis, a list of findings, and, most importantly, recommendations for corrective or future preventive action. Also included is a brief summary of safety actions taken—operating departments are not required to wait for a formal recommendation from the safety department prior to initiating corrective or preventive action.

Findings from safety investigations are derived from the facts and circumstances associated with the event and are based on the investigator's analysis. There is no effort to determine a cause or probable cause of an event. Findings are a list of fac-tors related to the event under investigation.

The recommendation process is probably the most important part of this safety investigation program, and it is important to understand how it is handled. The fol-lowing points should be noted when recommendations are formulated:

1. Recommendations should be derived from the findings of an investigation. In general, there will be a direct link between a finding and one or more recommendations.

2. Recommendations should be coordinated with the appropriate departments prior to issuance; however, the safety department should retain sole responsibility for the decisions to issue recommendations.

3. The receiving department should respond in writing as to the disposition of each recommendation. In the event that the recommendation is not adopted or an alternative safety action is taken instead, the rationale should be set forth.

4. Open recommendations should be tracked on a regular basis (30, 60, and 90 days) until completed.

The accident investigation process described applies to investigations of minor accidents. In the case of a major accident (hull loss, occupational death, or a major environmental spill), coordination will be required with external agencies (NTSB, FAA, OSHA, EPA). For example, in the event of an aircraft crash, it is important that an airline have a detailed emergency response plan that contains, among other things, detailed plans for a go-team who will participate in the accident investiga-tion. The go-team should be headed by a senior manager, who serves as the primary coordinator for all company activities related to an accident investigation. Appropriate technical personnel are designated as potential members of a go-team, the actual makeup of which is determined on the basis of known facts at the time of notification, most importantly, aircraft type and location. With regard to the latter, for any accident occurring outside of U.S. airspace, the company go-team reports to the U.S. accredited representative, in full accordance with International Civil Aviation Organization (ICAO) Annex 13.

To support the go-team, all necessary equipment and supplies, such as personal protective equipment, communications gear, and other material that may be necessary to conduct a major accident investigation, must be ready for instant shipment to the scene of an accident. All members of the go-team should be formally trained in accident investigation and must have the required (by OSHA) blood-borne pathogen training. The go-team roster should also identify other personnel who are tasked with providing administrative support to its technical members. The emergency response plan specifies the staffing of a command post by designated personnel from the affected airline. These people are responsible for the coordination of all activities related to the accident investigation, including the assembly of records, manuals, bulletins, and other necessary materials. Oversight and leadership of the entire effort should be under the direction of senior management, including the corporate safety officer. An annual emergency response drill is required to ensure that the plan is adequate and that the designated personnel are appropriately trained.

Although absolutely necessary, an emergency response plan and associated elements of the safety function are quite obviously something that no airline wants to activate for real. To meet this challenge, every airline company must have some form of independent, proactive accident prevention effort. The best means to accomplish this is to seek to identify hazards and risks consistently and then to eliminate these risks through accident prevention measures using SMS concepts.

On a final note, virtually all operational incidents will require a certain level of technical investigation and analysis to fully understand and identify the underlying cause factors. Within the airline corporate structure, investigative responsibility for flight safety incidents must be clearly assigned. Similarly, professional investigative methods must be consistently employed in the technical area. Use of investigative tools such as the digital flight data recorder (DFDR) requires a consistent objective and confidential method of analysis; often under the supervision of the NTSB if the Board assumes jurisdiction of the accident or incident. Also, since analysis methods require complex transcription methods, it is likely that DFDR analysis will occur at the maintenance/engineering department. However, DFDR information must be maintained in a strict confidential status with operational DFDR analysis performed by personnel familiar with current operational procedures.

AUDITING—A TOOL OF SAFETY ASSURANCE

Analyzing and addressing hazards after they have materialized (caused accidents and injuries) are commonly referred to as a *reactive approach* to safety. The preferred *proactive approach* to safety requires that hazards and hazardous conditions

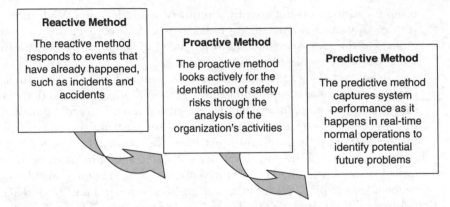

FIGURE 12-7   Safety management strategies: reactive, proactive, and predictive. (*Source: ICAO Safety Management Manual (2nd edition) pp. 3–11. www.icao.int*)

be identified before they cause accidents and injuries. Using modern Safety Management System strategies, the *predictive approach* to safety does not require a triggering event to take place. Instead, the predictive method aggressively seeks safety information to identify potential future problems (Fig. 12-7). The ICAO Safety Management Manual illustrates these three safety management strategies as follows:

One way of accomplishing safety assurance in an SMS environment is through the audit process. The words audit and inspection are often used interchangeably with some confusion about their definitions. The problem is compounded when the word survey is thrown into the mix. Most schools of thought (including the authors of this text), however, consider an audit to be more comprehensive and have a greater scope of work than an inspection. While inspections usually identify workplace hazards, audits are designed to evaluate programs and management issues that have resulted (or could result) in workplace hazards. Safety audits are an integral part of a safety program, and in addition to uncovering hazards, audits reveal the level of compliance with regulatory standards as well as measure the effectiveness of the safety program. Some of the common approaches to auditing an airline or airport facility are as follows:

- *Comprehensive audits*. These are extensive, detailed safety inspections/surveys (site conditions and programs) of the entire organization (an air carrier), a facility within an organization (an air carrier hub or maintenance facility), or an operation within a facility (fuel handling at the ramp). These audits are usually conducted on an annual basis. These audits are conducted by a team of safety professionals from within and outside the organization, facility, or site, and they are usually led by the company safety director or the divisional or site safety manager. This type of audit results in a formal written report of findings and recommendations and usually requires a response within an agreed time frame.

- *Self-audits.* These are informal daily, weekly, or monthly audits (more appropriately inspections) and form the crux of the internal inspection program (IEP). Additional information on IEP is presented in the next section. The frequency of checks depends on the application. For example, certain areas of airports have to be checked daily (or more often), and some operations in a hangar or on the ramp may require daily inspections. These audits are conducted by line management at the site or facility and usually result in a checklist of items that either are in acceptable condition or need fixing. No reports are normally generated as part of these audits. One of the functions of the director of safety is to help set up and provide expertise to local internal inspection teams.

- *Status audits.* These audits are intended to determine the status of compliance at the time of observation. These are usually subject-oriented audits and are conducted in areas of high risk and/or areas of known or perceived concern. A spot inspection of refueling for a randomly selected aircraft (or a group of aircrafts) and evaluating a welding operation are examples of status audits. These audits may generate reports if the findings are serious enough to require management's attention.

COMMUNICATIONS

Communication takes on many forms. The one addressed earlier in Chapter 7 concentrated on verbal/voice radio communications within the cockpit and between the pilot and the air traffic controller. It was further discussed how ineffective (inadequate or incorrect) communications in addition to diction, accents, and unique nuances (double meaning and confusing words) in the spoken (English) language can lead to aviation accidents and disasters. Other avenues of communication are now explored.

SAFETY, MAINTENANCE, AND FLIGHT COMMUNICATION

The airline safety, maintenance, and flight operations organizations are often mistakenly viewed as separate entities with little or no shared mutual interests, when in actuality the three organizations are closely interlinked in myriad complex relationships and objectives. While the maintenance/engineering department provides virtually all the technological expertise necessary to maintain the aircraft fleet, the flight and in-flight (flight attendant) organizations are considered the end user of the technical product and, therefore, must maintain a user's level of technical knowledge. The development and maintenance of this user requirement, whether for access of a complex onboard database or the operation of an evacuation system, define the first link in the flight/maintenance relationship.

The second link of the relationship is forged by the legal airworthiness concept. While the captain is responsible for ensuring the final airworthiness and safety of the aircraft, it is the maintenance engineering department that maintains or returns an aircraft to an airworthiness condition. Therefore, an active and formalized communication link between the two groups is necessary for mutual satisfaction of the end product, a flyable aircraft. A fundamental component of this link is the aircraft logbook, which serves to document the degradation and restoration of the aircraft between variable levels of serviceability, or airworthiness.

The third component of the flight–maintenance–safety relationship is the regulatory-procedural link. Both procedural and technical regulatory issues must be coordinated between the three departments. While regulatory requirements emanate from the FAA, they are often precipitated by the NTSB investigative findings, thereby necessitating direct communication with both agencies for effective implementation of evolving safety requirements.

In all the above-defined levels of interdivisional relationships, the concept of safety is prevalent, and dependent upon uninhibited access of information and communication between the flight, maintenance, and safety organizations. To enhance and facilitate the communication of time-critical safety information between the maintenance, flight operations, and safety departments, dedicated formal communicative links must be established.

FAA, NTSB, AND FLIGHTCREW COMMUNICATION

To ensure that complete and consistent support of NTSB and FAA technical-based investigations is maintained, it is contingent upon the safety department to forge strong communication links with the NTSB and FAA. Included in this process is the requirement for consistent analysis of the technical aspects of NTSB and FAA investigations. Because of the highly complex nature of modern aircraft, a means to quickly analyze and disseminate critical safety information is required. In addition, conduct of both major and minor technically oriented investigations will require in-depth and thorough coordination of NTSB recommendations and FAA requirements. This is also true with foreign regulatory agencies; however, the specific processes vary widely between nations.

Critical to the safety process is the ability to quickly communicate technical information with both pilot and flight attendant groups. Safety information must be continually reviewed for information pertinent to each airline's operations. For example, developments in technical detail or aircraft operational requirements/procedures must be provided to crewmembers in a timely manner. Included in the rapid communication process are issues related to cabin safety.

SAFETY TREND EVALUATION, ANALYSIS AND DATA EXCHANGE SYSTEM (STEADES)

STEADES is a service of the International Air Transport Association, which features the world's largest database of de-identified airline incident reports. It originated by adapting the British Airways Safety Information System (BASIS) into an international secure forum for the sharing of safety data consisting of over 120 airlines and over 100,000 pilot reports each year. In 2010, STEADES members received quarterly reports focusing on global safety trends with unbiased benchmarking analysis of key issues from aviation industry specialists.

In addition to the interactive benchmarking workbooks, in 2010 STEADES members also received safety analysis on the following subjects:

- Airport marking deficiencies
- Anti-icing and de-icing operational events
- Injuries caused by turbulence
- Fire warnings and cargo smoke incidents
- Foreign object damage
- Aircraft configuration warnings

This anonymous incident database adds great value to smaller airlines who can benefit from the feedback, lessons learned, and combined experience of many airline safety departments throughout the world.

COMMUNICATION MEDIA

Safety publications are another important component of an airline safety program. Most airlines publish regular internal safety documents to maintain high individual awareness of safety and risk management. Videos are also used to disseminate safety information to all employees. Communication and education are critical elements of any proactive airline safety program; there must be an effective mechanism in place to ensure the flow of critical safety information within the company.

In conjunction with the SRM process, the communication of pertinent information necessary for reduction of the risk must be provided to the safety user. Depending on the type of risk involved, the user may be a pilot, flight attendant, mechanic, and/or related support person. The primary recipient of risk identity depends on the type of failure cause. Recurring material failure causes (e.g., fuel pump failure) are generally the responsibility of the technical services or engineering organization, while human-factor–related causes are best communicated to employee groups (e.g., pilots, flight attendants, mechanics) as appropriate.

TRAINING

Training in general is a very important component of any business plan, and safety training is no exception. Safety training can help employees develop the knowledge and skills they need to comprehend workplace hazards and protect themselves. In fact, at least one major regulatory agency (OSHA) considers training to be one of the four major elements of its recommended safety and health program guidelines. Safety training is most effective when it is integrated into a company's performance and job practices requirements. The importance of training has been discussed previously in Chapters 6 and 7. While each regulatory agency sets different safety training requirements, the following important common issues should be noted and adhered to:

- The company's policy should clearly state the company's commitment to safety.
- The training should be conducted on paid work time.
- Training must be in the language that the employee understands and should be delivered at an appropriate pace and comprehension level.
- Employees and management should be involved in developing training programs.
- Training should be conducted based on identified needs.
- Training should have clearly defined goals.
- Training should follow good teaching pedagogues and should incorporate the following basic principles:
  - The attendee must understand the purpose of the instruction.
  - The information should be organized to maximize understanding.
  - Embed "hands-on" experiences whenever possible.
  - Design the training for different learning styles such as written instruction, audiovisual instruction, hands-on exercises, and group learning assignments. Straight lectures should be no more than 20 minutes long.
  - Repeat concepts as often as practical.
- There should be an evaluation component to training. Both the attendees and the training itself should be evaluated. Attendees can be evaluated through tests, quizzes, assignments, etc. Course evaluation questionnaires/surveys can be used to evaluate course content and its ability to achieve the stated goals.
- All employee training records should be documented and maintained. This will help ensure that every employee who needs training receives it, that refresher courses are provided when needed, that documentation is available when required to prove that training was conducted (a regulatory requirement in some cases), and that the training was appropriate. Documentation should include:
  - The name and signature of the trained employee
  - The training date
  - The topic, including a brief lesson plan
  - Evidence of the employee's successful completion
  - The name and signature of the trainer

ROLE OF ALPA IN AIRLINE SAFETY

Air safety is the primary responsibility of every airline pilot. As the oldest and largest airline pilots' union, the main role of the Air Line Pilots Association's (ALPA's) air safety structure is to provide channels of communication for line pilots to report air safety problems. An additional role is to stimulate safety awareness among individual pilots, to enable flightcrew members to be constructive critics of the airspace system. Finally, the air safety structure helps investigate airline accidents. Over the years, ALPA's air safety structure has contributed significantly to air safety.

ALPA's air safety structure can be compared to a three-sided pyramid, each side helping to support the others. That is, the air safety structure handles air safety problems from three perspectives:

1. By airline

2. By geographic area

3. By subject

ALPA Safety Structure

The basic unit of ALPA is the local council—the pilots of a single airline at a particular domicile. Each local council normally has an air safety committee, which processes local air safety problems. The committee is composed of not more than three members, headed by the local council air safety chairperson (LASC), who are appointed to 2-year terms by the local executive council (LEC) or the LEC chairperson.

An airline's central air safety committee is made up of each of the LASCs from that airline's various councils. The central air safety chairperson (CASC), appointed to a 2-year term by the master executive council (MEC) or MEC chairperson, presides over the committee, which handles problems unresolved at the local council level and those broader in scope than local issues.

ALPA's national safety structure is composed of five management groups, headed up by an Executive Air Safety Chairman, in these specific safety areas:

• Aircraft Design and Operations

• Airport and Ground Environment

• Air Traffic Services

• Human Factors and Training Group

• Accident Analysis

In 2010, ALPA named its top safety issues, which include the following topics:

• Modernization of the National Airspace System (NextGen)

• Retention of voluntary safety reporting systems—Flight Operation Quality Assurance (FOQA), and Aviation Safety Action Program (ASAP)

• Modernizing duty and rest regulations for pilots (combating pilot fatigue)

- Enhanced pilot training and professionalism
- Integration of unmanned aircraft systems (UAS) into the ATC structure

ACCIDENT INVESTIGATION

ALPA's Accident Investigation Board oversees investigation of all air carrier accidents by an appropriate ALPA subgroup. The board coordinates ALPA's participation in accident or incident investigations by the NTSB or the FAA and ensures that the appropriate ALPA subgroup determines the significant factors in each air carrier accident.

The 10 pilots on ALPA's Accident Investigation Board are scattered throughout the United States, so that at least one trained accident investigator can rapidly reach the scene of the accident no matter its location. ALPA's president appoints the board's chairperson. The executive central air safety chairperson, with the approval of ALPA's president, appoints the other nine members, and all 10 serve terms without time limits.

Under ALPA policy, each airline pilot group establishes an accident investigation team to function under the direction and responsibility of either the central air safety chairperson or the chief accident investigator, whichever that airline's MEC chooses to designate. (In most cases, the CASC heads the accident investigation team for his or her airline.) Either the CASC or the chief accident investigator is in charge of the technical aspects of any investigation involving one of his or her airline's aircraft. The CASC may request, however, that complete direction and operational control of an association accident investigation effort be assigned to a member of the ALPA Accident Investigation Board.

ALPA also maintains a worldwide accident/incident hotline for its pilots to call for help if involved in an aircraft critical safety event. If deemed necessary, ALPA will also employ its Critical Incident Response Program (CIRP), whose trained flight deck counselors work to mitigate the psychological effect of an accident before harmful stress occurs.

ALPA SAFETY MANAGEMENT SYSTEMS POLICY

Since 2008, ALPA has been a leader in supporting SMS principles as set forth in the ICAO Safety Management Manual. The ALPA policy statement is set as follows:

ALPA Supports SMS when it is developed and implemented in accordance with the following:

a. A documented, clearly defined commitment to the SMS from the CEO—a written SMS policy, signed by the CEO, which recognizes the business benefit of asset preservation and mishap prevention. The policy must show commitment to continuous improvement in the level of safety, management of risk and to a strong safety culture, and

b. Documented lines of safety accountability; and

c. Active involvement of the affected employees in a non-punitive reporting system and a commitment to a "just" safety culture, and

d. A documented, robust Safety Risk Management (SRM) program. The SRM program requires the participation of labor organization(s) as the representative of their employee groups in both the identification of hazards and in the development of risk mitigation strategies; and

e. A documented process for collecting and analyzing safety data and implementing corrective action plans; and

f. A documented method for continuous improvement of the SMS.

(Source: ALPA Administrative Manual, Section 80, Part 1 page 18, www.alpa.org)

Working to maintain the highest levels of airline safety is an enormous challenge. The depth and breadth of ALPA's air safety structure, and the critical role it plays in meeting that challenge, are unique in the air transportation industry.

FLIGHT SAFETY FOUNDATION

Another organization that plays a significant role in air safety is the Flight Safety Foundation (FSF). It was founded in 1947 by then-leaders of the aviation industry, who recognized the need for an independent body that would promote safety in aviation, anticipate flight safety problems, act as a clearinghouse on safety matters, and disseminate aviation safety information. Through the years, FSF has been responsible for the development of many aviation safety improvements that are taken for granted.

FSF doctrine is to anticipate and study flight safety problems and to collect and disseminate safety information for the benefit of all who fly. The most safety conscious airline shares the same airspace with the less-informed or even careless operator, so it is of benefit to invest in the education and awareness-raising of such operators. FSF, with more than 1200 member organizations in more than 150 countries, provides an information collection and feedback function that many less developed aviation industries rely on for aviation safety information.

As an apolitical, independent, nonprofit, and international organization, FSF benefits from a nonofficial status because it avoids a great many of the postured responses that many businesses are obliged to present to their peers, governments, and media. Because it has no enforcement authority, its task is friendly persuasion. Several aviation leaders have described FSF as the "safety conscience" for the industry. FSF has the support from major manufacturers and airlines (which have a sense of responsibility as well as an enlightened self-interest) to make the skies as safe as possible.

The agendas of FSF's annual safety seminars, held in locations throughout the world for the past 60 years, feature a strong program of accident prevention methodology presented by the best safety experts in industry, government, and academia. Their aim, of course, is to provide effective feedback to the aviation community about hazard identification, design, training, inspection, procedures, trend analysis, etc., to use collective knowledge for the prevention of accidents. Feedback occurs in other forums, such as industry association meetings, industry-government committees dealing with specific safety topics, meetings with other independent associations focusing on specific areas of safety improvement, and computer-based data exchanges.

Another means of obtaining information for feedback to the airline industry is FSF's confidential safety audits of corporate and airline operations. This is a valuable method of gaining firsthand information about how companies comply with their own operating standards, how they value safety, and how they manage risk. FSF shares this information on a non-attributable basis with its members through the regular publications it produces as well as its safety seminars. In addition, it completes the feedback loop by special workshops and conferences that focus on specific safety problems in various regions of the world.

FSF has helped the former Soviet Union to establish a Flight Safety Foundation in what is now the Commonwealth of Independent States (CIS). FSF is actively working through FSF-CIS to inculcate a safety-conscious culture in Aeroflot and the more than 60 emerging airlines in the CIS. Coordination of risk management information is a real challenge. For their part, the agencies of the former Soviet Union have been quite generous in sharing safety and accident information they have developed for their aviation operations. FSF, in turn, has shared these data with its worldwide membership.

SMS INFRASTRUCTURE AT A MAJOR U.S. AIRLINE

As noted earlier, in 2007, major U.S. Airlines were invited by the FAA to participate in the SMS Pilot Project to work closely with the FAA to establish a program of voluntary SMS implementation. This pilot project has been successful following the basic guidance of the ICAO Safety Management Manual (2nd edition) and the FAA Advisory Circular 120-92A. The following sections describe the journey of a major U.S. airline along the path to SMS Maturity Level Four. Level four is the final stage of SMS implementation where all SMS processes are in place.

The primary roadmap to implement SMS is the FAA Safety Management System Implementation Guide, Revision 3, dated June 1, 2010. (www.faa.gov)

THE PHASED SMS IMPLEMENTATION PROCESS

The FAA Implementation Guide provides a detailed process for implementation of the SMS functional expectations described in AC 120-92A. The overall objective of the guide is to assist aviation service providers (such as airlines) develop and implement a phased, integrated, comprehensive SMS for their entire organization.

The implementation roadmap is set forth in a four-phased process similar to that outlined in the ICAO Safety Management Manual. The phases of implementation are arranged in four levels of "maturity" similar to that developed in the proven capability maturity model for software engineering at Carnegie-Mellon University

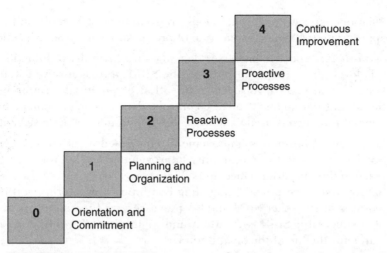

FIGURE 12-8   FAA—SMS maturity model. (*Source: www.faa.gov*)

in Pittsburgh, Pennsylvania. Figure 12-8, from the FAA Web site, is an FAA depiction of this maturity model, which is further described.

1. *Level zero:* Orientation and commitment is not so much a level as a status. It indicates that the Aviation Product/Service Provider has not started formal SMS development or implementation and includes the time period between an Aviation Product/Service Provider's first requests for information from the FAA on SMS implementation and when they commit to implementing an SMS.

2. *Level one:* Planning and organization. Level 1 begins when an Aviation Product/ Service Provider's Top Management commits to providing the resources necessary for full implementation of SMS throughout the organization. Two principal activities make up level one:

   a. *Gap analysis:* The first step in developing an SMS is for the organization to analyze its existing programs, systems, and activities with respect to the SMS functional expectations found in the SMS Framework. This analysis is a process and is called a "gap analysis," the "gaps" being those elements in the SMS Framework that are not already being performed by the Aviation Service Provider.

   b. *Implementation plan:* Once the gap analysis has been performed, an Implementation Plan is prepared. The Implementation Plan is simply a "road map" describing how the Aviation Service Provider intends to close the existing gaps by meeting the objectives and expectations in the SMS Framework.

3. *Level two:* Reactive process, basic risk management. At this level, the Aviation Service Provider develops and implements a basic SRM process. Information acquisition, processing, and analysis functions are implemented and a tracking system for risk control and corrective actions are established. At this phase, the

Aviation Service Provider develops an awareness of hazards and responds with appropriate systematic application of preventative or corrective actions.

4. *Level three:* Proactive processes, looking ahead. (Full-Up, Functioning SMS) At this level, the activities involved in the SRM process involve careful analysis of systems and tasks involved; identification of potential hazards in these functions, and development of risk controls. The risk management process developed at level two is used to analyze, document, and track these activities.

5. *Level four:* Continuous improvement, continued assurance. The final level of SMS maturity is the continuous improvement level. Processes have been in place and their performance and effectiveness have been verified. The complete Safety Assurance process, including continuous monitoring and the remaining features of the other SRM and SA processes are functioning. A major objective of a successful SMS is to attain and maintain this continuous improvement status for the life of the organization.

(Source: www.faa.gov)

Following this four step phased SMS process, the airline made a commitment to participate in this pilot project in 2008. Rather than attempt to tackle the whole airline at once, senior airline management decided to implement SMS into its new structure one department at a time. Since the flight operations department had mature safety systems already in place, this department was selected first and a Gap Analysis was conducted. The implementation plan proceeded smoothly by closing the existing gaps and meeting the objectives and expectations in the SMS framework, and in a three year period, the entire airline had reached SMS Maturity level four. A key element of strategy was to ensure that the operating departments remain responsible for departmental policy, procedural development, and SMS implementation in support of the company-wide SMS policy commitment. Although the Safety Department was responsible for company-wide management, the SMS process was carefully inserted into the operations manual of each operating department.

The CEO of the airline is the Accountable Executive. The company's Safety Management System is organized around the "four pillars" of safety management:

- *Policy*—Clearly defined policies, procedures, organizational structure, and accountabilities.

- *Risk management*—Formal system of hazard identification, risk assessment, risk management, resource allocation, and system monitoring.

- *Safety assurance*—Applies the processes of quality assurance, internal evaluation, continuous monitoring, employee safety event reporting, external oversight, data analysis, etc., to assure that risk controls perform their intended functions and remain effective at maintaining within acceptable levels.

- *Promotion*—Appropriate training for the level of safety responsibility and continuous communications to all employees of safety values and practices that support a sound safety culture.

The four pillars of SMS are communicated to the employees based upon the requirements of ICAO Annex 6 and FAA AC 120-92A. All employees receive initial and recurrent training appropriate for the person's SMS responsibilities. The airline has defined six broad categories for specific training requirements tailored to each employee's safety responsibilities. These categories for SMS training range from the CEO to the general employees as follows:

- Top management—CEO, President, and COO
- Line management—individuals, regardless of organizational level, responsible for overseeing daily operations and activities with decision or policy-making responsibility who may be responsible for risk management and risk mitigation within the work group.
- Safety manager—Vice President, Safety and Regulatory Compliance
- Safety staff—individuals with data collection, data analysis, or hazard identification responsibilities
- Front line, operational, or production employees—specific to job functions such as pilot, flight attendant, dispatcher, inspector, mechanic, engineer, maintenance controller, fleet service agent, customer service agent, and cargo agent are examples of this category. This list may not be all inclusive and is not intended to limit line management from assigning others within their work groups to this category.
- General employees—operations group employees who do not fall within one of the other five categories

A key position in this organization is the Vice President of Safety and Regulatory Compliance, who reports directly to the Chief Operating Officer. This position is responsible for the development, implementation, maintenance, and oversight of company SMS policy in accordance with FAA Advisory Circular 120-92A. As noted in Fig. 12-9, five directors report to this Vice President, who has broad responsibilities for safety matters throughout the corporation. Under each of these directors are safety managers who interface with company line management in their areas of expertise.

Among some of the significant responsibilities of the Vice President are to

- Ensure that SMS processes and outputs are documented, monitored, measured, and analyzed to determine where improvement is needed.
- Ensure clear and regular communication of safety policy, goals, objectives, standards, and performance to all employees of the organization.
- Maintain liaison with FAA, NTSB, DoD, and TSA authorities; interpret and evaluate regulatory and compliance requirements, and suggest appropriate courses of action to the CEO, President, and COO of the airline.

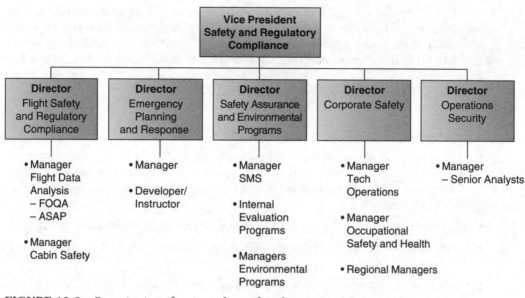

FIGURE 12-9   Organization of major airline safety department.

- Maintain a vigorous corporate-wide flight and ground safety program and ensure that all activities related to aircraft accidents, incidents, and personal injuries are handled in accordance with company procedures.

- Ensure that company environmental programs conform to regulatory standards and industry requirements.

- Monitor compliance and conformance through SMS Safety Assurance and the internal evaluation program.

- Ensure that safety lessons learned are documented and used in the development of corrective/preventive action plans and are publicized to employees as appropriate.

- Promote a positive safety culture and active safety awareness throughout the company.

The top level company SMS organization has four functional levels which are integrated into the company's line matrix organization. The operation of these SMS components can be shown as follows (Fig. 12-10).

Company Operations Standards Board

- Provides senior management a forum to discuss significant safety issues, review trends and SMS outputs, and identify opportunities for safety policy and culture reinforcement.

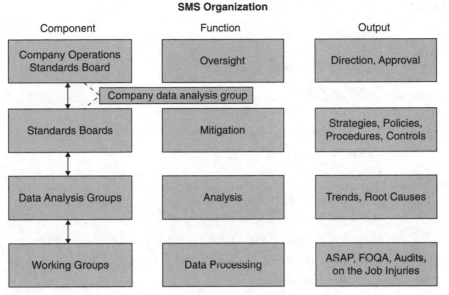

FIGURE 12-10   SMS functional organization in a major airline.

COMPANY DATA ANALYSIS GROUP

- Evaluates operational data from the standards boards to ensure consistency and provide a venue for the identification of system issues.
- Reviews operational data for divisional and cross-divisional inpact (cross-polination).
- Develops preliminary mitigation and recommendations for the COSB.

STANDARDS BOARDS

- Conduct formal assessment; reviews, creates, approves and coordinates policy and procedure changes; training and implementation of corrective action.

DATA ANALYSIS GROUPS

- Analyze trends and determine root causes to submit recommendations to the Standards Boards with a preliminary risk assessment.

WORKING GROUPS

- Identify hazards and process safety risk management data for further analysis.

It is anticipated that airline SMS will become regulatory in 2012 for FAR Part 121 carriers. The final chapter of this text will provide more details regarding aviation Safety Management System principles.

KEY TERMS

Safety Management Systems (SMS)

Four Pillars of SMS

    Safety Policy

    Safety Risk Management

    Safety Assurance

    Safety Promotion

ICAO Annex 6

FAA Advisory Circular 120-92A

ICAO Safety Management Manual (SMM)

Reactive, Proactive, Predictive approach to safety

Emergency Response Plan (ERP)

Audit Process

Airline Safety and FAA Extension Act of 2010 (PL 111-216)

Safety Trend Evaluation, Analysis and Data Exchange System (STEADES)

Airline Pilots Association (ALPA)

Flight Safety Foundatoin (FSF)

Accountable Executive

SMS Maturity Model

REVIEW QUESTIONS

1. Discuss the international development aspects of Airline Safety Management Systems (SMS).

2. List and discuss the salient features of the Four Pillars of SMS.

3. Explain the development of airline SMS in the United States.

4. What is the purpose of FAA Advisory Circular 120-92A?

5. Discuss the role of accident investigation and auditing in airline safety matters.

6. What does the title Accountable Executive mean?

7. What is the Safety Trend Evaluation, Analysis and Data Exchange System (STEADES)?

8. Describe the steps of the SMS maturity model.

9. Compare and contrast the reactive, proactive, and predictive approaches to safety management.

10. What is the role of the Airline Pilots Association (ALPA) in safety matters?

REFERENCES

Anthony, Thomas 2009, *SMS on Wheels* Aerosafety World, Flight Safety Foundation, September.

ICAO Safety Management Manual, 2nd Edition, 2009, Doc 9859, www.ICAO.int

Federal Aviation Administration Advisory Circular 120-92A, Subject: Safety Management Systems for Aviation Service Providers, dated 8/12/2010, www.faa.gov

Federal Aviation Administration, Flight Standards Service—SMS Program Office; Pilot Project and Voluntary Implementation of Organization SMS Programs, Revision 6, September 29, 2010.

_____. 2010 *Safety Management System (SMS) Assurance Guide,* Revision 3, June 1, 2010.

_____. 2010 *Safety Management System (SMS) Implementation Guide,* Revision 3, June 1, 2010.

_____. 2010 *Safety Management System (SMS) Pilot Project Participants and Voluntary Implementation of Organization SMS Programs, Revision 6,* September 29, 2010.

Ferrari, Jim and Orlady, Linda; 2008 *"What Will SMS Do For Me?"* Air Line Pilot, ALPA June/July.

Stolzer, Alan J., Halford, Carl D., and Goglia, John J. (2008), Safety Management Systems in Aviation. Burlington, VT: Ashgate Publishing.

U.S. Department of Transportation/FAA, Recommendations of the Aviation Safety Management System, Aviation Rulemaking Committee, March 31, 2010.

WEB REFERENCES

Air Line Pilots Association, www.ALPA.org

Flight Safety Foundation, www.flightsafety.org

ICAO publications and resources, www.icao.int

International Air Transportation Association, www.iata.org

Delta Airlines, www.delta.com

U.S. Airways, www.usairways.com

AVIATION SAFETY MANAGEMENT SYSTEMS

LEARNING OBJECTIVES

After completing this chapter, you should be able to

- Discuss the evolution (history) of SMS principles.
- List and describe the fundamental components of SMS—The Four Pillars.
- Compare and contrast the differences between a hazard and a safety risk.
- Discuss the significance of the Probability $\times$ Severity = Risk formula.
- Explain the purpose of the risk assessment 5×5 matrix in terms of severity and likelihood.
- Discuss the significance of reducing risk to a level as low as reasonable practicable (ALARP).
- Describe the three principle elements of the safety assurance process.
- Discuss the future of Safety Managements Systems in aviation.

INTRODUCTION

In many chapters of this book, the reader has been introduced to aviation Safety Management Systems (SMS) thinking and concepts. In fact, SMS has been presented as the future state of commercial aviation safety. This chapter will discuss the history of SMS principles, the fundamental concepts, the tools and techniques used by aviation safety professionals, and the future of SMS in aviation.

THE EVOLUTION OF SMS PRINCIPLES

It is clear that modern safety principles have significantly evolved in the years following World War II. Following the war, commercial aviation saw significant growth with the introduction of the B-47, jet engine, and swept back wing configuration. This "technical era" of aviation safety concentrated on the improvement of technical factors during the decades of the 1950s and 1960s.

In the 1970s, the Japanese gained world market share in the electronics and automobile industry using the "Total Quality Management" (TQM) concepts of two quality experts, W. Edwards Deming and Joseph Juran, among others. This movement in Japan focused on improving the quality of manufactured products through teamwork and employee involvement. In the aviation world, the focus shifted to the "human factors" era using such tools as crew resource management (CRM) and line-oriented flight training (LOFT), among others. As outlined by ICAO in the Safety Management Manual, the mid-1970s to the mid-1990s has been called the golden era of aviation human factors.

In the early 1990s, aviation safety thinking had shifted its focus to the "organizational era," which viewed safety from a systems perspective which included technical, human, and organization factors as well. Figure 13-1 reflects this evolution of safety thinking.

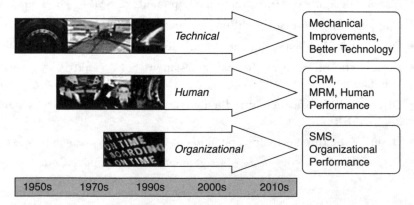

Evolution of Safety Thinking-Factors in Accidents

FIGURE 13-1   The evolution of safety thinking—factors in accidents. (*Source: www.faa.gov*)

As noted previously in Chapters 6 and 7, Dr. James Reason's Swiss Cheese model introduced the concept of the "organizational accident," which is a breakdown of organizational processes resulting in an accident. Using quality-based employee involvement and cultural empowerment concepts employed by the Japanese and adopted by the International Organization for Standardization (ISO), aviation Safety Management Systems theory has now evolved into a worldwide movement which should pay great safety dividends in the future.

THE FUNDAMENTAL COMPONENTS OF SMS—THE FOUR PILLARS

In the last chapter, the fundamental components were introduced based upon the ICAO SMS framework. These four components were further broken down into 12 specific elements necessary for an organization to have a fully functioning program. The FAA has adopted ICAO SMS theory in its Advisory Circular 120-92A, and explains each of the components on its Web site at this link: www.faa.gov

The four components of an SMS are as follows:

1. *Safety policy*—Establishes senior management's commitment to continually improve safety; defines the methods, processes, and organizational structure needed to meet safety goals.

2. *Safety Risk Management (SRM)*—Determines the need for, and adequacy of, new or revised risk controls based on the assessment of acceptable risk.

3. *Safety Assurance (SA)*—Evaluates the continued effectiveness of implemented risk control strategies; supports the identification of new hazards.

4. *Safety Promotion*—Includes training, communication, and other actions to create a positive safety culture within all levels of the workforce.

(Source: www.faa.gov) See Fig. 12-6.

As previously indicated, safety policy encompasses the leadership of the Accountable Executive to provide clear safety objectives, methods, processes, and the organizational structure needed to meet safety goals. Safety Promotion encompasses a nonpunitive, positive safety culture (also known as a "Just Culture") which empowers all employees to have a role in safety. This is accomplished through comprehensive SMS training, communication, and awareness techniques as discussed in detail in Chapter 12.

The heart of any SMS organization is the inter-relationship between Safety Risk Management and safety assurance. The next sections will focus on these two key functional processes. Figure 13-2 illustrates how the SRM process relates to the SA process at a summary, introductory level.

Basically, the SRM process provides for the identification of hazards and assessment of risk. Once it is assessed, risk should be controlled (mitigated) to a

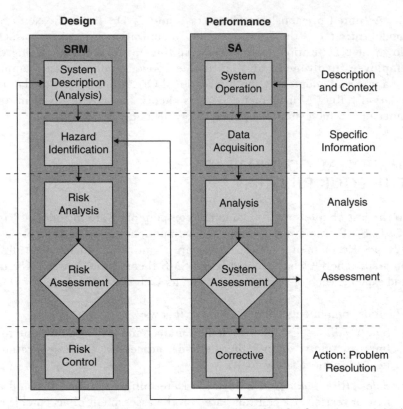

FIGURE 13-2   Safety Risk Management (SRM) and safety assurance (SA) processes. (*Source: FAA Advisory Circular 120-92A, p. 6 www.faa.gov*)

level "as low as reasonably practicable" (ALARP). At this point, the SA process takes over to assure that the risk controls are effective using system safety and quality management concepts of monitoring and measurement. The SA process also provides a means for new controls if there is a change in the operational environment by returning to the hazard identification step.

Now that the overall SMS relationships have been introduced, we will examine specific details of the relationship of the SRM and SA processes.

SAFETY RISK MANAGEMENT EXPLAINED

HAZARD OR RISK

Before a risk can be managed, a potential hazardous condition must be identified. These two concepts of hazard or risk can be easily confused. ICAO defines *hazard* as follows:

A condition, object or activity with the potential (emphasis added) of causing injuries to personnel, damage to equipment or structures, loss of material, or deduction of ability to perform a prescribed function.

(Source: ICAO Safety Management Manual, 2nd Edition, Section 4.2.3)

An example of a hazard used by ICAO, a crosswind of 15 knots blowing perpendicular to a runway is a hazard; it is not a risk. ICAO defines *risk* as follows:

The assessment, expressed in terms of predicted *probability* and *severity* (emphasis added) of the consequences of a hazard taking as reference the worst foreseeable situation.

Thus, risk can be expressed in a formula as:

$$P \times S = R$$

where,

- *P* is the probability that a given hazard may materialize (likelihood of a loss event per unit of time or activity).
- *S* is the severity of consequences (effects) that the materialized hazard can generate (loss per event).
- *R* is the expected risk (loss) per unit of time or activity.

So, returning to the previous example:

- A crosswind of 15 knots perpendicular to a runway is a hazard.
- The potential for running off the runway because the pilot might not be able to control the aircraft is one of the consequences of the hazard; however,
- The assessment of the consequences of literally running off the runway expressed in terms of *probability (P)* and *severity (S)* is the safety *risk (R)*.

Stated another way for clarification

- *P* = The probability that the aircraft would skid/run off the runway given that there is a 15 knot crosswind
- *S* = the severity of the above event which could vary from no aircraft damage or injuries, to total destruction or both (depending on variables such as the aircraft speed, attitude at time of runway contact, tire pressure, etc.).

Safety Risk Management (SRM) involves hazard identification and risk assessment to obtain mitigation of the safety risk to a level "as low as reasonable practicable," also

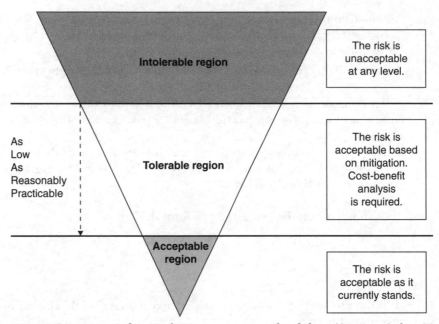

FIGURE 13-3   Safety Risk Management—tolerability. (*Source: Safety Management Manual, 2nd Edition, 5.3.2*)

known as ALARP. ICAO illustrates this risk tolerability as an inverted triangle as shown in Fig. 13-3.

To conduct a safety risk assessment, it is very helpful to plot it visually to clearly see the relationship between severity of consequences and likelihood of occurrence. The classic ICAO/FAA example of using an alphanumeric table and chart is provided in Table 13-1 and Fig. 13-4.

Note the coding of the risk matrix:

Dark gray—unacceptable risk

White—acceptable with mitigation

Light gray—acceptable

CONTROL STRATEGIES

The purpose of risk mitigation is to control hazards so they do not become risks, or alternatively, reduce the severity or likelihood of the risk to an acceptable level. The order of precedence for these controls is well established, and they are listed below in rank order from most effective to least effective:

1. *Elimination of the hazard.* The most effective is the engineering control strategy which eliminates the safety risk completely. As example would be providing interlocks to prevent thrust reverser activation in flight. Since this is a firm control solution, it is sometimes known as a "hard" control strategy.

TABLE 13-1 Sample Severity and Likelihood Criteria (*Source: www.faa.gov*)

SEVERITY OF CONSEQUENCES			LIKELIHOOD OF OCCURRENCE			
SECURITY LEVEL	DEFINITION	VALUE	LIKELIHOOD	DEFINITION	VALUE	
Catastrophic	Equipment destroyed, Multiple deaths	A	Frequent	Likely to occur many times	5	
Hazardous	Large reduction in Safety margins, Physical distress or a Workload such that Operators cannot be Relied upon to perform Their tasks accurately Or completely. Serious Injury or death. Major Equipment damage.	B	Occasional	Likely to occur Sometimes	4	
Major	Significant reduction In safety margins, Reduction in the ability of operators to cope with adverse operating conditions as a result of an increase in workload, or as result of conditions impairing their efficiency. Serious incident. Injury to persons.	C	Remote	Unlikely, but possible to occur	3	
Minor	Nuisance. Operating Limitations. Use of Emergency procedures. Minor incident.	D	Improbable	Very unlikely to occur	2	
Negligible	Little consequence	E	Extremely Improbable	Almost inconceivable that the event will occur	1	

Risk Likelihood		Risk Severity				
		Catastrophic A	Hazardous B	Major C	Minor D	Negligible E
Frequent	5	5A	5B	5C	5D	5E
Occasional	4	4A	4B	4C	4D	4E
Remote	3	3A	3B	3C	3D	3E
Improbable	2	2A	2B	2C	2D	2E
Extremely Improbable	1	1A	1B	1C	1D	1E

FIGURE 13-4   Risk assessment, 5 × 5 matrix. (*Source: FAA Advisory Circular 120-92A, App 3, pp. 1–3. www.faa.gov*)

2. *Reduction of the hazard level.* Reduction of the severity or likelihood of the event reduces its overall impact, perhaps moving its classification from unacceptable to acceptable on the risk assessment matrix.

3. *Employment of safety devices.* This control method is not fail-safe, but its mechanical nature (safety guards, rails, etc.) often prevents exposure to potential hazards.

4. *Warnings and alert methods.* These could be visual (warning light) or audible, such as an aircraft stall or landing gear warning horn. One of the control weaknesses is that after a warning notice is received, human action is necessary.

5. *Safety procedures.* These are known as administrative control strategies, and thus are considered a "soft" control measure. Training and regulations can be put into place, but human error can negate this type of control.

Control strategies, techniques, and equipment expressed in terms of the risk equation ($P \times S = R$) yield the following considerations:

- *Design and Engineering* will aim to reduce/eliminate the hazard and works on the probability (P) side of the risk equation which subsequently reduces/eliminates severity (S).

- *Safety devices,* which can be active or passive, reduce risk in a similar way to design and engineering but are considered less effective. Active safety devices require human action (ex. seat belts) and are not as effective as passive devices. Passive devices are those that sense and deploy automatically when needed (ex. airbags) and require no human action to activate.

- *Warning devices* increase the chances of one being aware of the potential hazard and hence reduce the probability of contact which reduces (P). Should the warning be ignored, the full extent of severity will be felt.

- *Procedures and training* are mainly intended to reduce (P), which subsequently reduces (S).

- *Personal protective equipment* (PPE) serves to reduce the severity (S) from a materialized hazard. Some may consider PPE as an active safety device. PPE should be considered the last line of defense as it requires human input to be worn correctly, and even if worn correctly, it can be uncomfortable/cumbersome to wear.

SAFETY ASSURANCE EXPLAINED

As indicated in Fig. 13-2, once the risk assessment and controls are in place and "ALARP" has been declared, the SMS process shifts to the Safety Assurance side of the ledger to provide feedback on how the Safety Risk Management process is performing. If performing well, the safety assurance process provides positive reinforcement that risks are properly managed. In Chapter 12, the three main elements of safety assurance were outlined in an airline safety context. Provided below is a further explanation of the subelements of safety assurance from FAA Advisory Circular 120-92A.

PERFORMANCE MONITORING AND MEASUREMENT

Eight subelements are performed in the Data Acquisition Process of Safety Assurance (Fig. 13-2):

Continuous monitoring comes from a variety of sources, but especially from line managers who are the technical experts in any organization.

Internal audits are conducted by operating departments of the organization.

Internal evaluation audits are conducted by people who are functionally independent of the process being evaluated.

External audits are conducted by the FAA, or by third-party organizations using tools such as the IOSA Standard of The International Air Transport Association.

Investigations of accidents, incidents, or other safety related events are conducted in a non-punitive fashion to determine why the safety problem happened.

An *employee reporting system* such as ASAP or other voluntary method is utilized.

Analysis of data is conducted, especially to identify root causes of any nonconformance and to identify potential new hazards.

System assessment is conducted to determine the safety performance and effectiveness of risk controls.

MANAGEMENT OF CHANGE—NEW HAZARDS TO CONSIDER

New hazards may inadvertently be introduced into an SMS system whenever a change occurs. A mature process which is ostensibly under control may become unsafe when faced with changes to the system. Also, a new operational procedure or system change needs to be processed through the SRM process to see whether it has an impact upon other processes. One of the more important aspects of safety assurance is an effective formal change management system.

CONTINUOUS IMPROVEMENT—GETTING BETTER THROUGH THE SMS PROCESS

The final element of Safety Assurance is very important; continuous improvement of the Safety Management System is a never ending goal. There are two subelements of continuous improvement:

Preventative/corrective action assignments must be tracked and managed. This is an active process requiring routine follow up and continuous assessment.

Top management review should be regularly conducted by the SMS Accountable Executive and other safety boards and committees at all levels of management. This review should include the inputs and outputs of SRM and the lessons learned from the safety assurance process.

THE FUTURE OF SMS IN AVIATION

Today, the transformation to aviation Safety Management Systems is in full swing on a global basis. Using quality principles as a guide, SMS programs will continue to employ new methods using empowerment and positive safety culture techniques to obtain safety levels never before achieved in commercial aviation.

For students of SMS processes, the future is very bright indeed. The next few years will bring increased international collaboration and employment opportunities. Aviation safety professionals will find that SMS is a journey of continuous improvement, not a final destination.

Welcome aboard and enjoy the flight!

KEY TERMS

The Four Pillars of SMS
    Safety Policy
    Safety Risk Management
    Safety Assurance
    Safety Promotion
Risk Assessment Matrix
Severity
Likelihood
Risk
ALARP
International Organization for Standardization (ISO)
Just Culture

REVIEW QUESTIONS

1. Discuss the evolution (history) of the SMS principles.
2. Describe the fundamental components and salient features of SMS—The Four Pillars.

3. Compare and contrast the differences between a hazard and a safety risk.

4. Discuss the significance of the $P \times S = R$ risk management formula.

5. What is the purpose of the risk assessment matrix?

6. Describe the relationship between Safety Risk Management and safety assurance.

7. What is the concept of ALARP and why is it significant?

8. Describe the three principle elements of the safety assurance process.

9. Discuss the future of Safety Management Systems in aviation to today's aviation safety professional.

REFERENCES

Federal Aviation Administration Advisory Circular 120-92A. 2010 *Introduction to Safety Management Systems for Air Operators* www.faa.gov

International Civil Aviation Organization "*SMS Training, ICAO SMS Module.*" 2011. and "*SMS Senior Management Briefing.*" www.icao.int

International Civil Aviation Organization *Safety Management Manual*, 2nd ed. 2009, www.icao.int

International Air Transport Association, 2011 www.iata.org

Stolzer, Alan J., Halford, Carl D., and Goglia, John J. 2008. *Safety Management Systems in Aviation*. Burlington, VT: Ashgate Publishing.

INDEX

Note: Page numbers followed by *f* denote figures; page numbers followed by *t* denote tables.